P9-DDA-102

Optimizing Quality in Electronics Assembly

Related Titles

ALVINO • *Plastics for Electronics*

BOSWELL • *Subcontracting Electronics*

BOSWELL and WICKAM • *Surface Mount Guidelines for Process Control, Quality, and Reliability*

BYERS • *Printed Circuit Board Design with Microcomputers*

CAPILLO • *Surface Mount Technology*

CHEN • *Computer Engineering Handbook*

CLASSON • *Surface Mount Technology for Concurrent Engineering and Manufacturing*

COOMBS • *Printed Circuits Handbook,* 4/e

DI GIACOMO • *Digital Bus Handbook*

FINK and CHRISTIANSEN • *Electronics Engineers' Handbook,* 3/e

GINSBERG • *Printed Circuits Design*

GINSBERG and SCHNORR • *Multichip Modules and Related Technologies*

HARPER • *Electronic Packaging and Interconnection Handbook*

HARPER and MILLER • *Electronic Packaging, Microelectronics, and Interconnection Dictionary*

HARPER and SAMPSON • *Electronic Materials and Processes Handbook,* 2/e

HWANG • *Modern Solders and Soldering for Competitive Electronics Manufacturing*

JURAN and GRYNA • *Juran's Quality Control Handbook*

LAU • *Ball Grid Array Technology*

LICARI • *Multichip Module Design, Fabrication, and Testing*

MANKO • *Solders and Soldering,* 3/e

RAO • *Multilevel Interconnect Technology*

SERGENT and HARPER • *Hybrid Microelectronics Handbook,* 2/e

SOLBERG • *Design Guidelines for Surface Mount Technology,* 2/e

STEARNS • *Flexible Printed Circuitry*

SZE • *VLSI Technology*

VAN ZANT • *Microchip Fabrication,* 3/e

Optimizing Quality in Electronics Assembly

A Heretical Approach

James Allen Smith
Frank B. Whitehall

McGraw-Hill

New York San Francisco Washington, D.C. Auckland Bogotá
Caracas Lisbon London Madrid Mexico City Milan
Montreal New Delhi San Juan Singapore
Sydney Tokyo Toronto

To Sonia, for the countless selfless sacrifices

JAS

Library of Congress Cataloging-in-Publication Data

Smith, James Allen, date.
Optimizing quality in electronics assembly : a heretical approach / James Allen Smith, Frank B. Whitehall.
p. cm.
Includes bibliographical references and index.
ISBN 0-07-059229-2 (hc)
1. Electronic industries—Quality control. I. Whitehall, Frank B. II. Title.
TK7836.S63 1996
621.381'068'5—dc20 96-29182
CIP

McGraw-Hill

A Division of The ***McGraw-Hill*** *Companies*

1 2 3 4 5 6 7 8 9 0 DOC/DOC 9 0 1 0 9 8 7 6

ISBN 0-07-059229-2

The sponsoring editor for this book was Stephen S. Chapman, the editing supervisor was Stephen M. Smith, and the production supervisor was Donald F. Schmidt. It was set in Palatino by Estelita F. Green of McGraw-Hill's Professional Book Group composition unit.

Printed and bound by R. R. Donnelley & Sons Company.

This book is printed on recycled, acid-free paper containing a minimum of 50% recycled de-inked fiber.

Contents

Acknowledgments

Optimizing Quality in Electronics Assembly would not have been possible without the unstinting support and inspiration provided over the years by our manifold professional associates, clients, and friends, only a small number of whom can be named here.

Maynard Eaves of Hewlett-Packard, the modest patriarch of North American printed circuit technology, has acted as James Allen Smith's mentor, inspiration, supporter, and stalwart confidant throughout the years.

Mike Critser of Reed Exhibition Companies provided numerous NEPCON platforms for the views presented in these pages. A more prudent or less principled conference director would have yielded to pressure from proponents of mainstream beliefs who tried earnestly to suppress our message.

Dr. Eli Goldratt and Wendy Donnelly of the Avraham Y. Goldratt Institute kept our intellects sharp and spirits high with generous encouragement and critiques. Core elements of this book rely on insights gained from Dr. Goldratt's work.

Maury Flynn of ITTA Motors & Actuators Worldwide, Keith Joyner of Lucas Aerospace, and Prof. John Roulston of GEC-Marconi Avionics graciously accepted the thankless and time-consuming task of reviewing the final manuscript. To all three our sincere thanks and apologies for disrupting their lives.

Neil B. Whitehall served as astute reviewer and critic of the manuscript, contributing no small amount of vital technical information. Neil's assistance with Chap. 11 was invaluable.

Frank B. Whitehall acknowledges a profound intellectual debt to Dr. J. M. Juran, with whom he studied and corresponded in the early 1970s. Dr. Juran has been a rich source of inspiration and motivation to us both over the years. Other powerful influences that shaped Frank's attitudes and thinking came from Dr. Charles Allen (deceased) of Ferranti; Alan Wesley (retired), also of Ferranti; Art M. Schneiderman of Analog Devices; and Dr. Charles C. Harwood, late of the Quality Improvement Company.

Steve Chapman, our editor at the McGraw-Hill Professional Book Group, guided the book from concept to reality, helped us find the necessary tone, and shielded us from the business pressures inflicted on him by our failure to meet the anticipated delivery dates for copy. If patience is truly a virtue, Steve approaches sainthood. Also within the McGraw-Hill organization, we are indebted to Donna Namorato for cheerfully and efficiently expediting the surprisingly large number of tedious logistical tasks that authors too often take for granted; to Steve Smith, who supervised the physical production without once complaining about our editorial nit-picking; and to Susan Sexton, whose eye for grammatical consistency and typographical errors can only be described as remarkable.

The sacrifices made on behalf of this work by our families can never be adequately conveyed on the printed page. The patience and steadfast encouragement of our wives—Sonia Smith and Elizabeth Whitehall—passes all understanding. From them, we learned the real meaning of quality.

James Allen Smith
Frank B. Whitehall

Authors' Note

This book began as a three-month project to put on paper our collective knowledge of quality in the electronics assembly industry. Three months of full-time effort seemed about right considering our backgrounds. We are, after all, electronics assembly veterans with roughly six decades of hands-on experience and untold thousands of lecture hours between us. Most of that experience involves quality management, either directly or through the various activities—including design and production—that determine quality. Our views have helped shape many industry standards and procedures. In short, we expected that a compilation of our past work and knowledge would be entirely satisfactory.

We miscalculated. The project ended up taking more than two years of full-time writing and research. There were times when we referred to it as our American tricentennial project. At other times, we thought that deadline—the year 2076—might be unrealistically short. If the project did not take a lifetime, there were times when it seemed that way.

What went wrong? Actually, nothing. The problem involved what went right. We began to notice disturbing patterns in prominent contemporary quality literature. The literature regularly cited earlier works in ways that made us wonder how many of the modern authorities had actually read the referenced material. Ultimately, those concerns about possible revisionism caused us to reread the original works—many of which have been out of print for half a century or more. It turned out that misquotation and misinterpretation are much more common than we had previously believed.

Discovering the historical inaccuracies in so many influential quality writings then required intensive reappraisal of popular modern quality attitudes. The results were vastly more intriguing, consequential, and—at times—disturbing than we had imagined at the start of the project. When we first sat down at the word processor, our views about quality were not profoundly atypical of most contemporary quality professionals. Our findings and experiences shook us out of that complacency. Certainly, we will never again regard the meaning, significance, and methods of achieving quality as we did two years ago. We suspect that readers may find themselves thinking of quality in quite different ways as well.

Introduction

After lavishing attention on quality in recent years, many executives today are moving on to new interests such as reengineering. This shift in emphasis is not entirely undesirable, since much of the professed commitment to quality was superficial and cynical; those who stay the quality course may be smaller in number but are infinitely more loyal and knowledgeable than the masses still looking for a managerial magic bullet.

While there are certainly compensations in seeing the decline of quality as a fad, the movement away from quality is sadly premature. The large numbers of American companies that discovered astounding new vitality through quality improvement programs since 1980 proved the potency of quality management as a competitive tool. But vastly more companies have failed to take advantage of the benefits that they would enjoy from better quality.

Two important reasons for the adoption of new operational strategies in place of quality improvement should be familiar to any student of management history. First, business leaders are notoriously impatient, constantly seeking quick fixes to satisfy investors. Second, the executive community as a whole consistently demonstrates pronounced herd mentality; any well-publicized flirtation with a new mode of operation by a few industry leaders normally creates a stampede away from the current strategies toward the new ideology.

It can be said in all fairness as the infatuation with quality wanes among American businesses that neither the meaning of quality nor the reasons that it should be improved have been understood by more than a handful of individuals in an even smaller number of companies. The "quality" sought in most cases was not quality at all, and the expectations for rewards were grossly exaggerated. No industry has demonstrated greater confusion than electronics assembly about the what, why, and how of quality.

The consequences of this misunderstanding are often serious and occasionally catastrophic. In attempting to become stronger through better quality, electronics assembly companies all too frequently act on unfounded beliefs and value judgments rather than rigorous business logic. Emotions have overruled reason. And, as it turns out, the road to quality hell is truly paved with the best of intentions.

For a combined total of more than half a century, we—Smith and Whitehall—have been active global participants in development, implementation, and appraisal of quality systems for electronics assembly. The dark ages when quality activities were held in lower esteem than janitorial tasks remain vividly etched in our minds. We have also lived through the heady days when the electronics industry initially recognized quality improvement as one of the most—if not *the* most—important operational strategies. And we had the good fortune to participate in the development of several highly regarded electronics assembly quality systems.

Our career choices and experiences mean that we are neither unbiased nor dispassionate in our attitudes about quality. Quite simply, we believe quality must be among the paramount concerns of any electronics assembly company. Companies that relegate quality to secondary stature will ultimately find their very survival threatened. However, the costs of pursuing an unsuitable quality improvement strategy can be equally ruinous. Electronics assembly companies that unquestioningly adopt the most fashionable recent quality ideology generally end up worse off than if they had done nothing.

Quality improvements should be pursued. But the planning and implementation of that pursuit must be undertaken in a rigorously scientific manner quite unlike the strategies most frequently advocated today. Not all approaches work equally well. And methods that produce impressive gains for some companies can perform poorly in others.

Crucial Questions

Choosing to embark on a quest for greater quality can therefore be one of the most rewarding decisions an electronics company will ever make—or one of the most disastrous. The outcome will be rewarding only for those companies that (1) fully comprehend the true meaning of quality, (2) are realistic in their expectations of benefits they hope to obtain from better quality, and (3) plan their quality improvement investments accordingly.

The questions that must be asked *and answered* even before deciding whether to invest in quality improvement are formidable. Just a few of those questions include:

- What is the meaning of "quality" in the context of electronics assembly?
- Should quality investments be subject to the same financial assessment standards as other investment decisions? If so, how can those evaluations be made?
- Does the endless pursuit of ever-better quality necessarily lead to improvements in the company's prosperity and security? If not, how can the ideal level of investment in quality improvement be established?
- Can some quality tools cause more damage than improvement to a company's health?
- Are newer quality improvement tools truly better than the older tools they replaced?
- Should quality improvement always be the company's foremost strategic tool?

If the decision is taken to proceed, the questions become even more numerous and challenging.

Modern quality orthodoxy makes no allowance for such questions. Conventional quality wisdom insists that a company can never have enough quality and any improvement in quality will result in benefits exceeding the costs of obtaining that improvement. We, too, would like to believe that greater quality is always better for every company. However, our experiences in recent years do not support that conclusion. By the standards of current quality conventions, therefore, we could be considered quality heretics.

During the last quarter of 1994, Cambridge Management Sciences, Inc., with which we are affiliated, conducted a study of electronics assembly companies to determine whether our experiences—and, therefore, our views—were unique or whether they reflect prevalent industry conditions that have not been widely reported. On the basis of our work with clients in the Americas and Europe, we had anticipated the survey results would be less enthusiastic than what is normally heard or read about the various systems. However, we were not at all prepared for the staggering level of disenchantment surrounding essentially *every* program. In particular, the results showed that:

- With rare short-term exceptions or only isolated instances of success, no system introduced in the past 10 years to improve profitability, market share, or quality in electronics assembly companies has earned more than it cost.
- Every year, more than 30 percent of American electronics assembly companies replace their strategies for improving profitability, market share, or quality with new approaches that also fail in large numbers.
- The frequency with which new systems are developed, adopted, and abandoned has been accelerating; in the past decade, the life expectancy of new systems has averaged less than 5 years.
- Quality directors increasingly find themselves and/or the programs they administer (often pushed on them in the first place by other corporate executives) under attack from managers of other departments and, ever more frequently, by senior executives.
- New programs fail for exactly the same reasons that the strategies they replace failed.
- Most strategies, however, do possess inherent strengths and would make important positive contributions to user companies *if a relatively small but lethal number of pitfalls could be avoided.*

The Cambridge Management survey results coincide with other recent findings. In particular, sources that had been strong advocates of either or both TQM and reengineering have publicly stated that those systems are not working. First *Business Week* magazine, long an aggressive champion of quality improvement, published a feature article titled "Quality: How to Make It Pay"[1] that said bluntly, "...at too many companies, it turns out, the push for quality can be as badly misguided as it is well-intended...it's wasted effort and expense." Then James Champy[2] conceded that reengineering also is ailing. "Reengineering is in trouble. It's not easy for me to make this admission. I was one of the two people who introduced the concept."[3]

Considering these bleak experiences, it should not be surprising that unhappiness about the results of quality investments when compared to the promises is already common among general management—and spreading

---

[1]*Business Week*, Aug. 8, 1994, pp. 54–59.

[2][Champy 1995], p. 1.

[3]See [Hammer and Champy 1993].

rapidly. However, the disillusion of those nonquality managers with quality programs has been largely overlooked by the quality profession. A very real danger exists that continued failure by quality personnel to recognize and adapt to changing expectations may bring about reversal of the profession's hard-won gains made over the past decade. Sweeping changes must be made in the promises and practices of the quality profession—and they must be made quickly. Mainstream quality practices simply are not working well—at least when seen through the eyes of other departments.

Making Quality Pay

This is a book about why quality programs often deliver less than satisfactory results for electronics assembly companies. More important, however, it is a book about how to achieve success with quality programs: How to ensure that the quality programs best suited to this industry and the reader's company will always be chosen; how to avoid the strategies with high probability of failure; how to produce not just positive net benefits but the *maximum* possible benefits. In these pages will be found detailed blueprints for implementing a powerful and highly cost-effective quality program specifically designed for the special needs of the electronics assembly industry.

For personnel involved in any aspect of electronics assembly—including but by no means limited to quality personnel—the techniques presented here offer several advantages over orthodox quality programs. Those advantages include the following:

- The tools and program contained in these pages were developed *within* the electronics assembly industry specifically *for* the electronics assembly industry; they are not the products of academics lacking hands-on experience in this industry.
- All have been field tested—most for more than a decade—in roughly 100 plants in the Americas and Europe; invariably the annual return on investment has exceeded 100 percent without including intangible benefits such as greater customer satisfaction.
- They are relatively inexpensive to install and, when applied according to the guidelines, almost invariably correct the most serious quality and productivity problems of any electronics assembly company within a matter of months.
- The systems continue to work almost indefinitely with only minimal maintenance; companies where early versions of these systems were implemented a decade ago still employ the tools and program with great success.

Using This Book to Maximum Advantage

This book is organized in a carefully determined sequence of material that:

1. Begins with historical and theoretical foundations of quality practice.
2. Progresses through obscure yet crucial scientific foundations of those processes unique to the electronics assembly industry.
3. Presents a variety of useful hands-on nuts-and-bolts techniques for achieving highest quality at lowest cost. These tools are neither as glamorous nor as fashionable as the more lavish "total quality" programs that dominate contemporary quality attitudes. They do, however, offer an altogether more compelling reason than total quality for inclusion in any quality improvement program: They work!
4. Casts a critical eye on several of the most popular quality fads of the past decade, including various flavors of "total quality" such as ISO 9000, the Malcolm Baldrige National Quality Award, Six Sigma™ and benchmarking.
5. Details quality optimization, a simple yet profound technique for managing quality investments.
6. Concludes with a proven effective step-by-step program for achieving fast and lasting quality improvements in the electronics assembly industry.

The reader may be tempted to skip the preliminaries and go directly to the program proposal in the concluding chapter. While impatience is understandable, the urge to start at the end must be resisted. The program depends on material contained in the preceding chapters. Since most of that material is either presented in print for the first time or differs substantially from standard interpretations, the significance of the program—its challenges as well as its opportunities—will inevitably be lost on anyone who missed laying the intellectual foundation. Taking the linear route from beginning to end will not necessarily be easy, but we are confident that the journey will be rewarding for novice and veteran members of the electronics assembly industry.

1 Introduction to the Theory of Quality Optimization

1.1 Chapter Objectives

Quality professionals frequently believe that attempts to quantify the benefits of quality programs are incompatible with the commitment to quality. Today, however, many corporations require their quality departments to demonstrate that benefits of quality programs exceed costs. This chapter explains that:

- The ultimate goal of every company must be maximization of long-term profits.
- Quality improvement cannot be the goal, only a tool to help the company achieve the goal.
- Every company activity—quality operations included—must be assessed according to benefits received for funds expended.
- Return on quality (ROQ) is increasingly popular among nonquality executives for assessing the effectiveness of quality investments.

- The new emphasis on cost-benefit analysis of quality initiatives typically strengthens both the company and the quality department.
- Measuring the net benefits of every quality improvement project always advances the quality cause.
- The most effective method of managing quality is by the technique of quality optimization.

1.2 Quality circa 1996

These are not the worst of times for quality. Nor, unfortunately, are they the best. Superficially, the indications point to a healthy environment for quality management—more than ever before, companies today claim they are solidly committed to something called quality. At the same time, however, signs of cynicism and bewilderment are evident and growing. Increasingly, managers say "quality" when what they really mean is "cost reductions," "greater efficiency," "shorter cycle time," "participative management," and "higher profits." While most or all of those objectives may be eminently desirable and attainable through quality improvement, none is the same as quality.

Meanwhile, the role of the quality professional has rarely been less clear. New functions ranging from inspiring personnel to determining broad company operating policies to assessing customer satisfaction have entered the realm of the quality department's responsibilities. Often these added tasks demand as much attention from the quality manager as do the traditional challenges of quality assurance themselves.

Broadening the quality function carries considerable risks. The most notable hazard is the very real possibility of stretching the quality department's resources beyond all reasonable limits. Far too many quality departments today are overworked and understaffed, with the inevitable result that all quality tasks suffer. "Quality" is increasingly applied to everything and therefore means nothing.

Moreover, prospects for improvement are not encouraging. The latest ISO management standard, ISO 14000 (designated BS 7750 in the United Kingdom), for example, calls for an environmental management system structured similarly to the ubiquitous ISO 9000 (also known as BS 5750 in the United Kingdom). Companies are expected to base new environmental management documentation on the existing documentation and auditing procedures developed under ISO 9000 and BS 5750. The next step, already under way, is establishment of an ISO 9000 type of accreditation system for bodies intending to certify compliance with the new standard. Although environmental issues clearly have no correlation to quality, the use of similar methods for managing both ISO 14000 and ISO 9000 means that auditing the company's environmental compliance will also become the responsibility of quality managers already carrying excessive workloads. This trend is already well established in Britain and, early in 1996, was rapidly gaining momentum in North America.

The ISO 14000 issue is just one example of how quality professionals today are being saddled with tasks that have nothing to do with quality. As will become apparent in these pages, numerous other activities unrelated to quality have also been placed on the quality department's shoulders. At least three seriously negative consequences of the added workload are obvious:

1. Diminished ability to find time for true quality matters
2. Reduced understanding throughout the company of what constitutes quality
3. Less respect by other parts of the organization for the importance of true quality management

These trends toward dilution and distortion of quality management threaten to reverse many or all of the gains made by the profession over the past 15 years. In our view, only one possibility exists for preventing this regression. The quality profession must adopt the rigorous analytical management techniques that sustain the influence of other departments such as finance and sales. The new approach may not be easy or even palatable for many quality practitioners, but it is absolutely essential unless quality managers are willing to accept a return to second-class status in the management hierarchy.

The propositions to be developed in these pages begin with what we trust is an uncontroversial position: that quality management plays a key role in the success of any business. However, the currently prevailing belief that a company can never have too much quality is valid only if several unlikely conditions prevail. Further, much "quality improvement" activity has little to do with the real meaning of quality. Ultimately, all investments in quality improvement must be governed by a rigorous set of economic criteria that we call quality optimization.

1.3 The Need for Quality Optimization

Quality—despite well-known claims to the contrary—is not necessarily free.[1] For that matter, some quality initiatives never produce benefits equal to or greater than their costs. Moreover, quality improvement actions that provide excellent benefits in relation to cost at one plant may be more costly than beneficial at others. While the natural inclination of any quality manager is to push boldly ahead on measures to improve the company, there are also times when the most prudent course is the physicians' creed: "First do no harm." The company is the quality professional's patient and must be cared for with more than blind faith.

The possibility that quality programs could cost more than they return is not a thought unanimously endorsed by the quality profession. Until recently, in fact, the overwhelming majority of quality professionals insisted that better quality always translates into a stronger company. Increasingly, though, there is recognition that quality improvement efforts—like almost all investments—suffer from diminishing returns. In other words, at some point in a quality program's life, it will begin to produce fewer and fewer gains for every additional dollar spent on it. Eventually, the costs may easily exceed the benefits.

Rational managers monitor quality improvement investments carefully and terminate programs when the return on the next dollar invested will be less than a dollar. Accordingly, we have devised a process of managing quality improvement investments for maximum benefit to the company. We call this process *quality optimization.*

Quality optimization is explained in Chap. 18. Before quality optimization can be considered, however, much groundwork needs to be laid. Specifically, it is necessary to examine:

- The origins of current attitudes about quality
- The meaning of "quality"
- The effectiveness of various quality improvement tools
- The validity of claims about benefits derived elsewhere in companies and countries that pursued quality improvement aggressively

Quality optimization requires quantifying and comparing both costs and benefits of each quality investment. Therefore, it is best to begin by

---

[1]"Quality is free" is grammatically incorrect. "Quality improvement" should be substituted whenever "quality" is used in this context.

noting that many quality professionals absolutely reject the validity of any management system in which the monetary aspects of quality programs are measured.

1.4 The Case against Quantifying Quality Gains

For analytical purposes, perspectives about quality can conveniently be divided into two eras: Before Deming and After Deming. Despite some commentators' opinions,[2] the distinction between pre- and post-Deming periods by no means implies Dr. W. Edwards Deming discovered quality or was even personally responsible for the quality revolution.[3] Rather, the rise of Demingism marks a watershed in quality thinking.

Prior to Deming, quality improvements were sought because of Philip B. Crosby's thesis that quality improvements reduce company expenses by more than the cost of generating the improvements.[4] While Crosby did not invent the hypothesis that quality improvements earn more than they cost, he was certainly the most articulate and persuasive spokesman for a decades-old quality doctrine.

Deming transformed Western thinking with the argument that money should not even enter into the quality equation. The clearest and strongest of Deming's warnings about the dangers of breaking out the costs and benefits of quality improvement projects is contained in *Out of the Crisis*.[5] "He that expects to quantify in dollars the gains that will accrue to a company year by year for a program of improvement of quality by principles expounded in this book will suffer delusion," Deming argued. "He should know before he starts that he will be able to quantify only a trivial part of the gain....

"As the outlook for a company grows bleaker and bleaker, management falls heavier and heavier on the comptroller for management by figures. In the absence of knowledge about the problems of production, the comptroller can only watch the bottom line, squeeze down on costs of materials purchased, including the costs of tools, machinery, mainte-

---

[2]See, for example, [Gabor 1990].

[3]The single most important factor in corporate America embracing quality as a vital strategic weapon most likely was the publication of *Quality Is Free* [Crosby 1979]. Discussion of contributions by leading quality thinkers may be found in Chap. 2.

[4][Crosby 1979].

[5][Deming 1986], p. 123.

nance, supplies. Neglect of the more important invisible figures, unknown and unknowable, on the total costs of these moves, causes further shrinkage of profit from whatever business remains."

1.4.1 Demingism as Conventional Wisdom

Few quality professionals today disagree with Deming's contentions—which for the sake of brevity will be referred to here as *Demingism*—that (1) too much quality can never exist and (2) quality's contributions to the company must never be subjected to monetary considerations. In essence, Demingism has become the quality profession's version of what economist John Kenneth Galbraith termed *conventional wisdom.*

"[E]conomic and social behavior are complex, and to comprehend their character is mentally tiring," Galbraith wrote.[6] "Therefore we adhere, as though to a raft, to those ideas which represent our understanding. This is a prime manifestation of vested interest. For a vested interest in understanding is more preciously guarded than any other treasure. It is why men react, not infrequently with something akin to religious passion, to the defense of what they have so laboriously learned. Familiarity may breed contempt in some areas of human behavior, but in the field of social ideas it is the touchstone of acceptability.

"Because familiarity is such an important test of acceptability, the acceptable ideas have great stability. They are highly predictable. It will be convenient to have a name for the ideas which are esteemed at any time for their acceptability, and it should be a term that emphasizes this predictability. I shall refer to these ideas henceforth as the conventional wisdom."

Galbraith goes on to explain how beliefs that become part of conventional wisdom continue to evolve. "Moreover, with time and aided by debate, the accepted ideas become increasingly elaborate. They have a large literature, even a mystique. The defenders are able to say that the challengers of the conventional wisdom have not mastered their intricacies."[7]

Once a position becomes conventional wisdom, considerable energy is expended to overcome any suggestions of holes in its logical or factual fabric. Or, in Galbraith's words, "There are many reasons why people like to hear articulated that which they approve. It serves the ego: the individual has the satisfaction of knowing that other and more

[6][Galbraith 1958], pp. 7–8. These page references refer to the fourth revised edition of 1984.

[7]p. 9.

famous people share his conclusions. To hear what he believes is also a source of reassurance. The individual knows that he is supported in his thoughts—that he has not been left behind and alone. Further, to hear what one approves serves the evangelizing instinct. It means that others are also hearing and are thereby in the process of being persuaded.

"In some measure, the articulation of the conventional wisdom is a religious rite. It is an act of affirmation like reading aloud from the Scriptures or going to church. The business executive listening to a luncheon address on the immutable virtues of free enterprise is already persuaded, and so are his fellow listeners, and all are secure in their convictions. Indeed, although a display of rapt attention is required, the executive may not feel it necessary to listen. But he does placate the gods by participating in the ritual."[8]

Therefore, Galbraith concludes, "The conventional wisdom accommodates itself not to the world that it is meant to interpret, but to the audience's view of the world. Since the latter remains with the comfortable and the familiar, while the world moves on, the conventional wisdom is always in danger of obsolescence."[9]

All too often today, quality thinking embodies all the characteristics of Galbraith's conventional wisdom. Within that realm of conventional wisdom, Demingism has become the most prominent example of the "comfortable and familiar." Inevitably, therefore, time is passing Demingism by.

1.4.2 Return on Quality

While Demingism remains the conventional wisdom of the quality world, its stature among company officials outside the quality department is increasingly under attack. *Out of the Crisis* converted executives of all types to Demingism so that, until recently, every company of consequence has at least claimed that quality issues dominate their business strategies. More and more, however, commitment outside the quality department to quality improvement involves opinions that would not have met with Deming's approval. In particular, companies have begun to question the wisdom of following Deming's advice against measuring quality's costs and benefits.

The pre-Deming belief that quality, like other company activities, must be assessed on the basis of its contribution to earnings is undergoing a strong revival. Interestingly, early converts to comparison of tan-

[8] p. 10.

[9] pp. 11–12.

gible gains from quality against expenses for quality include several of the last decade's most enthusiastic advocates and practitioners of Deming-style quality. Two prominent advocates of the new quality thinking—AT&T and Federal Express—are highly regarded past winners of the Malcolm Baldrige National Quality Award, which epitomizes Demingism.

The new standard by which companies increasingly choose to evaluate their quality programs has been labeled *return on quality* (ROQ).[10] ROQ is the quality-specific term for the same manner of cost-benefit analysis that has long applied to almost every other business operation from production to marketing. When the return on investment in any given quality initiative fails to meet the company's minimum return on investment requirements, ROQ policy requires the termination of that initiative. Advanced practitioners of ROQ would also require that proposed quality projects pass a cost-benefit analysis before being adopted.

The implications of ROQ are clear. Quality managers who cannot prove conclusively that their departments produce meaningful net gains in their companies' performance will be subjected to traumatic reductions in their department budgets, their personal influence within the company, and—in many cases—their very job security.

Deviation from conventional wisdom is always unsettling. The sudden prominence of ROQ therefore gives rise to two urgent questions:

First: Does the rise of ROQ signify return of the pre-Deming corporate mentality?

Second: Have the post-Deming lessons been lost?

The same answer applies to both questions: not necessarily. Asking for measurement of quality's contributions by no means requires reversion to the practices of what we have called the pre-Deming era. Important progress has been made in the years since 1985 in techniques for identifying and approximating the financial value of formerly intangible quality benefits (the "unknown and unknowable" of Deming's argument).

Deming's philosophy of quality was a product of his time and experiences. With few exceptions, business leaders during Deming's formative years regarded quality entirely as cost, never as investment, and constrained spending on quality improvement accordingly. Even the more enlightened companies that Deming encountered emphasized

[10]"Quality: How to Make It Pay," *Business Week,* Aug. 8, 1994, p. 55. The article quotes Vanderbilt University professor of management Rolan Rust's describing ROQ as, "If we're not going to make money off it, we're not going to do it."

quality control—that is, attempting to catch and repair defects after production—rather than the defect prevention approach known as quality assurance.[11] Indeed, doing the job right the first time was essentially unknown, in large measure because few managers (including quality management) believed such an objective possible and/or feasible. The universe of quality benefits "unknown and unknowable" was vast. In the presence of such tremendous antiquality bias—accompanied by gross ignorance of what and how to measure quality's contributions to the company's financial health—allowing quality to be subject to cost-benefit analysis merely strengthened the hand of those quality agnostics opposed to any spending whatever on quality.

1.5 The Case for Quantifying Quality Gains

While Deming's stand was appropriate to conditions of his time, the world constantly evolves. Executive attitudes toward quality—together with understanding of the less obvious gains provided by quality improvement—have changed radically (and, from the perspective of the quality department, for the better) since the early 1980s. Today's senior managers overwhelmingly support quality improvement; their quality departments enjoy powers and funds far beyond Deming's dreams. Total quality, the Malcolm Baldrige National Quality Award, ISO 9000, worker teams, and Six Sigma™ are part of the normal business environment. In the presence of an executive mindset that knows quality can be a positive contributor to the company's well-being rather than a necessary evil—in other words, in the modern corporate world—quality effectiveness measurements no longer necessarily threaten quality advances. The time has come to reexamine the quality profession's distrust of measuring the cost-effectiveness of quality improvement efforts.

British professor and business consultant Charles Handy summarizes quality's conventional wisdom particularly well. "In the long-term, Deming argued, you stay competitive and in business by being the best there is, not necessarily the cheapest, by taking the customer seriously and giving him or her what they want and need. The product comes before the money."[12] Moreover, "Short-term profit at the expense of

[11]Terminology is important, and quality control is often confused with quality assurance. The distinction is explained in Sec. 3.8. For the purposes of this book, quality control will be considered postproduction policing of output while quality assurance—the practice advocated in these pages—is the science of doing the job right the first time.

[12][Handy 1989], p. 114.

quality will lead to short-term lives. In that sense, quality is, to my mind, the organizational equivalent of truth. Quality like truth, will count in the end....Profit is increasingly recognized as what it always should have been, a means and not an end in itself."[13]

1.5.1 Goldratt and The Goal

Wrapping quality in the trappings of truth and moral superiority makes the Handy/Deming argument almost irresistible to any member of the quality profession. After all, being on the side of the angels certainly enhances job satisfaction. However, Dr. Eliyahu M. Goldratt presents a much stronger case for the opposite position that profit is the end while quality improvement is only a means of achieving that end. Goldratt[14] refers to profit as The Goal in his book of the same title.

A deceptively simple novel (or, more accurately, extended parable), *The Goal* looks at business realities through the eyes of a plant manager and his subordinates in a money-losing plant. If the manager and his staff cannot reverse the plant's dismal financial performance within 90 days, the parent company intends to sell or close the facility. Either action by the parent would cost most of the plant's management team their jobs.

The principal lesson learned by the plant manager and his staff is that companies have only one reason for existence: to earn the maximum possible profits. "Because what happens if a company doesn't make money?" the manager asks. "If the company doesn't make money by producing and selling products, or by maintenance contracts, or by selling some of its assets, or by some other means...the company is finished. It will cease to function. Money must be the goal. Nothing else works in its place."[15]

1.5.2 The Case for Maximizing Profits

The consequences of failing to maximize profits are substantial. Without profits, there can be no internally generated capital. Outside sources of capital such as investors and banks also disappear. A company harried by capital shortages finds itself unable to carry out the activities required for survival, including:

---

[13] p. 115.

[14] [Goldratt 1984]. A revised edition appeared in 1986 and an expanded edition in 1992.

[15] p. 41.

- Creation of desirable new products to keep existing customers and open additional markets
- Acquiring additional capital equipment capable of producing more sophisticated and reliable products at lower cost
- Conducting the sales campaigns to make customers aware of its products
- Attracting the "best and the brightest" personnel at all levels of the company
- Providing secure, rewarding jobs for employees

Or, as Goldratt's plant manager puts it, "I make a list of all the items people think of as being goals: cost-effective purchasing, employing good people, high technology, producing products, producing quality products, selling quality products, capturing market share. I even add some others like communications and customer satisfaction.

"All of these are essential to running the business successfully. What do they all do? They enable the company to make money. But they are not the goals themselves; they're just the means of achieving the goal." Throughout this book, we will refer to the "means" as *tools.*

Goldratt's concept of the goal was articulated succinctly by *Star Trek*'s Mr. Spock in his famous salute, "Live long and prosper."

Comparison with Goldratt's unemotional scientific logic[16] reveals serious flaws in the quality conventional wisdom. Consider, for example, Handy's summation of Demingism: "To make 'profits' or 'the bottom line' is not, by itself, a useful way to describe the purpose behind an organization. It does not begin to tell you what to do or what to be. It is akin to an individual saying that he or she wants to be happy. Of course, happiness and profitability is [sic] a state devoutly to be wished for but it [sic] is not a purpose. If anything, profits are a means and not an end."[17]

Handy's elaboration on Deming's thesis, though noble and eloquent, can be valid only if it is legitimate to apply individual human values to an inanimate object such as the company. But human values cannot apply to the corporation. Companies are not human beings; they exist only on paper and in the eyes of the law. By its very nature, a company has no soul. While employees of the company may well find themselves questioning their destinies or personal fulfillments, the company feels nothing. Therefore, profits cannot be equated to happiness any more

---

[16]Goldratt holds a doctorate in physics.

[17][Handy 1989], p. 181.

than breathing or any other life-sustaining human act can be called happiness. Survival may be preferable to the alternative but cannot be considered a synonym for happiness. Moreover, the for-profit company has no reason to search for meaning.[18] Its purpose in life is clear: to survive prosperously so as to compensate investors.[19] This is not a moral question, only a statement of fact.

It helps to consider both sides of the argument from another angle. Suppose that the company labors long, hard, and successfully to achieve the highest quality in its industry. What do the employees and shareholders hope to achieve in return for their considerable efforts and risk? In other words, what motivated the various human providers of capital and labor to strive for greater quality? The employees can—and likely will—take pride in their workmanship. The shareholders may feel more virtuous knowing their financial backing has enabled new standards of excellence to be obtained. Above all, however, both employees and shareholders expect to be financially rewarded. And those financial rewards are made possible exclusively by profits.

1.5.3 Closing the Communication Gap

Recognizing that profit is the goal and quality improvement only a tool helps explain a chronic problem afflicting the quality profession: why nonquality executives so often frustrate quality personnel. For decades, quality personnel have complained that nonquality management doesn't understand quality and the company will never achieve "world-class" quality without an executive epiphany. Goldratt's goal makes it clear that a lack of understanding does exist—but the cause of the problem differs from the quality profession's contention. While the properly focused senior management are oriented toward the goal of greater profita-

---

[18]A nonprofit organization must be distinguished from the for-profit businesses for which this book is intended. The standard by which a nonprofit organization should be judged is delivery of the service for which it was founded. In such cases, revenues are the means while both quantity and quality of delivered service are the ends.

[19]The position is often taken that the company's role is also to provide security and satisfaction for all employees. See, for example, [Goldratt 1994], p. 273. While such aspirations are commendable, security for workers can be ensured only by healthy long-term profits for the company. A company without profits is a company unable to offer employment security. Conversely, a company with insecure, unhappy employees is also unlikely to achieve satisfactory profits in the long term; thus achieving happy, productive employees can properly be considered yet another tool that must be employed by all companies hoping to achieve the goal. A third "goal" introduced in [Goldratt 1994]—"Provide satisfaction to the market now as well as in the future"—upon similar analysis turns out to be a tool rather than a goal.

bility,[20] the quality personnel—secure in their belief that quality is intrinsically good and therefore not subject to financial analysis—too often fail to provide the necessary data that would bring about executive support. And what is that necessary data? Evidence that investment in quality improvement constitutes a wise use of the company's funds. After all, when Crosby declared "quality is free," CEOs lined up at his door.

In other words, the principal deterrent to communication between the quality department and the executive suite can generally be attributed to inability—or refusal—of quality personnel to operate and communicate on the same terms as the rest of the company. Clearly there can be no universal answer; the level of quality insight among nonquality managers varies from company to company, as does the quality department's ability and willingness to measure and present its accomplishments according to the same standards as other departments. There is, however, one certain universal truth: the quality department serves its own and the company's interests best by willingly and competently justifying its activities in the financial terms that constitute the standard frames of reference in the rest of the company.

As will be shown in Chap. 3, success of any quality program depends on effective communication with every participant from the highest decision maker to the most junior operator. Effective communication, in turn, occurs only when every party to the activity or decision employs the same vocabulary and the words in that vocabulary have identical meaning to everyone speaking, hearing, writing, or reading them. Thus a quality department that speaks a different language from the rest of the company and judges its successes or failures according to its own rules cannot function effectively and harmoniously with the rest of the company. Since every other department in the company operates according to return on investment—and it is unrealistic to expect the rest of the company to change—the onus to adapt falls on the quality department.

1.5.4 Cost-Benefit Analysis Helps Quality Management

Taking this reasoning to the next logical step, it follows that quality personnel should impose upon themselves and their departments the discipline of measuring return on resources used in pursuit of higher quality. Quality departments that quantify their achievements are equipped to satisfy their company's most senior decision makers that further

[20]Of course, their profit goal may be strictly short-term, achieved by cutting investments for new product development and similar endeavors that are vital to the firm's long-term profits.

investments in quality improvement are fiscally responsible and therefore desirable—presuming, of course, that the achievements are real and not figments of the quality department's imagination. Selecting and managing quality programs according to their effects on the company's financial health will invariably contribute to the quality department's well-being. This will be true even if senior management does not explicitly demand clear evidence that funds spent on quality programs deliver satisfactory payback. The management committee's question "What has your department done for the company lately?" is never far away; quality managers armed with the quantified, verifiable answers can sleep better at night.

The natural concern of the typical quality professional toward return on quality centers on one pivotal question: what happens if the ROQ turns out to be less than required or, worse yet, negative? Quality managers are, after all, humans and consequently subject to the normal human fears of the unknown. Not having tried to establish whether quality's benefits exceed quality's costs, many quality managers naturally distrust the consequences of such an exercise. Moreover, every quality professional knows that some of the most important benefits of quality efforts—notably greater customer satisfaction translating into higher sales and profits—cannot be measured in terms of contribution to profit. Therefore, the quality professional is inevitably haunted by suspicion that the measurable returns will sufficiently exceed costs to satisfy management. The implicit but overwhelming appeal of Demingism to quality personnel lies in their well-founded suspicion that, if required but unable to prove their financial worth, the company will probably reduce or eliminate its demand for quality programs. These suspicions have intensified in the current corporate infatuation with layoffs.

The irony of such fears is that any quality activity properly conceived, managed, and monitored will show benefits far in excess of costs. The quality program for electronics assembly companies explained in Chap. 19, for example, should deliver readily quantifiable first year benefits that exceed all the costs of the program in perpetuity. The "unquantifiable benefits" need not even be taken into account, although it will be shown in subsequent chapters that they can be closely approximated and therefore should be counted.

1.5.5 Short- vs. Long-Term Thinking

Quality personnel harbor another great fear: that financial justifications bias planning time horizons toward the very short term while real quality improvement requires longer to produce impressive results. Handy was right in his statement that short-term thinking leads to short-term lives. At the same time, he chose a common misconception among quality observers that profits are necessarily a short-term consideration. That is not the case, although it can be fairly argued that American companies have all too often maximized immediate earnings at the expense of longer-term health.[21] The goal of profit maximization established in *The Goal* is definitely intended to be for the long term with short-term considerations restricted to providing sufficient cash flow for survival into the long term.

Short-term profits can be maximized in many ways that are incompatible with both quality and long-term survival (to say nothing of long-term prosperity). Just a few common ways in which companies harm their long-term prospects for the sake of more immediate earnings include:

- Using inferior (and less expensive) parts
- Ignoring preventive maintenance of plant and equipment
- Taking quality assurance and quality control shortcuts
- Reducing warranty coverage
- Eliminating R&D activity

Short-term thinking contradicts the entire point of quality management.

1.5.6 The Goal and Quality Improvement

Rejecting the thesis that quality improvement per se is the company's goal does not negate the importance of quality improvement to the company's prosperity. However, adjustments must be made in view of why quality programs are important. Once again, Goldratt shows the way. "If the goal is to make money," Goldratt writes, "then...an action that moves us toward making money is productive. And an action that takes

[21]The short-term focus can be seen in the frenzy of production activity at the end of each fiscal year. In addition, major production peaks characterize the end of the quarter. Crunches at the end of the month are also common. Other times are only hectic.

away from making money is non-productive." Therefore, every corporate activity must be designed to maximize profits.[22]

"Every corporate activity" clearly includes quality programs. Accordingly, the new thinking must be that quality improvement is a tool that can assist the company in its journey to the goal of greater profitability. The conventional wisdom that quality is inherently good and should be pursued without concern for the costs must be discarded. Moreover, it must be clearly recognized that *higher quality may be a means to greater profits but higher quality does not guarantee higher profit.* Tools that help us reach the goal must not be confused with the goal itself.

Quality improvement is not the only tool available to the company, though in the current competitive environment it tends to be one of the more important. Among the other tools can be found delivery time, inventory control, planned maintenance, and facilities management. Beyond the shop itself lie tools such as marketing (including price, value,[23] and features), sales, and even that most loathed of all corporate activities, accounting.

Quality investments should be undertaken only when they will move the company farther along the road of greater long-term profits. If a direct link between additional spending on quality and greater profits cannot be shown, the quality activity cannot be justified. Other measures of keeping score, such as reports of greater customer satisfaction, are irrelevant unless the monetary value of that result can be determined in a reasonable fashion.

1.6 Quality Optimization

Useful though ROQ is, in our experience the financial analysis must be taken a step farther into activities that we collectively term *quality optimization.* The most important differences between ROQ and quality optimization are twofold:

1. Rather than merely collecting data and considering it in aggregate, quality optimization considers the net return obtained from each

---

[22]One qualification to the profit criterion is raised by Goldratt. The company must not allow itself to be put into a fatal cash flow squeeze even if receivables far exceed payables.

[23]Value may be defined by the classic value engineering and analysis equation: performance ÷ cost = value. Chapter 3 examines the value concept in greater detail.

incremental increase in resources invested in the program and responds according to changes at the margin.

2. The measurements of marginal return on quality are integrated with corresponding measurements for all other possible uses of the company's financial resources. Funds are directed to whatever tool provides the highest marginal rate of return until the best available return from any additional spending equals the cost to the company of acquiring the additional funds. At the point where the company can no longer raise additional funds to invest, the marginal cost of capital becomes infinite—a rate of return that no investment can provide.

Quality optimization can be seen as the litmus test that determines (1) whether to initiate a quality improvement project and (2) the point at which the project should be terminated. If the project moves the company closer to the goal of greater profits, it is good. If not, the company applies its resources to other activities that will. If no investments can be found that yield positive returns after taking account of the cost of capital, quality optimization does not allow further investment.

In the normal course of events, a project delivers greatest profits in the early stages. Returns decline as the project matures and ultimately drift into negative returns. Therefore, a project that makes financial sense in the early stages can eventually become a burden on the company's earnings. Quality optimization monitors each project and ascertains the point at which a project should be canceled or rescaled.

ROQ can prove useful as an approximate indicator of whether a quality project was properly conceived and, to a point, implemented. However, ROQ cannot help the company allocate capital in the face of too many opportunities in which return on each meets the company's minimum required return on capital. In such circumstances, quality optimization is again best suited to the task.

How could any company possibly have "too many" opportunities where returns will meet the company's requirement? The answer lies in capital constraints. After making payments on debt, dividends to shareholders, costs of existing projects, and contingency savings (to cite just four of many demands on profits), every company will have a finite amount of capital left for new projects. All opportunities, from quality improvement to new sales materials, must be evaluated according to the probable return on investment and listed in descending order of return. Keeping a running total of costs while moving down the list, the company commits to projects until the total cost commitment equals the total available capital. Other less profitable investments must be excluded, even though they meet ROQ's less stringent requirement of enhancing profits.[24]

Again, this form of project analysis shows that quality improvement is only one of many paths to greater profitability. Since the funds available to the company for all uses will be finite, expenditures will always be ranked according to:

1. The anticipated rate of return if the activity is new
2. The latest rate of return for those projects already under way

Investments are steered to those new opportunities and existing programs in descending order of return until the point is reached where the return from the next most attractive investment falls below the cost of providing funds for that investment. In a quality optimization environment, quality improvement projects compete for funds not just among themselves but also against every other investment opportunity open to the company.

Quality optimization is explained in greater detail in Chap. 18.

1.7 Quality Is Relative

One seldom appreciated aspect of quality is that it has little meaning in isolation. As will be seen in Chap. 3, quality can and must be defined in a rigorously objective fashion that allows quality characteristics to be quantified. Without reference to levels of quality at competing companies, however, our own company's level of quality tells us little of value. In particular, regarding our own quality in a vacuum gives us no insights into how our product is likely to be perceived relative to competing products by the marketplace. While we can and should measure improvements or declines in our quality, those measurements become useful only when we know the customer's options from other suppliers. The only practical way for a manufacturer to consider quality is by comparing its own level of quality to that of competitors and *potential* competitors. Whenever one speaks of a company's level of quality, therefore, the immediate question to be posed is "level of quality compared to what?"

Similarly, while a manufacturer can measure changes in its own quality, "quality improvement" matters only when changes in competitors'

[24]The marginal cost of the next dollar to be spent after all the available capital has been committed is infinite. The investment that can provide an infinite return does not exist.

quality levels are also taken into account. If a competitor's quality improves while ours stays the same, we have effectively suffered reduced quality. It follows from this realization that even the decision not to pursue higher quality will bring about changes in our quality level if competitors improve their quality. Therefore, a company's quality level must be assessed in a market context.

Unfortunately, it is not always possible to measure even our own company's changes in quality let alone changes taking place in competitors' products. The modern quality manager must be continually attempting to assess the company's quality in relative as well as absolute levels. The company's level of quality relative to real and prospective competitors figures prominently in quality optimization.

1.8 Summary

- The company can have only one goal: maximization of long-term profits.
- A subsidiary requirement must also be met: sustaining sufficient short-term cash flow to survive into the long term.
- The basic theory can be expressed as "Live long and prosper."
- Quality improvement is a tool that may help the company reach the goal; quality improvement must not be mistaken for the goal.
- Regardless of whether quality personnel approve, today's corporate management expects the quality department to quantify the net financial outcomes of all quality initiatives.
- The quality department will be stronger and more influential if it measures and reports its progress according to the same financial yardstick used by the rest of the company.
- ROQ is a useful but suboptimal way of evaluating the quality department's financial performance.

2 A Concise History of Quality Thinking

2.1 Chapter Objectives

Many of the most contentious quality issues are not new. More often than not, the issues that trouble us today have been raised previously and, in at least a few important respects, already answered. Knowing what others have done helps us avoid repeating their efforts, maximizes the probability of capitalizing on their successes, and minimizes the chances of repeating their mistakes.

In this chapter, an overview of quality history results in several significant benefits. Studying the work of earlier quality theorists and practitioners makes it easier—and, in many cases, possible—for the contemporary quality practitioner to:

- Raise questions that might be overlooked. Devising the answers is often easier than framing the questions.
- Comprehend the complexity of core quality issues. Most of today's most troublesome quality questions also plagued previous generations of quality professionals.

- Learn from past successes. If quality theorists and practitioners in the past effectively answered any of the perennial questions about quality that continue to be asked today, the trap of "reinventing the wheel" can be avoided by taking advantage of that earlier work.
- Profit from past mistakes. Recognizing where and how others failed makes it easier to avoid repeating those errors ("reinventing the Edsel").
- Better understand both the conflicts and similarities in the positions of leading quality figures. Even among the most prominent figures of quality thinking, great rifts in beliefs exist.
- Discover that seemingly objective analytical quality doctrines are actually subjective and based on anecdotal "proof."

2.2 Early American Quality Practices

In the familiar words of George Santayana, "Those who cannot remember the past are condemned to repeat it."[1] In other words, to know our journey's objective and how to get there, it helps to know where we have been. Of course, knowing where we have been does not in itself ensure that we will safely reach our goal any more than steering a boat by watching the wake guarantees a successful journey. Nonetheless, when in unfamiliar waters that have been charted previously, taking time to check the navigational chart is more prudent than chancing unpleasant encounters with hidden shoals.

The history of quality practices presented in this chapter is deliberately selective. For example, although Greek philosophers—most notably Socrates, Plato, and Aristotle—hotly debated the meaning of quality roughly 2500 years ago and reached conclusions that continue to be studied to this day, their work is not included here. The outline deals primarily with the years after World War II and concentrates on the four most influential American quality figures of this period.

Our survey begins with a speech delivered 30 years before the birth of W. Edwards Deming by the famous essayist Ralph Waldo Emerson. Despite its age, the advice could have been taken from almost any modern quality symposium. "Build a better mousetrap," Emerson said,

---

[1]Santayana, George. 1906. *Life of Reason*. Chap. 12, "Reason in Common Sense."

"and the world will beat a path to your door."[2] Today's "customer-driven quality" is really nothing more than Emerson warmed over. Moreover, Emerson was not unique in his thinking. While Emerson was speaking about quality in general, transportation mogul Collis P. Huntington using similar words in a real-world context at his Newport News Shipbuilding and Drydock Company: "We shall build good ships here; at a profit if we can, at a loss if we must, but always good ships."[3]

In the years immediately preceding World War I, quality thinking had evolved to the point where companies included quality as part of their formal statements of objective (what has more recently become known as the "mission statement"). For example, in 1911, Joseph O. Eaton, president of the Torbenson Gear and Axle Company (soon to become Eaton Corporation), defined his company's purpose as "Producing the highest quality products at costs which make them economically practical in the most competitively priced markets."[4] North of the border, a Canadian department store chain operated by another but unrelated Eaton—Timothy—promised, "Goods satisfactory or money refunded."[5]

2.3 Dr. Walter A. Shewhart

Between the two world wars, a tool destined to change all facets of industry—quality at least as much as any other function—appeared: statistics. Statistical methodology did not originate in quality management. The basic principles had been practiced by agronomists for more than a decade before the first attempts to utilize statistical process control in manufacturing. But it is worth noting that the first meaningful applications of statistical quality (and process) control for manufacturing[6] were developed, employed, and promoted by the electronics assembly industry. Specifically, many of the statistical methods later associated with Deming were devised and refined by Dr. Walter A. Shewhart, initially at Western Electric's Hawthorne telephone assembly

---

[2]Emerson is credited with including those words in an 1871 speech. The complete text actually read: "If a man can write a better book, preach a better sermon, or make a better mousetrap than his neighbor, though he build his house in the woods, the world will make a beaten path to his door."

[3]Quoted in [Dobyns and Crawford-Mason 1991].

[4]*The History of the Eaton Corporation 1911–1985,* an in-house publication of Eaton Corporation.

[5]From a single Toronto-based store opened by Timothy Eaton in 1870, the family-owned company developed into a chain stretching from Canada's Atlantic coast to the Pacific. The company continues to be owned and operated by Eaton descendants.

[6]The pioneering work in statistical design was carried out in British agriculture during the second decade of this century.

plant in Chicago and subsequently at Bell Laboratories. Accordingly, Shewhart deserves recognition as the father of modern statistical quality control (SQC) while the electronics assembly industry provided the nursery.

Shewhart[7] defined quality as "the management of variation." His definition held up so well that a leading American business publication could still write more than 60 years later: "It's not that quality is hard to define; it's simply the absence of variation."[8] In an ideal world, no variation can be measured from unit to unit of any given product. In practice, however, some variation does exist. Shewhart's management of variation should be interpreted as keeping deviation from wandering outside specific maximum tolerances; products in which the amount of variation extends beyond the upper or lower tolerance limits are defective.[9]

Shewhart divided the causes of variation into two groups: a steady component inherent in the process and an intermittent component. He attributed the inherent variation—now generally known as "random variation"—to change and undiscoverable causes while intermittent variation results from assignable causes. Shewhart considered the distinction between the two types of variation important for statistical and economic reasons. Assignable causes of variation can be discovered and controlled by plant personnel in an economical manner. Inherent (random) causes, on the other hand, are unpredictable, and the costs of preventing such variation would be uneconomical.

Because preventing intermittent variation would cost more than could be saved, Shewhart believed some level of defects to be inevitable and acceptable. Later quality professionals, especially Philip B. Crosby, strongly disagreed with that thinking.[10] However, it seems fair to judge that Shewhart spoke of the world as it is, Crosby as it should be. The roots of quality optimization fit well into Shewhart's universe.

---

[7][Shewhart 1931].

[8][*Business Week* Editors and Green 1994], p. 3. The strengths and weaknesses of this statement are examined again in Chap. 3.

[9]Many—perhaps most—products do not fit into the upper and lower tolerance limits model to which most statistical quality addresses itself. See Chap. 14 for more information about special characteristics of so-called go/no-go situations.

[10][Crosby 1984], p. 75, contains the statement: "I have listened for years," he writes, "as otherwise reasonable people explained and explained how Zero Defects was an impossible goal. Yet, in their own companies, there were departments that routinely had no defects....Check the payroll department and see how often an error pops up. Whenever a problem comes about in someone's pay, it is usually because the individual, the supervisor, or the personnel department did something....Payroll doesn't make mistakes." A more cynical observer might question whether it matters how the mistakes arise, and Shewhart would have regarded payroll errors (even if they did result from mistakes in other departments) as intermittent variation.

Shewhart's influence on quality procedures included application of scientific rigor. Noting that scientific methodology consisted of three stages—frame the hypothesis, run the experiment, and test the hypothesis against the experiment's results—Shewhart devised a three-stage procedure for quality management. The three stages (which make up a "cycle") consist of:

1. Specify what the required outcome must be.
2. Carry out production.
3. Inspect the product to ensure that actual outcome meets the requirements.

The emphasis on inspection is worth noting. In modern terms, Shewhart was definitely a proponent of quality control rather than quality assurance.

While the actual Shewhart three-stage control protocol is little known today, a direct descendant of that cycle is almost universally known by quality professionals. During his lectures to the Japanese, Deming taught his own version of the Shewhart cycle. The Deming cycle has four stages rather than three—plan, do, check, and act—from which derives the name PDCA cycle.

In the final analysis, however, Shewhart's most durable contribution to production management is undoubtedly the process control chart. More than 70 years after Shewhart first demonstrated the technique at Western Electric, the graphing of production characteristics against upper and lower acceptance limits remains a familiar sight in the world's factories. Control charts are considered in Chap. 14.

2.4 Four Quality Mentors

The American Society for Quality Control has a membership roster numbering into the tens of thousands. Each of these members can be considered a quality professional. However, the concept of quality in the last half century has been dominated by those four men whom we term "mentors." Our terminology differs from most other writers, who often refer to Crosby, Deming, Feigenbaum, and Juran as "gurus." Although the guru designation is doubtless intended to show respect as in the context of trusted adviser, the word shares its linguistic roots with less flattering terms: "grave," "grief," "aggravate," "brute," and even "blitzkrieg." Consequently, mentor[11] seems to be a more suitable title than guru.

While the differences among the four mentors are more striking than their similarities, the four did share three notable characteristics:

1. All were well into late middle age or even elderly before being "discovered" by the general business community.
2. All achieved something never previously experienced in the quality business: fame. In the space of just a few years, all four men earned unprecedentedly large incomes. In particular, Crosby—the youngest of the four and the only one not to have a doctorate—became enormously wealthy.
3. None of them apparently intended to work in quality. Crosby studied podiatry and failed to recognize until after completing his degree that treating foot problems was not the way he wanted to spend the rest of his life. Deming studied engineering as an undergraduate, then turned to physics for his doctorate. Juran is an engineer. Feigenbaum began as an apprentice toolmaker.

2.4.1 Philip B. Crosby

Though an environment making revolution possible may evolve over many years, social upheavals ultimately reach critical mass through a single pivotal event. The spark igniting America's quality revolution was Crosby's 1979 book *Quality Is Free.*

As we have seen, a well-established quality profession existed throughout most of this century. From around 1945 until 1979, however, interest in quality matters outside the quality department was at best

[11]Softkey International's *American Heritage Dictionary* version 3.5 (1994) explains, "The word *mentor* is an example of the way in which the great works of literature live on without our knowing it. The word has recently gained currency in the professional world, where it is thought to be a good idea to have a mentor, a wise and trusted counselor, guiding one's career, preferably in the upper reaches of the organization. We owe this word to the more heroic age of Homer, in whose *Odyssey* Mentor is the trusted friend of Odysseus left in charge of the household during Odysseus's absence. More important for our usage of the word *mentor,* Athena disguised as Mentor guides Odysseus's son Telemachus in his search for his father. Fénelon in his romance *Télémaque* (1699) emphasized Mentor as a character, and so it was that *mentor,* going back through Latin to a Greek name, became a common noun meaning 'wise counselor,' first recorded in 1750. *Mentor* is an appropriate name for such a person because it probably meant 'adviser' in Greek and comes from the Indo-European root *men-,* meaning 'to think.'"

negligible and more usually either nonexistent or hostile. Corporate America saw quality as a cost of doing business, a necessary evil tolerated but rarely embraced. Crosby convinced America's most powerful companies that quality was not only free but also warranting unequivocal personal endorsement and involvement by the company's most senior executives.

Prior to Crosby, quality professionals argued that quality efforts should be supported essentially because quality was synonymous with good.[12] Crosby knew that nonquality managers and executives required more tangible reasons before they would become supporters. So he made quality synonymous with the most basic business tangible of all—profit.

"Quality is not only free, it is an honest-to-everything profit maker,"[13] Crosby said. "Every penny you don't spend on doing things wrong, over, or instead becomes half a penny right on the bottom line."[14]

Where science had failed in the past to elevate the importance of quality activities, Crosby's unabashed salesmanship succeeded. Crosby's systems contain just as much evangelical fervor but no more science than Dale Carnegie's motivational approach to self-improvement.[15]

2.4.1.1 Zero Defects. In 1961, while employed as a quality control engineer at the Martin (subsequently Martin-Marietta and now Lockheed-Martin) Company's missile production plant in Orlando, Fla., Crosby was reluctantly forced to head a new company project to determine whether intensified inspection would result in the ability to ship missiles completely free of defects.[16] Like all such bureaucratic undertakings, it required an identifying name; the company settled on "Z-D"—for "Zero Defects"—which they marketed with the same quality-as-sales-tool zeal that Motorola has shown more recently with Six Sigma.™

Z-D was successful within the restrictive definition of "success" that the company had laid out for itself—the intense postproduction inspec-

---

[12]The most notable published exception to this philosophy was Juran, who by the mid-1950s was already integrating quality into the corporate financial structure. Deming, of course, had been making much the same case but was invisible and unheard in America.

[13][Crosby 1979], p. 1.

[14]The statement is true only in the presence of a 50 percent corporate tax rate—and the company is already paying taxes on every extra dollar earned.

[15]Of his many books, Carnegie is best known for *How to Win Friends and Influence People.*

[16][Juran, Gryna, and Bingham 1974], pp. 18–50. The true nature of the Martin program depends on the source. Juran says Martin "exhorted its inspectors to find all discrepancies so that perfect missiles could be delivered." *Time* magazine, however, described the approach as prevention *before* inspection. Since very little of consequence in American quality programs of that vintage occurred without careful scrutiny by Juran, we would be more inclined to trust Juran's account than the impressions of a general-purpose newsmagazine reporter.

tion pared the number of defects shipped to a small fraction of what had previously been normal and accepted. On the other hand, there is no evidence that the number of first-pass defects fell, only that greater inspection reduced the number of defects reaching the customer. In quality assurance terms, therefore, Z-D provided no gain whatever; the entire exercise was a definitive demonstration of quality control's need for massive end-of-line screening.

For a brief time, Z-D attracted considerable attention from government, other industries, and the press. In the summer of 1964, the U.S. Army Missile Command presented a series of workshops and seminars about how companies could achieve Z-D. Not to be outdone, the Department of Defense held seminars on the same topic later in the year.[17] Coinciding with the DoD seminars, *Time* magazine devoted a page to the subject.[18] At that point, the *Time* article claimed, 200 "major" U.S. companies had "adopted" Z-D. As so often happens with management fads, the Zero Defect wave had crested by the time it was discovered by the press, and early users of the concept were already defecting.

More than 30 years later, that *Time* article provides fascinating insights into the development of Crosby the quality evangelist. "A stern word from the boss started the whole Z.D. idea three years ago," the article says. "At Martin Marietta's plant in Orlando, Fla., a quality control engineer named Philip Crosby had succeeded in cutting defects on Pershing missiles to half the acceptable level—but his boss complained that this was still too high. Incensed at first, Crosby soon began to agree: 'If management tolerates a low standard, people work to that standard. Well, why not a no defects job?' He persuaded workers in his department to sign no-defects pledges, soon surprising the Army by delivering a Pershing missile two weeks ahead of schedule with no detectable defects among its 25,000 parts."

Although he had started out being the instrument of an unhappy (and, by the standards of the time, visionary) boss, Crosby leveraged himself into a position where, at least in quality circles, his name

---

[17][Juran, Gryna, and Bingham 1974], pp. 18–50.

[18]"Let's See Z.D.," *Time,* Nov. 6, 1964, pp. 93–94. The article notes that the Zero Defect "system…reaches past the inspector and goes back to the man on the line, emphasizing prevention instead of cure. This uniquely simple system is part common sense, part snow job, and wholly successful." Just another of the endless "successful" quality programs that over the decades have turned out to be more fruitful as public relations exercises than as methods of improving operations. The passage is notable for two other points. First, the "emphasis on prevention" really means more rigorous inspection, commonly known today as quality control. Second, the reference to "common sense" helps explain why the "wholly successful" program became extinct; common sense is to rigorous logic what a comic book is to great literature.

became synonymous with zero defects. In a 1964 article,[19] themes appear which form the cornerstone of Crosby's ideology to this day. "We are going to talk about preventing errors," he wrote. "*Preventing,* not detecting. Detecting costs a lot of money and only saves you future grief. It can't do a thing about the problems with which you have already been blessed." No rational person, then or now, would take issue with the desirability of preventing defects rather than trying to find them after the fact. But his words clash with the actions of Z-D. Crosby relied on intensive postproduction inspection to achieve the defect-free condition of Martin's shipped product. In text, Crosby was arguing for what is now thought of as quality assurance, but in practice his operation still relied on traditional quality control.

Crosby's solution to the creation of defects depended on attitude adjustments. "Mistakes are caused by two things," he argued in the same article,[20] "lack of knowledge or lack of attention. You can measure lack of knowledge and fix it by tried and true means. But lack of attention is an attitude problem and must be repaired by the person himself. He must develop a conscious desire to perform the job right the first time." Norman Vincent Peale, famous for *The Power of Positive Thinking,* would have agreed.

Even if he had not really managed to move his own department from the finding / fixing mentality (a Herculean labor for any employee of the defense industry at any time and probably impossible three decades ago), Crosby deserves credit for promoting the idea of defect prevention rather than detection.

Crosby's zero defects article culminated in a caustic rejoinder by J. M. Juran titled "Quality Problems, Remedies and Nostrums."[21] The abstract accompanying the main paper sums up Juran's feelings about zero defect programs rather well. "In all essential respects, the effectiveness of the ZD movement is grossly exaggerated; the unsuccessful programs have been more numerous than the successful; motivational programs have a narrow, not a broad range of application; the premises underlying the ZD programs are suspect; the main purpose behind the

---

[19]Crosby, Philip B. "Z Is for Zero Defects," *Industrial Quality Control,* October 1964, pp. 182–185, reprinted from the Apr. 13, 1964, issue of *American Machinist/Metalworking Manufacturing.*

[20]"Z Is for Zero Defects," p. 183.

[21]Juran, J. M. "Quality Problems, Remedies and Nostrums." *Industrial Quality Control,* June 1966, pp. 647–653.

movement has probably been customer relations, not quality improvement."[22]

2.4.1.2 Conformance to Requirements. The problem with "zero defects" is that it requires an explicit definition of exactly what constitutes a defect. Crosby concluded that a defect is any aspect of the product that fails to conform to requirements. Accordingly, if defects equal failure of the product to conform to requirements, "...quality means conformance. Nonquality is nonconformance."[23]

Crosby is adamant that conformance to requirements is not synonymous with the "fitness for use" notion of quality generally associated with Juran. When someone wants to ship a product that is "good enough" but does not meet the company's requirements, Crosby argues that the individual will contend that the product is fit for use.[24] In reality, Crosby and Juran were closer than Crosby realized; he had simply misinterpreted the meaning of "fitness for use" intended by Juran. The nature of that misunderstanding is addressed in Sec. 2.4.4.2.

Crosby's position against shipping product that fails to meet the requirements (however those requirements were reached) cannot be attacked. If the product is fit to use even though it fails to meet certain specifications, the specifications are clearly meaningless and should be scrapped or amended. Defense industry practices show that a product can be fit for use but not meet requirements. Substantial numbers (if not the majority) of MIL Specs no longer make any sense but must be met if the company hopes to keep the contract.[25] The product may not meet requirements but may actually be more reliable and fit for use than will be the case after reworking to meet the requirements. The solution to this dilemma lies not in waivers, however; instead, the requirements must be redrawn so they no longer call for undesirable production and evaluation procedures.

Unfortunately, the seemingly simple concept of "conformance to requirements" also falls apart under closer scrutiny. How, for example,

---

[22]The same statement, without altering a single word, could be applied to the "pseudoquality" programs examined in Chap. 16. Despite Juran's solidly based objections, however, Crosby rode Zero Defects to great personal wealth. Which, come to think of it, is exactly what the pseudoquality providers are doing today.

[23][Crosby 1979], p. 39.

[24][Crosby 1989], pp. 74–75.

[25]The tragic heritage left to the electronics assembly industry by NASA standards setters and later adopted by the Department of Defense is examined more thoroughly in Chap. 9.

should requirements be set and who does the setting? Without specific guidelines for establishing those requirements, designing a program to eliminate those nonconformances makes no sense. Juran's thoughts on these questions—summarized in Sec. 2.4.4.2—are important.

On the fundamental question of who determines requirements, Crosby has never provided an answer that we could accept. The issue is completely ignored in *Quality Is Free*. Five years later, in his next book,[26] Crosby announced that the requirements can only be established by management. "Management really has three basic tasks to perform: (1) establish the requirements that employees are to meet, (2) supply the wherewithal that the employees need in order to meet those requirements, and (3) spend all its time encouraging and helping the employees to meet those requirements." But who precisely are "management" and how can they know what requirements are needed and/or desirable? Crosby ignored those meaningful issues.

Six years later, Crosby provided an alternate solution. "All requirements come from the customer in one form or another, because with no customers there is no business," he decided.[27] Herein is found the core of the "quality is what the customer says it is" school of thought so popular in the 1990s. As Chap. 3 shows, however, customer-defined quality is generally unworkable. Crosby himself inadvertently provides some reasons why customer-defined requirements are unlikely to work well in practice: "[T]here are many customers besides the actual user. There is the Internal Revenue Service, which has its requirements; the SEC; the bank; the union; the landlord; and so on. All of these want conformance to requirements from us. They do not appreciate deviations." Other books from Crosby suffer the same failing of defining quality in ways that offer no foundation on which to construct a solid program of corrective action.

2.4.1.3 Cost of Quality. While Crosby may be weak on concrete solutions, he is almost certainly the master of selling commitment to quality. Not surprisingly, therefore, he found the sales abilities of quality departments sadly underdeveloped. "Quality is free," he wrote,[28] "but no one is ever going to know it, if there isn't some sort of agreed-on system of measurement...in spite of the fact that such a method was developed by

[26][Crosby 1984], p. 59.

[27][Crosby 1989], p. 75.

[28][Crosby 1979], pp. 102–103.

General Electric in the 1950s as a tool for determining the need for corrective action on a specific product line.

"The quality profession...clings to the very management concepts that allow them to be inadequate, so cost-of-quality measurement was never really implemented except by a radical here and there....

"All you really need is enough information to show your management that reducing the cost of quality is in fact an opportunity to increase profits without raising sales, buying new equipment or hiring new people....It is normal to obtain only one-third of the real cost the first time you try it."

To increase the proportion of real cost obtained the first time measurements are taken, Crosby broke down cost of quality into three broad categories: prevention costs, appraisal costs, and failure costs. He then identified several subcategories for each. His list of prevention costs, for example, includes:

- Design reviews
- Product qualifications
- Drawing checking
- Engineering quality orientation
- Make Certain program[29]
- Supplier evaluations
- Supplier quality seminars
- Specification review
- Process capability studies
- Tool control
- Operation training
- Quality orientation
- Acceptance planning
- Zero Defects program
- Quality audits
- Preventive maintenance[30]

---

[29]"Make Certain" was the name Crosby applied to his brand of employee involvement activities.

[30][Crosby 1979], p. 105.

For appraisal costs, Crosby specifies:

- Prototype inspection and test
- Production specification and conformance analysis
- Supplier surveillance
- Receiving inspection and test
- Product acceptance
- Process control acceptance
- Packaging inspection
- Status measurement and reporting[31]

Finally Crosby's failure costs (a figure representing the costs of "lost customer credibility") include:

- Consumer affairs
- Redesign
- Engineering change order
- Purchasing change order
- Corrective action costs
- Rework
- Scrap
- Warranty
- Service after service
- Product liability[32]

Obviously several of the costs in each category are specific to Crosby's practices; the reader can add or subtract costs according to conditions within the company concerned.

The measurement task may seem daunting, but Crosby—quite rightly, in our view—is adamant that the work must be done and the results brought to the attention of all personnel, not just those employed by the quality department. "Measurements should be established both for measuring the overall cost of quality and for determining the current status of specific product or procedure compliance. These measure-

[31]pp. 105–106.

[32]p. 106.

ments should be displayed for all to see, for they provide visible proof of improvement and recognition of achievement. Measurement is very important. People like to *see* results."[33]

The most important contributions by Crosby, therefore, have little to do with concrete tools; he is, above all, a theorist rather than an implementer. All the same, the theories he advanced and the missionary work he carried out made quality improvement appealing to the highest-ranking officers of most meaningful American companies. Crosby may have lulled many naive readers into a false sense of how to make quality pay, but he also brought quality functions into the mainstream of corporate life.

2.4.2 W. Edwards Deming

Depending on one's point of view, Deming is either a tragic figure of American quality or the patron saint. Totally a product of the twentieth century, he was born in 1900 but went almost unnoticed in the United States until just a fraction of his life remained. He was 81 before many people in his homeland cared about his opinions. Yet, in the space of just a dozen years before his death in 1993, he converted hundreds of thousands of Americans to an ideology that probably merits the title "Demingism."

In more ways than one, Deming was a product of Shewhart. Deming's only personal experience working in a factory consisted of part-time work during the mid-1920s under Shewhart at Western Electric's Hawthorne plant.[34] He could not have been in a better place at a better time, however. Shewhart was in the middle of his most exciting period, defining and testing the principles that would form the basis of his pioneering book on quality management.[35] Through his work at Western Electric, Deming was exposed to Shewhart's three-stage cycle, powerful tools for teaching statistical methodology (including the famous red and white beads experiment) and the concept of quality being expressed as a function of variation. His time at Western Electric also spawned an

[33] p. 16.

[34] Much of Deming's disdain for corporate management stems from his work experiences at Western Electric. A massive plant employing more than 5000 workers, the Hawthorne facility operated according to the "scientific management" techniques of F. W. Taylor, which Deming believed stressed output over quality. The modern quality department began as a centralized inspection department in companies following Taylor's guidelines. Deming found the rigidity of the plant structure too confining and the enormous amount of inspection repugnant. He never worked in another manufacturing plant.

[35] [Shewhart 1931].

often heated rivalry with another future giant of quality management named Joseph Juran, an employee at the Hawthorne plant from 1924 although not working directly for Shewhart.

Widely believed to have received his doctorate in statistics, Deming actually holds no formal degrees in the field for a very good reason—the science of statistics did not emerge until after Deming finished school. Like Shewhart, Deming's doctorate was in physics. Also like Shewhart, Deming mastered statistics in the best possible way: from hands-on participation. Between 1926 and 1940, whenever a significant advance occurred in America's statistical knowledge, Deming generally played some sort of role. Working under Shewhart, Deming was more student than teacher; but a decade later, having become widely regarded as the country's leading authority on sampling techniques, it was Deming who edited and wrote the foreword to Shewhart's classic *Statistical Sampling from the Viewpoint of Quality Control.* Sampling methods for the 1940 census—generally considered the most important statistical project to that time—were prepared and supervised by Deming.

Considering his statistical management credentials, therefore, it is unsettling to realize that Deming's primary contributions to the understanding of quality are almost exclusively philosophical, based on personal prejudice and anecdotal evidence. Many commandments of Demingism do not withstand rigorous objective analysis, and it is quite possible that Deming would have arrived at very different conclusions if he had subjected his rhetoric to scientific scrutiny. In particular, his insistence on continual improvement would have benefited from more thorough research. (Deming insisted that his method requires *continual*—not *continuous*—improvement.[36] His reasoning was that continuous, the word used by most Americans, means "unbroken, never-ending" while continual means "occurring on a frequent or regular basis." The importance placed by Deming on this semantic hair splitting has not been fully appreciated by some of his admirers.)[37]

2.4.2.1 The 14 Points for Management. The heart of Demingism consists of his well-known 14 Points for Management. In Deming's words, "Adoption and action on the 14 points are a [sic] signal that the manage-

[36][Dobyns and Crawford-Mason 1994], p. xxii.

[37]In the spring of 1994, a prospective client informed a consultant who needed to visit the plant that he was worried the outsider could destroy the plant's quality fabric. His main fear: "We have built a special culture here and I don't want you to bring in a lot of new words to confuse the people." He then proceeded to describe the company's commitment to "continuous improvement" at the same time as he extolled the virtues of Deming.

ment intend to stay in business and aim to protect investors and jobs."[38] He lists the 14 points as:

1. Create constancy of purpose toward improvement of product and service, with the aim to become competitive and to stay in business, and to provide jobs.
2. Adopt the new philosophy. We are in a new economic age. Western management must awaken to the challenge, must learn their responsibilities, and must take on leadership for change.
3. Cease dependence on inspection to achieve quality. Eliminate the need for inspection by building quality into the product in the first place.
4. End the practice of awarding business on the basis of price tag. Instead, minimize total cost. Move toward a single supplier for any one item, on a long-term relationship of loyalty and trust.
5. Improve constantly and forever the system of production and service, to improve quality and productivity, and thus constantly decrease costs.
6. Institute training on the job.
7. Institute leadership. The aim of supervision should be to help people and machines and gadgets do a better job. Supervision of management is in need of overhaul as well as supervision of production workers.
8. Drive out fear, so that everyone may work effectively for the company.
9. Break down barriers between departments. People in research, design, sales, and production must work as a team, to foresee problems of production and in use that may be encountered with the product or service.
10. Eliminate slogans, exhortations, and targets for the work force asking for zero defects and new levels of productivity. Such exhortations only create adversarial relationships, as the bulk of the causes of low quality and low productivity belong to the system and thus lie beyond the power of the workforce.

11*a*. Eliminate work standards (quotas) on the factory floor. Substitute leadership.

b. Eliminate management by objective. Eliminate management by numbers, numerical goals. Substitute leadership.

12*a*. Remove barriers that rob the hourly worker of his right to pride of workmanship. The responsibility of supervisors must be changed from sheer numbers to quality.

---

[38][Deming 1986], p. 23.

b. Remove barriers that rob people in management and in engineering of their right to pride of workmanship. This means, inter alia, abolishment of the annual or merit rating and of management by objective.
13. Institute a vigorous program of education and self-improvement.
14. Put everyone in the company to work to accomplish the transformation. The transformation is everybody's job.[39]

Over the years, we have observed the results of employing Deming's 14 points in scores of electronics assembly companies. The best results have always been obtained by those companies that did not blindly follow the Deming template but discarded the rules that struck them as unworkable. In our experience, companies adhering to Deming's Point 11*b*—eliminating management by numbers—generally wished they had been more cautious in adopting Deming's principles.

2.4.2.2 The Seven Deadly Diseases. The flip side of the 14 Points for Management consists of seven "deadly diseases (that) afflict most companies in the Western world. An esteemed economist (Carolyn A. Emigh) remarked that cure of the Deadly Diseases will require total reconstruction of Western management."[40] Deming's Seven Deadly Diseases are:

1. Lack of constancy of purpose to plan product and service that will have a market and keep the company in business, and provide jobs.
2. Emphasis on short-term profits: short-term thinking (just the opposite from constancy of purpose to stay in business), fed by fear of unfriendly takeover, and by push from bankers and owners for dividends.
3. Evaluation of performance, merit rating, or annual review.
4. Mobility of management; job hopping.
5. Management by use only of visible figures. With little or no consideration of figures that are unknown or unknowable.
6. Excessive medical costs.
7. Excessive costs of liability, swelled by lawyers that work on contingency fees.[41]

Several of the "Deadly Diseases" are nothing more than failure to follow one or more of Deming's 14 Points for Management; accordingly,

---

[39]pp. 23–24.

[40]p. 97.

[41]pp. 97–98.

acceptance of the disease requires belief in the validity of the corresponding Point. Whether adopting the entire package of Points serves the company's best interests is by no means certain, however. For example, some of Deming's Points conflict with the position adopted in the previous chapter of this book that the company's purpose is not merely to stay in business (or, for that matter, provide jobs) but to maximize long-term profits while sustaining sufficient short-term cash flow for survival.

Additionally, despite Deming's opposition, some form of ongoing job evaluation is generally desirable, at least at the management level; failure to keep the employee advised of perceived performance can be disastrous, especially if the employee is performing well but feels insecure.

Further, management turnover is not always harmful. When new management personnel introduce positive new ideas or end complacency that has been holding the company back, the change in management is desirable. Deming's warnings are most valid when new management discards existing programs that are working well in favor of less effective programs with which the newcomer is familiar or replaces vigorous career company personnel with its cronies. The dilemma facing the company when weighing the merits and demerits of restructuring management ranks is that the consequences will only be apparent after the fact; in some cases, the outcomes may never be clear. If no changes are made and the company continues to prosper, can anyone know definitively that greater progress might not have taken place with new management? Or, if management changes do occur and the company languishes, how much of the decline should be attributed to new leadership and how much is attributable to mistakes in judgment by previous management?

2.4.2.3 The Points and Deadly Diseases Revisited. All of Deming's 14 Points and 7 Deadly Diseases can be arranged into just six categories in total:[42]

1. *The purpose and mission of business organizations.* Deming argued that American companies suffer because they lack constancy of purpose (Point 1 and Deadly Disease 1) and emphasize short-term profits (Deadly Disease 2). These ideas did not originate with Deming, how-

[42]Duncan, Jack W., and Van Matra, Joseph G. "The Gospel According to Deming: Is It Really New?" *Business Horizons,* July–August 1990, pp. 3–9.

ever; the same messages can be found in many business texts[43] published and widely taught decades before Deming produced his theory of management.

2. *Philosophy instead of technique.* Many of Deming's points and Deadly Diseases concern philosophy. Point 2 says to adopt the new philosophy; Point 3 says to cease dependence on inspection to achieve quality; Point 5 says to improve constantly and forever the system of production and service; and Point 14 says to put everyone in the company to work to accomplish the transformation. Ironically, the same sentiments can be found in Frederick W. Taylor's 1947 book on scientific management[44] that Deming's followers believe to be responsible for the worst of American management practices.

3. *Instruction and training.* Point 6 advocates on-the-job training while Point 13 calls for zealous pursuit of education and self-improvement. Again, overtones of Taylor are at work here. Taylor contended that scientific management includes training and instructing workers in how to do the job properly.

4. *Cooperation-competition-conflict.* Point 4 warns against buying strictly according to price. Point 9 insists on breaking down barriers between departments. The latter point appeared in a well-known 1931 book[45] that admonished management to integrate and coordinate the disparate departments and personnel found within the typical organization.

5. *Manager-worker relations.* Point 7 (institute leadership), Point 8 (drive out fear), Point 12 (remove barriers that rob the hourly worker of his right to pride of workmanship), and Deadly Disease 4 (mobility of management) all logically fall into this category. All these ideas were well known and respected long before Deming's first visit to Japan.

6. *Quantitative goal setting.* Deming refers through several points and Deadly Diseases to reasons why quantitative goals should be eliminated from the company. Deming opposed slogans, exhortations, and targets for the workforce (Point 10), work standards and management by objectives (Point 11), evaluations of employee performance, merit ratings, and annual reviews (Deadly Disease 3), and management according to visible figures only (Deadly Disease 5). This category rep-

---

[43]See, for example, Barnard, Chester I. 1938. *The Functions of the Executive.* Cambridge, Mass.: Harvard University Press, and Drucker, Peter F. 1954. *The Practice of Management.* New York: Harper.

[44][Taylor 1947].

[45]Mooney, J. D., and Reilley, A. C. 1931. *Onward Industry.* New York: Harper.

resents Deming's singular—and definitive—break with Taylor and standard management practices. In this one area, Demingism breaks new ground. However, it is also the least defensible of his positions and one which Japanese companies ignored.[46] In much the same way that a journey requires a destination, management requires specific numerical targets for the company as a whole, departments, and employees.

Perhaps the most important question raised by Deming's 14 Points and Seven Deadly Diseases is whether they relate to quality at all. Is it possible that they actually focus on productivity and general management? The difference between quality and productivity is real and significant. As Chap. 3 will emphasize, too many companies freely label every activity a "quality improvement program" when the objectives really concern issues totally divorced from quality. By blurring the distinction between quality improvement actions and productivity issues, Deming created serious confusion that gives rise to the rampant misuse of the "quality" label.

2.4.2.4 The Shewhart Factor. Deming was not solely a management philosopher. Indeed, it was not until late in his life that discourses on management style replaced statistical process control as his main theme. His seminars relied heavily on the red and white beads demonstration of random probability[47] right to the end. However, Deming's statistical contributions never came close to his influence on attitudes. Indeed, in the field of statistics, he was little more than a recycler of Shewhart methodologies (the beads experiment included), perhaps because he believed there was little room for improvement on the principles of the man he termed "the master."[48]

Deming is often credited with developing tools that actually originated with Shewhart. The Japanese, for example, renamed the Shewhart cycle the Deming cycle, and that was the name under which it became known to most Americans. Although Deming did improve on the original concepts—Shewhart's three-step cycle, for example, became four steps in the Deming version—the differences are primarily cosmetic.[49]

---

[46][Fucini and Fucini 1990] describes the rigorous application of goals and standards at Mazda's Flat Rock, Mich., plant.

[47][Gabor 1990], pp. 60–63.

[48][Deming 1986], p. 168.

[49]A footnote on p. 88 of [Deming 1986] reads: "I [Deming] called it in Japan in 1950 and onward the Shewhart cycle. It went into immediate use in Japan under the name of the Deming cycle, and so it has been called there ever since."

Shewhart's cycle consisted of:

1. Specify what the required outcome must be.
2. Carry out production.
3. Inspect the product to ensure that actual outcome meets the requirements.

In the Deming version, the steps become:

1. Plan. (Plan a change or test. Decide how to use the observations.)
2. Do. (Carry out the change or test decided upon, preferably on a small scale.)
3. Check. (Observe the effects of the change or test.)
4. Act. (Make the necessary changes identified by the results of step 3.)

Terminology aside, there are no real differences between the first three steps in the Shewhart and Deming cycles. Deming, however, took the concept one crucial step farther by explicitly stating that corrective action is required after the outcome is determined. The unstated imperative of both cycles is that the process of planning, producing, verifying, and changing is continual. That is, after changes are made, the quality engineer returns to step 1 and repeats the operations in sequence.[50]

Shewhart's emphasis on quality as control of variation also figures prominently in Deming's work. Deming's terminology—common causes and special causes—differs from Shewhart's inherent and assignable causes, but as Deming acknowledges,[51] in other respects the concepts are the same.

2.4.3 A. V. Feigenbaum

During World War II, Feigenbaum, a former apprentice toolmaker at GE, was assigned to the company's development program for jet airplane engines. With the jet propulsion engine still in its infancy, failures were frequent and dramatic: failure of a jet engine generally meant an explosion. Using statistical tools, Feigenbaum was able to determine the causes of failure. Feigenbaum would later say, "I realized that here was a body of knowledge that needed to be developed. It was as important as electronics."[52] The result was Total Quality Control, defined by

[50] [Deming 1986], p. 88.

[51] p. 310.

[52] [Dobyns and Crawford-Mason 1991], p. 55.

Feigenbaum as "An effective system for integrating the quality-development, quality-maintenance, and quality-improvement efforts of the various groups in an organization so as to enable production and service at the most economical levels which allow for full customer satisfaction."[53] In other words, Feigenbaum's Total Quality Control is a systems approach to quality improvement.

But what is the quality in Total Quality Control? In Feigenbaum's words, "In the phrase 'quality control,' the word *quality* does not have any popular meaning of 'best' in any absolute sense. It means `best for certain customer conditions.' These conditions are (i) the actual use and (ii) the selling price of the product. Product quality cannot be thought of apart from product cost."

By the third edition in 1983, although Feigenbaum continued to insist that the "quality" in quality control "does not have the popular meaning of 'best' in any abstract sense,"[54] other important determinants had been added. "Quality is a customer determination," Feigenbaum stated, "not a marketing determination or a general management determination. It is based upon the customer's actual experience with the product or service, measured against his or her *requirements*—stated or unstated, conscious or merely sensed, technically operational or entirely subjective—and always representing a moving target in a competitive marketplace."[55]

"Product and service quality can be defined as: The total composite product and service characteristics of marketing, engineering, manufacture, and maintenance through which the product and service in use will meet the expectations of the customer.

"The purpose of most quality measurements is to determine and evaluate the *degree* or *level* to which the product or service approaches this total composite.

"Some other terms, such as *reliability, serviceability,* and *maintainability,* have sometimes been used[56] as definitions for product quality. These terms are of course individual *characteristics* which make up the composite of product and service quality.

"It is important to recognize this fact because the key requirement for establishing what is to be the 'quality' of a given product requires the economic balancing-off of these various individual quality characteristics....[In addition to the previous three characteristics] the product must have appearance suitable to customer requirements, so it must

---

[53][Feigenbaum 1961], p. 1.

[54][Feigenbaum 1983, 1991], p. 9.

[55]p. 7.

[56][Juran and Gryna 1988], pp. 2.9–2.11.

have *attractability.* When all the other product characteristics are balanced in, the 'right' quality becomes that composite which provides the intended functions with the greatest overall economy, considering among other things product and service obsolescence—and it is the *total customer-satisfaction*-oriented concept of 'quality' that must be controlled."[57]

Feigenbaum believed that his "right" quality combination changes with time. "Moreover, this balance can change as the product or service itself changes. For example, each of the four stages of the maturity cycle through which many products pass demands a somewhat different quality balance." Those stages consist of:

1. The *innovation* phase in which the product is so novel that the characteristics alone determine the sale.
2. The *conspicuous consumption* phase when the core features are augmented by enhancements to appearance and "attractability."
3. By the *widespread use* phase, the product has become an integral part of everyday life and purchase decisions are based on consistent product performance and serviceability.
4. Finally, the product enters the *commodity* stage where the only distinguishing features between competing brands are reliability and economy.

Thus Feigenbaum eliminated some of the subjectivity that undermines the works of both Crosby and Deming. Feigenbaum's reasoning is employed extensively in the development of *quality optimization* principles in this book.

Feigenbaum's Total Quality Control is not accepted in Japan as being the same as Japanese TQC, although Ishikawa[58] acknowledges that "the concept of 'total quality control' was originated by" Feigenbaum. The Japanese objections include an erroneous belief that Feigenbaum felt quality should be administered by a professional quality department. Or, in Ishikawa's words, "Fearing that quality which is everybody's job in a business can become nobody's job, Feigenbaum suggested that TQC be buttressed and serviced by a well-organized management function whose only area of specialization is product quality and whose only area of operation is in the quality control jobs. His Western-style professionalism led him to advocate TQC conducted essentially by QC specialists.

---

[57][Feigenbaum 1983, 1991], p. 7.

[58][Ishikawa 1985], p. 90.

"The Japanese approach has differed from Dr. Feigenbaum's approach. Since 1949 [the Japanese] have insisted on having all divisions and all employees become involved in studying and promoting QC."[59] To avoid confusion with Feigenbaum's approach, Ishikawa and many other Japanese quality professionals call their technique "company-wide quality control."

The only problem with the Japanese position is that Feigenbaum was an avid advocate of involving *all* company personnel in the quality function. In Feigenbaum's words, "One essential contribution of total-quality programs today is the establishment of customer-oriented quality disciplines in the marketing and engineering functions as well as in production. Thus, every employee of an organization, from top management to the production line worker, will be personally involved in quality control."[60]

More than any other Western quality professional, Feigenbaum defined the Japanese approach to quality. For his contributions, he was wrongly criticized and unrewarded.

2.4.4 J. M. Juran

The dust jacket of [Juran 1964] provides the following brief biography. "J. M. Juran has pursued a varied career in management—as industrial executive, government administrator, university professor, impartial labor arbitrator, and management consultant. A holder of degrees in engineering and law, Dr. Juran is the author of seven books, several filmed lectures and over a hundred published papers on various management subjects. His lecture schedule has taken him literally around the world, as well as to the advanced management courses of many American companies and universities.

"Currently a free lance author and lecturer in management, Dr. Juran also serves a number of industrial companies as a member of the Board of Directors or as a consultant." Remarkably, at the time that brief biography was written, Juran still had three decades of productive work—undoubtedly the most important years of his career—ahead of him.

Contemporary American views of quality history have not been kind to Juran. For the first three and a half decades after World War II, Juran was the unquestioned American quality star. He wrote prolifically and with a grace found all too rarely in technical literature. He devised or

---

[59] pp. 90–91.

[60] [Feigenbaum 1991], p. 13.

refined most of the techniques now associated with the best of what the quality profession ever offered. His work in Japan is widely regarded there as more important to the country's development than Deming's lectures.[61] His *Quality Control Handbook* remains the most important work in the quality library after 45 years and three revisions. But in the last 15 years, Deming abruptly eclipsed Juran in the public eye.

While Deming—so closely associated with Shewhart's statistical process control achievements—worked only two summers at the Western Electric Hawthorne Plant, Juran had been a full-time employee there for 2 years before Deming's arrival and remained with the company well after Deming's departure. Juran too admired Shewhart, though he worked in a different department, and became a strong advocate of statistical methodology. In all respects, including statistics, however, he tended to take a middle road, recognizing that no one approach or set of tools could work well under all conditions. For example, he wrote of statistical quality control people: "They literally plastered the factory walls with Shewhart charts while the company's main quality problems—unrealistic expectations, vendor relations, clear definition of responsibility, unwise economics, vague standards, etc.—went on and on....There is nothing ill-intentioned or evil about it. It is just in the nature of things. Tool-oriented people approach problems with so heavy a bias that they should not be given the sole responsibility of choosing where they are to direct their efforts."[62]

2.4.4.1 Fitness for Use. Beginning with the third edition of the *Quality Control Handbook,*[63] Juran defined quality as "fitness for use." In the fourth edition, Juran described Parameters of Fitness for Use: "Beyond those product features which bear directly on product satisfaction there are additional aspects of the product which also contribute to fitness for use....For products which are consumed promptly (food, fuel, many services) fitness for use is determined by (1) the adequacy of the product

[61]An indication of Juran's stature in Japan is found in the Japan Quality Control Medal (often abbreviated to Japan Prize). Introduced in 1969 as a best-of-the-best competition open to Deming Prize winners who continued to show substantial improvement 5 years later, the Japan Prize was to have been called the Juran Medal by the Japan Union of Scientists and Engineers. When Juran was informed of the honor, he sent back a lukewarm response which caused the Japanese—who interpreted the letter as a rejection notice—to choose the alternative name.

[62][Juran 1964], pp. 88–89.

[63][Juran, Gryna, and Bingham 1974], p. 2.2.

design and (2) the extent to which the product originally conforms to that design. For long-lived products, some new time-oriented factors come into play: availability, reliability, and maintainability. These abilities are closely interrelated and are vital to fitness for use."[64]

Availability is the proportion of time that the product works. It can be expressed as

$$(\text{Uptime}) \div (\text{uptime} + \text{downtime})$$

where uptime = total time the product is in an operative state (in service)
downtime = total time the product is in an inoperative state (out of service)

Alternatively, availability can be expressed as the ratio

$$(\text{MTBF}) \div (\text{MTBF} + \text{MTTR})$$

where MTBF = mean time between failures
MTTR = mean time to repair

Reliability means "freedom from failure." The attainable reliability is determined by the design, but the achieved reliability will be less than the attainable reliability because of (1) transient impact by environmental factors during use, (2) poor maintenance, or (3) damage during production. Achieved reliability is also known as "operational reliability."

Juran defined maintainability as the ease of servicing the product. It has two components: (1) preventive or scheduled maintenance to prevent failure, and (2) unscheduled maintenance to restore service in the event of failure. The time required to maintain the product is important because other productive resources such as labor are idle while the product is out of service. Maintainability is affected by (1) ease of access to and modular replacement of defective or worn parts, (2) effectiveness of diagnostic tools, (3) clarity of and ease of access to service documentation, and (4) availability of spare parts. The most important measures of maintainability are:

- Mean time to repair
- Probability of restoring service in the time specified

[64][Juran and Gryna 1988], pp. 2.9–2.11.

- Mean time for scheduled maintenance (often further divided between the time to inspect and the time to service)

It is useful to consider maintainability as two separate categories: (1) "serviceability," which is the ease of carrying out regularly scheduled maintenance, and (2) "repairability," which is the ease of restoring the product to working condition after a failure.

2.4.4.2 Fitness for Use vs. Conformance to Requirements. Crosby—who defined quality as conformance to requirements—insisted that his definition differed substantially from fitness for use. "'Fitness for use has a nice ring, and in fact has a lot of meaning particularly when you are talking about the design concept of a product....But that (design) isn't what we are doing. We are buying and measuring things. The design and concept decisions were made long ago. When our program manager gets an order from us, he only wants to know what we want, how many and when. He is not too concerned about how we are going to use it or the 'fitness for use' evaluation."[65]

Ten years later, Crosby had evidently arrived at new conclusions about fitness for use. He apparently felt that fitness for use extended beyond the design stage, but a new concern—that the terminology might be misunderstood on the shop floor—had taken hold. "When some talk about 'fitness for use,'" Crosby wrote, "they want to use that phrase to permit them to make judgments on products or services. They want to use it as permission to make waivers or deviations. Yet that was not its intent. It merely means that the output should do the job for the customer and have requirements that spell it out clearly so it is 'fit for use'."[66]

Juran would not have agreed with Crosby's interpretation of fitness for use. In particular, Juran believed that, except for simple products, workers and supervisors in the plant cannot determine fitness for use. They are unable to do so because they lack full knowledge of the needs of internal and external customers. Accordingly, they are able to act only in terms of conformance, the parameters of which are defined by the few company individuals who are familiar with the customers' needs.

Therefore, conformance to requirements would seem to be an operational subset of fitness for use. The conformance requirements must be established for workers and supervisors by those few employees who do know the customers' needs that determine fitness for use. In our

---

[65][Crosby 1979], p. 38.

[66][Crosby 1989], pp. 74–75.

experiences, parameters of fitness for use and thus the standards by which conformance to requirements can be judged are normally determined jointly by the customer's experts and the company's experts. In the absence of a single large customer or handful of customers, the company's experts alone must interpret the needs of the customer universe; this will be shown in Chap. 3 to be a marketing function.

2.4.4.3 Requirements, Fitness, and Specifications. If Crosby and Juran were so close philosophically, how did they happen to differ so vehemently in language? Perhaps the answer lies in confusing "requirements" with "specifications." The former are what the customer needs, the latter what the company standards setters dictate. It is entirely possible for a product to conform to all specifications yet not meet the real requirements.

Juran clearly meant requirements—and only requirements—when he referred to "fitness for use." He was not concerned with specifications or customer wants. His only criterion was whether the product would meet the customer's needs.

Crosby is harder to follow because he saw "requirements" as interchangeable with "specifications."[67] On the other hand, by contending that specifications which do not relate to customer requirements should be dropped, Crosby effectively reduced requirements and specifications to a single entity. Therefore, Juran and Crosby were both concerned with customer needs and only customer needs.

2.4.4.4 Juran as Management Philosopher. Although Juran was very much the pragmatist who put enormous effort into determining quantifiable aspects of quality, he also possessed a strong philosophy about how a company should be managed. His philosophy differed sharply from those of Deming and Crosby. Unlike Deming, Juran saw managers as potential assets to improvement rather than a constant drag on performance. And, in contrast to Crosby, Juran recognized the folly of exhorting employees to work harder or more carefully.

In contrast to the other three quality mentors, Juran was too prolific and diversified to allow for easy summation. He was not given to anecdotal evidence in the manner of Deming or glib salesmanship like Crosby. Perhaps the only way to fully grasp the enormity of his intellect is by studying his exceptional book *Managerial Breakthrough*[68]—always

---

[67]For evidence, see Crosby's attacks on "fitness for use" as an excuse for shipping nonconforming units.

[68][Juran 1964].

bearing in mind that he was 60 years old on the publication date and would remain active in quality management for another three decades.

2.5 Summary

- The widespread belief that American quality has developed primarily in the last 15 years is wrong. The most useful theories and applications existed well before the start of the "quality revolution."
- The basics of statistical quality control were developed by Dr. Walter A. Shewhart of Western Electric (now AT&T) during the late 1920s.
- The four leading American quality "mentors" of the last 50 years use very different language but in some respects have much in common.
- Some ideas—particularly those of Deming—are presented as fact but have never been proved.
- None of the leading approaches to quality is entirely satisfactory; although most of the central questions have been asked, the answers provided have not always been satisfactory. In large part this is because the developers of those approaches have tried to find a universally applicable concept of quality and, from that, a set of quality tools that will work for any industry, whether manufacturing or service.
- American TQC is quite different from Japanese TQC.

3
Defining Quality

3.1 Chapter Objectives

Quality improvement programs have value to the company if—and only if—they increase the company's long-term profits without reducing short-term cash flow below survival levels. Responsible project management therefore requires that costs and benefits of every quality initiative be measured. However, there can be no quantification of a project's value if the meaning of "quality" is not absolutely concrete and clear to everyone involved. This chapter explains why:

- Quality theory results in quality improvements only when the theory can be translated into actions. "Quality" must be applicable, not purely theoretical.
- "Quality" must be defined. Without an objective definition that allows quality to be unambiguously measured and judged, quality has no meaning and cannot be applied.
- Many quality authorities feel that quality cannot be defined in quantitative terms. If that is true, quality can never be an effective business tool.
- Although definitions of quality have been proposed, the best known are flawed and/or subjective.
- "Quality" may have different meanings for different types of products and services. Rather than looking for a universally applicable definition of quality, we must determine a definition specific to our industry.

- The customer equates the quality of electronic products with dependability.
- Warranty length and terms of coverage are important—and underutilized—sales tools.
- Manufacturers can and should measure the extent to which their own actions during production and quality evaluation increase product failure rates.

A definition for "quality" that meets the requirements of objectivity and quantifiability in the context of electronics assembly is developed and presented here.

3.2 Consequences of Failure to Define "Quality"

Quality improvement has attracted great attention from the business world because it is believed that better quality must lead to better company performance. But just what is better company performance? Referring to Goldratt again, it is clear "better company performance" can have only one meaning: maximization of long-term profits while maintaining adequate short-term cash flow to remain in business. Therefore, the only reason for pursuing improved quality must be to increase long-term profits without reducing cash flow to critical levels.

The company will act according to its interpretation of what "quality" means. If its concept of quality is wrong, the company is unlikely to take optimal actions. While possessing an accurate definition of quality does not in itself guarantee that ideal action plans will be chosen, it profoundly improves the probability of choosing the optimal strategies.

A definition is like a map; the more accurate the map, the higher the probability of ending up at the desired destination. An accurate map does not ensure success—the skills of the navigator and the driver's familiarity with the region also affect the outcome—but it beats guesswork. On the other hand, guesswork may well be preferable to reliance on a faulty map.

In *The Goal,*[1] Goldratt set out to establish the ultimate objective of every for-profit company. The result was a definition that left no room for uncertainty about the reason for the company's existence. The definition of the goal made it possible for management to quantitatively evaluate every action, whether ongoing or planned. Moreover, although

[1] [Goldratt 1984].

the personal objectives of any individual employee may not coincide with the company's goal—indeed, the individual's goals may be inimical to the company's goal—the unambiguous statement of the company's ultimate function made those conflicts easier to identify and resolve.

The need to define quality is inescapable. In Deming's words, "Misunderstandings between companies and between departments within a company about alleged defective materials, or alleged malfunctioning of apparatus, often have their roots in failure on both sides to state in advance in meaningful terms the specifications of an item, or the specifications for performance, and failure to understand the problems of measurement."[2]

The absolute necessity for key words to have clearly understood meaning can be seen time and again in company discussions, meetings, and problem-solving exercises. In the absence of definitions, it is unlikely that everyone in the company shares an identical understanding of what a pivotal word means. Indeed, every participant in the discussion may have a personal, unique, and implicit concept of the key word's meaning.

In a definitional vacuum, decisions will be made with various members of the decision-making team or committee having differing perceptions of what was decided. For example, suppose individual A believes that quality means "free from visual imperfections," B believes quality means "the product will work without failure for a reasonable period of time in the customer's hands," and C believes that quality means "meeting customer demand."[3] The conclusions reached during any meeting about quality will therefore be interpreted differently by each of A, B, and C. Decisions reached by committees in which each member operates according to implicit assumptions unique to that member are destined to fail and the company's decision-making abilities will be paralyzed.

3.2.1 Consensus May Not Mean True Agreement

In too many cases, critical decisions are made before the participants explicitly agree on a single, precise, objective definition for each and every term. The consequences of those decisions inevitably disappoint

---

[2][Deming 1986], pp. 277–278.

[3]These are only three examples of the many meanings put forth by various quality authorities.

some or all of the individuals involved. Disappointments cannot be avoided if all parties to the decision anticipate that the activity will bring the company closer to their own singular objectives, which (without their awareness) may contradict the personal objectives of others involved in the decision. When the terms of reference are not explicitly stated and resolved, consensus is meaningless; there is no way of knowing that all participants have agreed on the same items.

For example, suppose individuals A and B are both concerned about the company's "quality" and share responsibility for the company's quality improvement. There is no formal company definition of "quality." In the absence of a clear company definition, A and B each act on their own implicit understandings of quality. A, for example, wants fewer solder defects when product leaves the plant. B, on the other hand, cares only about field performance. Without realizing that their individual objectives differ, A and B jointly agree to increase inspection, test, and touchup. They both believe that the chosen course of action will improve the company's quality, as they understand the meaning of the word. Also they are unaware that their concepts of quality differ. The outcome may well satisfy A, but B will soon find that failure rates actually increase. Reasons why the chosen course of action will increase failure rates can be found in Chap. 6.

Formal, explicit agreement on a single, companywide definition of each key issue—and quality assuredly ranks among any company's key issues—must always be reached before any decisions relating to that topic take place. Note that the company must restrict itself to a *single* definition of quality.[4] More than one "definition" effectively means the term has not been defined at all. Different committees working with different definitions can easily end up working at cross-purposes. Moreover—as in the example of the decision to implement more inspection, test, and touchup—lack of an explicit formal statement of quality's meaning can result in members of the same team or committee wrongly believing that they are in agreement on objectives and means.

Therefore, the quality department must overcome a fundamental hurdle: to provide a rigorous, unambiguous, and *applicable* definition of quality itself. This definition must be readily understood in its entirety by (1) every employee, (2) every customer (or, where the customer is

[4]More appropriately, a single definition of quality is required for each different type of product and service in the company. Discussions of total quality later in this book will examine the reasons why every distinct product and service category requires its own definition of quality. In these pages, for example, we will be developing a definition of quality as it applies exclusively to electronics assemblies.

another company, the customer's personnel who have direct or indirect[5] contact with our company), and (3) all supplier personnel who likewise have direct or indirect contact with our company.

3.2.2 The Better Mousetrap Reconsidered

There is no room for subjectivity in the definition; subjectivity is the enemy of accuracy. Yet most "definitions" of quality are based on enormously subjective language. Consider, for example, Emerson's statement that the world will "beat a path to your door" if you just build a "better" mousetrap. While the better mousetrap argument has long provided exceptional emotional support for the quality movement, closer analysis shows it to be an excellent example of subjectivity masquerading as concrete objectivity. The subjective nature of the statement undermines its usefulness as a decision-making tool.

The problems with Emerson's "better" mousetrap analogy begin with the contention that customers will seek out the better product even if barriers to access are placed in their way. In reality, development and production of the "better" product in themselves are unlikely to greatly alter the company's fortunes. The company must also:

- Build the "better" product at a cost that will earn profits at whatever price customers prove willing to pay.
- Vigorously market the virtues of its product.
- Maintain a strong sales force.

The need to take these additional actions to capitalize on having the "better" product comes from the findings of Chap. 1 that quality improvement is only one of several possible tools for improving the company's profitability. It should be apparent that those tools generally interact rather than working in isolation.

Accordingly, Emerson's reliance on "better" makes his proposition useless for all but the most abstract quality decision making. In order to take the proper action—that is, to become "better" and have the world beating a path to our door—we must be able to say *exactly* what "better" means. Unfortunately, "better" does not lend itself to exact interpreta-

[5]A customer's purchasing agent would have direct contact with our company. A customer's product designer would typically have indirect contact, although it is certainly not unknown for designers to meet with supplier representatives directly to learn about specific properties of the supplier's products.

tions. A representative dictionary[6] defines better as meaning "of superior quality or excellence." And that, in turn, forces the question: What is the meaning of "quality?" To which the same dictionary responds that quality is "character with respect to excellence, fineness, etc. or grade of excellence." The "definition" has become circular.

The impossibility of applying Emerson's better mousetrap position to the needs of our specific industry shows that any useful quality definition must answer two critical questions:

1. *What* does "better" (i.e., "higher quality") mean?
2. *How* can we measure it?

If we succeed in answering those two questions, we must then answer two additional questions on which quality optimization depends:

1. *Why* would we want more quality?
2. *How much* more quality—if any—would we want?

3.2.3 The Operational Definition

Our own painful search for the meaning of quality over many years certainly makes us sympathetic to anyone who finds defining quality difficult. Nonetheless, the hard truth is that only two choices exist. The first is to accept that quality is intangible and undefinable, in which case—because the unknowable cannot be applied—we must abandon the entire field of quality improvement as a strategic tool. Or we can pursue a workable definition in the hope that such a thing exists and therefore allows us, once the meaning of "quality" has been found, to add quality improvement to our list of profit enhancement agents.

Deming, who himself never defined quality in a satisfactory manner, called the required definition an "operational definition." In his words, "An operational definition puts communicable meaning into a concept. Adjectives like good, reliable, uniform, round, tired, safe, unsafe, unemployed have no communicable meaning until they are expressed in operational terms of sampling, test and criterion....An operational definition is one that reasonable men can agree on."

Furthermore, Deming wrote, "An operational definition is one that people can do business with. An operational definition of safe, round,

[6]*Webster's Encyclopedic Unabridged Dictionary of the English Language.* 1989. New York: Grammercy Press.

reliable or any other quality must be communicable, with the same meaning to vendor as to purchaser, same meaning yesterday and today to the production worker. Example:

1. A specific test of a piece of material or an assembly
2. A criterion (or criteria) for judgment
3. Decision: yes or no, the object of the material did or did not meet the criterion (or criteria)

If such an operational definition does not exist or cannot be found, the entire "quality" function must be discarded as an applicable management tool.

3.3 Challenges of Defining Quality

Can an operational definition of quality be found? Indeed, does such a definition even exist? We do know from the experiences of others that quality is not a simple concept and the search for its meaning will involve great difficulty. One good indication of how arduous the task will be is, as we saw in Chap. 2, the inability of the four quality mentors to agree on the meaning of quality. Indeed, "quality" means so many different things to various people that some authorities doubt that it can be defined at all.

Deming paraphrased Shewhart to describe the challenges involved in establishing a definition of quality: "The difficulty in defining quality is to translate future needs of the user into measurable characteristics, so that a product can be designed and turned out to give satisfaction at a price that the user will pay. This is not easy, and as soon as one feels fairly successful in the endeavor, he finds that the needs of the consumer have changed, competitors have moved in, there are new materials to work with, some better than the old ones, some worse; some cheaper than the old ones, some dearer."[7]

3.3.1 Can Quality Be Defined?

Sixty-five years later, doubts still exist about whether quality can be defined. For example, *Quality or Else* contains this quotation from McKinsey and Company director John Stewart: "There is no one definition of quality....Quality is a sense of appreciation that something is

[7][Deming 1986], p. 168.

better than something else. It changes in a lifetime, and it changes generation to generation, and it varies by facets of human activity."[8] The same book,[9] paraphrasing psychologist and consultant Michael Maccoby, argues, "The difficulty with defining quality is that the definition changes as industry does."

Not everyone feels that defining quality is impossible. Some researchers even believe that defining quality is a relatively simple matter. *Business Week* claims: "It's not that quality is hard to define; it's simply the absence of variation. In manufacturing circles, it has long been accepted that quality has nothing to do with how expensive or fancy or complicated a product is—a quality product simply does what it's supposed to do, over and over again. Thus, a Chevrolet can have just as much quality as a Rolls-Royce, and the service at a discount store can be equally 'good'—free of variations—as at Bergdorf Goodman. Of course, even a perfect product can't do more than it was designed to do: Don't count on a Chevy to perform like a Rolls."[10] This, however, raises another thorny issue: Is the quality of a product that invariably fails to work higher than the quality of a product that may or may not work? There is no obvious merit in a product that is invariably bad.

3.3.2 "I Know Quality When I See It"

On the other hand, some quality theorists believe that, even if quality is difficult to label, it is so obvious that it doesn't matter. A classic example of the "I can't define quality but I know it when I see it" school of thought was articulated by author Robert Persig in his book *Zen and the Art of Motorcycle Maintenance.* "If you want to build a factory, or fix a motorcycle, or set a nation right without getting stuck," Persig wrote, "then classical, structured dualistic subject-object knowledge, although necessary, isn't enough. You need to have some feeling for the quality of the work. You have to have a sense of what is good. *That* is what carries you forward. This sense isn't just something you *are* born with, although you are born with it. It's also something you can develop. It's not just 'intuition,' not just unexplainable 'skill' or 'talent.' It's the direct result of contact with the basic *reality,* Quality, which dualistic reason in the past tended to conceal."[11]

---

[8] [Dobyns and Crawford-Mason 1991], p. 21.

[9] p. 21.

[10] [*Business Week* Editors and Green 1994], p. 3.

[11] p. 255.

While Persig's interpretation of "quality" sounds sufficiently accurate that the book has remained a best-seller for more than two decades, there are treacherous shoals just beneath the surface of those placid verbal waters. Persig himself experienced the acute frustration inevitable in trying to apply abstract semantics to the reality of daily life. The ongoing theme of *Zen* is how Persig's search for the meaning of quality drove him to the verge of insanity.[12] The consequences for anyone in the electronics assembly industry hoping to employ Persig's totally subjective test of quality will be equally traumatic.

3.3.3 An Operational Definition Exists

Fortunately, quality—at least as it applies to the electronics assembly industry—can be defined in the objective, quantifiable manner that avoids a Persig-like fate. In other words, the "operational definition" which Deming says is "the one that people can do business with" exists and can be specified. The solution lies in eliminating the numerous immaterial concerns that permeate quality thinking. Above all, we must constantly remind ourselves that we do not need to generate a universally applicable definition that applies equally to every product or service; we need address only this one specific industry—electronics assembly.

3.3.4 Excluding Mechanics from Consideration

We must further narrow the definition of what this book means by "electronics assembly." Electronic components are employed in diverse ways. The purest form of electronic product consists entirely of electronic components. Such pure electronic products are relatively rare; even PCB assemblies generally involve some mechanical elements such as fasteners. As subassemblies merge into finished goods, the importance of nonelectronic elements increases. Cables enter the picture, as do cabinets. The number of fastener types increases, as does their complexity. When electromechanical systems are added to the mix, the

[12]Some readers of *Zen and the Art of Motorcycle Maintenance* wrongly believe that Persig's pursuit of the meaning of quality drove him insane. See, for example, Holbrook, Morris B., and Corfman, Kim P. "Quality and Value in the Consumption Experience: Phaedrus Rides Again." [Jacoby and Olson 1985], p. 31.

mechanical content of the unit can easily outweigh the electronics. Many plants that think of themselves as part of the electronics assembly industry also contain some or all of machine shops, sheet metal fabrication, paint facilities, and a variety of mechanical system assembly operations.

While the electronic content may be less significant than the nonelectronic content in the finished system, the scope of this book is restricted to pure electronic components and assemblies. The reason for this apparently narrow definition is that the quality requirements of electronic components and systems have little or nothing in common with the quality requirements of mechanical hardware. As will become clear in the following pages, the most important constituents of quality in the electronics world are not visible and sometimes not even measurable. Mechanical parts, on the other hand, are evaluated using techniques quite different from those most suitable for assessing electronics quality. Cosmetic issues that are overemphasized in the pure electronics world do have considerable relevance to mechanical parts such as cabinets; a scratch on a PCB, for example, is generally of little or no consequence but would certainly be a quality issue on the exterior of a cabinet.

So, while the importance of mechanical parts in the composition of an electronics assembly is not trivial, taking them into account here would make an already substantial book impossibly unwieldy. The discussions here should be presumed to refer exclusively to pure electronics unless the text explicitly states otherwise.

3.4 The Customer in Quality

As we saw in Chap. 2, prominent quality authorities have long subscribed to the belief that quality depends on the customer's opinion. Feigenbaum's position with respect to the relationship between a quality definition and the customer is not unusual: "Quality is a customer determination, not a marketing determination or a general management determination. It is based upon the customer's actual experience with the product or service, measured against his or her *requirements*—stated or unstated, conscious or merely sensed, technically operational or entirely subjective—and always representing a moving target in a competitive marketplace."[13]

---

[13]p. 7.

Deming put the case more succinctly and in a rather different context when he concluded that, "A product or service possesses quality if it helps somebody and enjoys a good and sustaining market."[14] On the other hand, brevity in this instance does not lead to usefulness; no manufacturer can construct a strategy around results that will not be known until long after the strategy has been implemented. Moreover, as Sec. 3.5 explains, many factors other than quality determine demand for a product.

In recent years, the opinion that only the customer knows quality has gained a wide following. *Electronic Business Today* put it bluntly: "Most experts agree that quality really means customer satisfaction. Consequently, many electronics managers have made customer satisfaction the No. 1 quality metric."[15] Jack West, chairman of the American Society for Quality Control, went even farther: "Quality is what the consumer says it is. It's not what the quality professional says it is. It's not what the engineer says it is. It's what the customer says it is."[16]

There is no doubt that rational ill-treated customers will take their patronage elsewhere, provided an alternative source exists for whatever they need or want. Lack of courtesy is one of the quickest ways to lose a customer. Similarly, false representations—whether about a product's performance capabilities, price, warranty, or delivery time—will also damage any company that faces competition (or encourage the entry of competitors that otherwise would have stayed out of the market). All these issues, of course, are negatives; they are reasons why a customer will be lost.

But before customers can be lost, they must first be attracted. That is where the positive issues enter the picture. Among the positives for influencing customers will be found conscientious and persuasive marketing, packaging, ease of making the purchase, price, and many other elements.

The ultimate question is not whether a happier customer is good for the company. If keeping the customer happy is feasible while still selling at a profit and if customers are not inherently impossible to please, then the company will be better off with happier customers than with unhappy ones. There is no doubt that gaining and keeping happy customers is better than the alternative *if all other factors are the same.*

Does this mean the company should not strive to please its cus-

---

[14][Deming 1993], p. 2.

[15]Young, Lewis H. "Exactly How Much Quality Is Enough?" *Electronics Business Today,* October 1995, p. 56.

[16]Armand Feigenbaum has used this definition on occasion. See, for example, Karbatsos, Nancy. "Quality in Transition, Part Two: Narrative for the Nineties," *Quality Progress,* January 1990, p. 23.

tomers? Of course not. Unhappy customers equal financial trouble for companies. But what—if any—correlation exists between happy customers and quality? Or, to rephrase the issue, how many of the factors that make for a happy (or unhappy) customer can properly be considered product quality and how many should be ascribed to other actions? An operational definition cannot be established before this question is answered.

3.4.1 The Customer Is Not Always Right

As Crosby pointed out,[17] "with no customers there is no business." So the customer must be treated with respect. However, it does not always—or even generally—follow that the customer is always right. Indeed, there are at least three excellent reasons why letting the customer define "quality" will be harmful to the supplier and, not infrequently, the customer: (1) inability to know what the customer wants, (2) conflict between what the customer *wants* and what the customer *needs,* and (3) unrealistic customer demands.

3.4.1.1 Customer Desires Cannot Be Identified. For companies whose customer base contains only a handful of names, discovering what the customer wants is fairly straightforward. Such manufacturers normally produce to order following specifications provided by the customer. Most manufacturing companies sell to the mass market, however, and the only way such companies can gain any insights into what prospective customers want is by market research techniques such as focus groups.

In fact, it will often be impossible for any company—no matter how small the customer base—to know the customer(s)' attitudes. The inability to know what the customer wants results from the fact that customers' desires are often unknown and unknowable, even to the customers themselves. This is particularly true of new products that represent a break with the customers' experiences. An enormous number of the world's most important electronics products would not exist today if their original backers had followed stated customer preferences when those products were introduced. The telephone (who needed an in-house alternative to the telegraph?), the photocopier (IBM turned down an investment opportunity in a very young Xerox), the Walkman (Sony chairman Akio Morita claims to have personally pushed the project

[17][Crosby 1989], p. 75.

through over objections from most company personnel), and the microwave oven (consumers were nervous about both the new way of cooking and possible dangers from microwave radiation) are just four examples of enormously successful electronics products that would never have proceeded beyond the prototype stage if the customers had been defining "quality."

A larger customer base—existing or potential—increases the difficulty of explaining new ideas to prospective buyers. Knowing what the customer wants is generally more difficult for manufacturers of finished goods than for manufacturers of components or subassemblies, whose products are more likely to be purchased according to technical specifications. Above all, the lack of customer knowledge presents special challenges to manufacturers of "new concept" products as opposed to companies that are merely repackaging existing accepted products. The home videotape recorder was a new concept product 20 years ago and faced considerable consumer skepticism; today, "new" VCRs face a marketplace of experienced and better-informed customers.

The fundamental problem with defining quality as "what the customer says it is" concerns how the customer's opinion can be determined in a manner that enables the manufacturer to act accurately on that opinion. Since the customer often has no idea of a product's desirability before it is actually available for purchase, sales volume is the only test of whether the product met the customer's approval. This information, of course, comes too late to help the manufacturer. Moreover, the customer may discover after the purchase that the product is not as good as it seemed before the acquisition. Above all, sales are not based solely on quality as it should be defined in the context of electronic products. Each of these points is developed further in subsequent sections of this chapter.

In the final analysis, defining quality as "what the customer says it is" means that:

1. Products that sell well will be said to have high quality while products that sell poorly are given low marks for quality.
2. The information necessary to determine "quality" will be available only after the product has been designed, produced, and sold. In other words, the customer's opinion arrives too late to be of predictive use to the company and is therefore worthless for planning.

Thus defining quality in terms of demand violates the requirement that a definition lead directly to an unambiguous course of action for the manufacturer.

3.4.1.2 Customer Wants Differ from Customer Needs. Customers frequently know what they want but fail to realize that what they want is not what they need and may even be dangerous. For example, customers of contract electronics assembly shops typically provide the contractors with workmanship guidelines dictating how solder connections should look but imposing no restrictions on how those cosmetically acceptable joints must be achieved. The assembly shop can achieve the cosmetically perfect connections by employing a dangerously active flux and/or relying on visual inspection and touchup to catch and "fix" cosmetically imperfect connections. Both strategies will meet the customer's definition of quality—attractive solder connections—but both also cause excessive failure rates.[18] The damage resulting from the customer's imposition of wrongheaded standards may be sufficiently severe to force the customer out of business. In addition to undermining the overall strength of the electronics industry, demise of the customer can hardly be considered beneficial to the personnel and owners of the contract manufacturer.

Customers are more inclined to blame the manufacturer than to accept personal blame for failing to have properly identified their real needs. Again, it will be the manufacturer who suffers by generating customer ill will. Yet in both cases, the company followed the advice of providing what the customer wanted.

Often, the seller knows that the customer is wrong but agrees when the customer claims that the product is defective. Again, the seller is simply obeying the common quality adage that the "customer is always right." However, acquiescing even though the customer is clearly wrong does a disservice to both supplier and customer. The supplier loses two ways:

1. By incurring the cost of reworking or replacing a perfectly acceptable product.
2. By reinforcing the customer's belief that the supplier is incompetent. More often than most suppliers realize, the customer walks away believing that the supplier would never have found the (nonexistent) problem without the customer's help.

Thus taking the position that "the customer is always right" can very well end up costing the company customers. The attitude of "we'll 'fix' it if you don't like it" has prevailed in defense contracting for decades and is a consequence of neither side knowing what the specification really calls for—or taking the trouble to find out.

[18]See Chap. 7.

Of course, many times the customer is right. A customer who is displeased with a product that fails to perform as the manufacturer promised, for a reasonable time span, is entirely justified in that displeasure. If the manufacturer then fails to revive or replace the failed product without delay, the customer again has the right to be displeased. A company that ignores customers' reactions does so at its peril. Ultimately, however, acquiescing to improper customer demands is counterproductive. The only course of action that maximizes the probability of gain for both customer and company is to educate the customer. And before we can educate the customer we must ourselves have complete knowledge.

3.4.1.3 Unrealistic Customer Demands. The manager of a large national electronics components distributor recently told us what every customer really wants. In his words: "They want what you don't have. They want it yesterday. And, above all, they want it for free." Obviously, he was not entirely serious. On the other hand, he was not entirely joking, either. If the customer has unrealistic expectations, no company can feasibly meet those requirements.

3.4.1.4 Who Is the Customer? As every economist knows, demand increases as price decreases. Which "customer," therefore, defines "quality": the customer who is willing to buy at any price, the customer who will only buy at a very low price, or a customer who falls somewhere between the two extremes? The answer is far from clear. But how can the customer define quality if the customer's identity is uncertain?

3.4.2 Contractual Obligations

When contracts exist between manufacturer and customer, the rights of the customer take on dimensions that do not apply under normal market conditions. Obviously, the customer enjoys certain legal rights under the contract. In addition, however, the customer is entitled to moral rights that otherwise would not be relevant. When contracts exist, the quality department is responsible for ensuring conformance to the contract requirements. Even when the contract requirements seem unreasonable to the production, engineering, or quality personnel themselves, the quality department's responsibility—to ensure compliance with the contract terms—is clear. This role is quite different from the quality department's function when all requirements are determined internally by the manufacturer.

3.4.2.1 Disposition of Nonconforming Product. The contract normally includes specifications of product performance under specific conditions. Other contract conditions can include product characteristics such as appearance or feel that are entirely cosmetic. Some contracts—especially those involving Department of Defense organizations—even dictate the manufacturing process parameters.

Whenever cosmetic requirements are part of a contract, there will almost certainly be instances when a dependably functional unit does not meet all the fit and finish standards. In other words, appearance does not affect performance and the blemished product works as well as if the product was cosmetically perfect. In such instances, the quality department is typically called upon to decide whether:

- The blemished unit can be shipped to the customer without remedial action.
- The unit must be reworked to conform to the exact cosmetic specifications.
- A waiver for that unit will be sought from the customer.

These are the situations that create friction between the production department that wants to ship and the quality department that insists on adhering to the letter of the contract. What is the conscientious quality department to do?

The answer is straightforward: If a manufacturer promises to supply products possessing specific attributes, it is morally bound to meet all the attributes on every unit shipped. This truth holds even if the contract requirements are unreasonable in the eyes of the production department and even if the price of meeting those cosmetic requirements involves compromising the unit's dependability (as, for example, by altering the appearance of solder connections through touchup or rework). Even if the manufacturer will lose money on the contract because meeting the visual criteria is costly, the only moral stance is for the manufacturer to absorb the losses and provide the cosmetically perfect product that it contracted to supply.

In the avionics industry, when delays occur in developing electronic systems that meet all the contracted specifications, it is not unusual for manufacturers of electronic systems to go one step farther in accepting responsibility for the costs of nonconforming product. Rather than hold up delivery of an aircraft worth many tens of millions of dollars while the avionics system in question is perfected, systems of lesser ability are provided at the manufacturer's expense until fully conforming product can be provided. The temporary systems provide all the functions necessary for safe operation of the aircraft but lack some of the specified "bells and whistles." The aircraft is accepted by the customer with the knowledge that the instrument in question does not offer all the performance features promised but will be refitted at a future date.

3.4.2.2 Consultations before Contract Finalization. The time to debate requirements is during contract negotiations, not after the agreement has been signed. Tragically, sales and marketing personnel rarely seek out participation from the production and quality departments during preparation of a bid. If they are not consulted during drafting of the contract, the plant personnel, including members of the quality department, become aware of illogical requirements only after the contract has been executed.

The burdens placed on the production and quality departments when they have been excluded from decisions before the contract is signed can be heavy indeed. Often, the company may be unable to meet the contract requirements without suffering losses—losses that could have been avoided through contract decision input from the departments that must make and evaluate the product. Faced with the realization that they have committed to unfavorable contract terms, it is not uncommon for manufacturers to ask the customers for contract revisions. But the customer should not be expected to pay for the manufacturer's inept management practices. Requests for waivers clearly demonstrate flaws in the manufacturer's sales and / or plant operations.

Contract proposals determine the profitability of any job. The contracting process is therefore too important to be left entirely in the hands of sales personnel. Every new contract proposal requires careful review by the plant personnel who will bear responsibility for meeting the product specifications. If the customer's expectations cannot be met at the price quoted, the conflicts must be dealt with before the contract is finalized.

Since the practicalities of providing product that meets the contract terms fall to the design and production departments, the quality management need not be involved in preparation of the contract proposal. Once the contract has been signed, however, the quality department's responsibility is clear: to ensure that the product conforms to all the promises made to the customer. Whenever contracts enter the picture, Crosby's dictum that quality means conformance to requirements applies absolutely. Shipping blemished product—where "blemished" means violating an appearance criterion of the contract terms—or seeking waivers are not ethical options.

3.4.2.3 Design Constraints. Contracts are often signed while the product involved is still in the conceptual stage and no working models exist. If design engineering subsequently proves unable to develop a product that meets the operating specifications of the contract, what is the quality department's role? In theory, quality personnel should never be involved in such problems because the product's deficiencies should be recognized before production begins. In practice, however, some design failures do not become apparent until units have been manufactured and reach quality evaluation.

Quality management cannot authorize shipment of units that do not conform to the contract requirements. A resolution—changes in the specifications, price revisions, legal action, and so on—must be decided in meetings between representatives of the customer and contractor. Quality personnel should never be expected to negotiate with the customer for acceptance of nonconforming product.

3.4.2.4 Exceeding Requirements. Just as the quality department must rigorously ensure that the product meets all contractual requirements, the department must be equally careful to avoid imposing criteria that exceed the customer's requirements. Certainly there is nothing wrong with exceeding requirements when no additional expense is involved (for example, when free process refinements lead to perfect rather than merely acceptable output). But rejecting acceptable product in a misguided search for subjective perfection is irresponsible. The quality department's function is to ensure adherence to requirements, not to arbitrarily revise those requirements upward.

Rejection of acceptable product is almost inevitable when the manufacturer issues workmanship guidelines that include more than one level of conformance (specifically, the ubiquitous picture books showing "acceptable preferred," "acceptable," and "acceptable minimum").

3.5 Defining Quality for the Electronics Assembly Industry

While many goods and services may not lend themselves to an applicable quality framework, a workable definition of quality in the electronics assembly industry can be formulated. To derive that definition, it is necessary first to recognize that customers make purchases for five reasons, some of which are interdependent:

1. *Features.* "Features" is defined as all characteristics of the product excluding quality and price. Features range from objective functions (e.g., speed of operation) through to subjective criteria (e.g., color). The customer will translate features into a perceived suitability of the product to meet the customer's needs.

2. *Dependability.* The customer needs to know whether the product will deliver all the features without breakdowns of any kind for a predictable period of time when operated in explicitly understood environments. Therefore, "dependability" is defined as "the degree to which the product will provide all the features promised by the manufacturer without interruption for a specified period of time under specified conditions of use."[19]

3. *Performance.* The product's usefulness to the customer is determined by the features and the percentage of time that the features will be available for use by the customer. In other words, "performance" is defined as "features plus dependability." This relationship is shown in Fig. 3.1.

4. *Value.* The customer may be faced by many different combinations of features and dependability when reviewing the attributes of several competing products. The choice will be based on which product provides the most attractive combination of features, dependability, and price. This combination, shown in Fig. 3.2, is known as "value" and is defined as

$$\text{Value} = \text{performance} \div \text{price}$$

5. *Affordability.* The final consideration is whether the customer has the necessary funds with which to make the purchase (i.e., the budget).

---

[19]Dependability, though related to reliability, has broader meaning. Juran, as noted in Chap. 2, defined quality as "fitness for use." Fitness for use, in turn, incorporates what we have termed "features" plus three other characteristics: availability (the percentage of time that the system is in service), reliability (freedom from failure under given conditions for a specified period of time), and maintainability (the ease and speed with which maintenance can be carried out). "Dependability," as employed in these pages, consists of Juran's availability and reliability functions together with some elements of maintainability (specifically the time required to put the system back in service in the event of failure). See [Juran 1988], pp. 2.8–2.10.

If a product performs according to its specifications (which include the "conditions of use" stated by Juran as a determinant of what he terms "reliability") for a period of time specified at the outset by the manufacturer, the product can be said to have dependability. Failure to perform all functions for the specified life expectancy means the product lacks dependability. Matters become more complicated when the manufacturer's specified period of use is set for unreasonably short duration. This complication is addressed in Sec. 3.9 (Warranties).

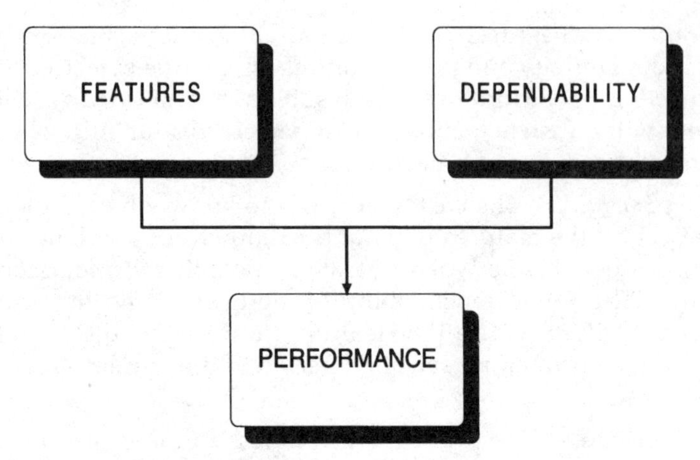

Figure 3.1. Determinants of performance.

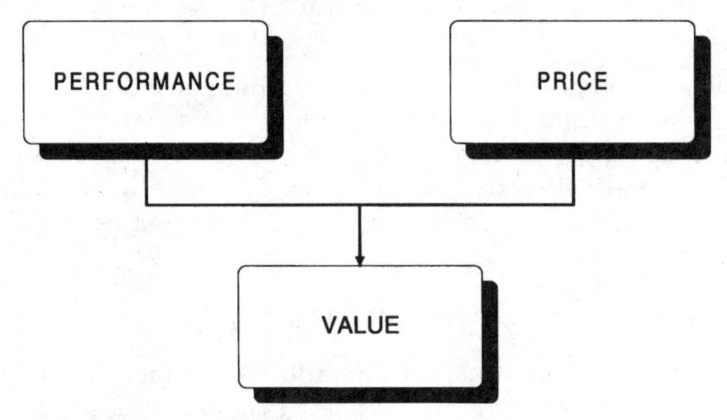

Figure 3.2. Determinants of value.

If the price is more than the customer has available to spend, no other considerations matter. "Affordability" is defined as "price in relation to budget." Affordability varies directly with price.

3.5.1 Characteristics Mistaken for Quality

All of these characteristics—features, dependability, performance, value, and affordability—are commonly believed to be characteristics of quality. With the notable exception of dependability, however, they are not, at least in the sense that they relate to the quality department's range of activities. Companies have sales and marketing departments to establish features and value.

Customers can choose among product features to find the combination that comes closest to what they want and fits their budget constraints. Those sales and marketing considerations lie beyond the authority of the quality department. (If the company defines the quality function to include sales and marketing, then features, price, and value would become part of the quality management's job description. However, these characteristics would not become elements of quality even if they happened to be overseen by the quality department. In any event, we have yet to see a company where the quality department's mandate includes sales and marketing.)

Of course, the efforts of the quality department can affect the company's cost structure and thus its ability to earn acceptable profits at lower selling prices. However, price should always be determined independent of costs. The sales and marketing department should set actual price in response to market forces such as demand and competition. Companies that set prices on the basis of costs plus a set profit margin are ignoring the realities of a market economy. The price established on the basis of costs rather than demand will equal only by accident the price best suited for the market. More often, the price based on costs will be:

- Too high, so that potential customers are frozen out of the market and competitors are encouraged to enter the market
- Too low, so that attainable profits are sacrificed.

3.5.2 The Subjective Nature of Value

Although value is given here as an equation, it is not a scientific constant. Value is largely perceptual and will be different for each prospective customer. This rather abstract nature of value is inevitable because the importance of any given combination of features varies from one person to another. Only the price is identical for all customers.[20] If a customer sees two products as equivalent in all respects except price, the product with the lower price will have greater value. As explained in

[20]This statement, while satisfactory for our purposes, is not always true in economics. In economic analysis, a seller capable of segregating customers by desires and purchasing power can charge different prices in different markets. So price need not be the same for all customers.

the next chapter, Japanese automobiles were inaccurately said to offer better quality than similarly priced American cars during the 1970s and 1980s; in fact, because of factors such as an artificially low value for the yen, prices for Japanese products were less than their real cost, so the Japanese car represented better value.

Affordability is not the same as value, although the two concepts are related. The affordable product may not be the best value if the customer is free to choose without budget constraints. Affordability simply places a limit on what the customer can pay. Affordability can also be somewhat elastic; if the customer finds performance sufficiently attractive, the customer may be willing to forgo purchases of other products to free up the necessary extra funds.

Dependability is independent of the other product characteristics. It relates to features only in the sense that any failure of the product to provide a feature or features to the customer equals a loss of dependability. Value relates to dependability by virtue of the fact that costs incurred through lack of dependability—ranging from inconvenience through to monetary losses caused by loss of product use—inversely affect value. Affordability (price), as we have seen, should be determined by market conditions, though again higher dependability may translate into greater demand and thus allow for increased selling prices.

In the final analysis, dependability and only dependability can be considered the critical product attribute that constitutes "quality." For the purposes of the electronics assembly industry, dependability and quality are synonymous and fully interchangeable.

3.5.3 Dependability in the Demand Equation

Dependability, features, price, and value all affect demand. Therefore, dependability is only one of the factors in the demand equation. Higher dependability typically—but not always and not necessarily predictably—leads to greater demand by increasing the amount of repeat and referral business.[21] At the same time, all the dependability in the world will be of scant comfort to the company if customers do not buy. That bleak outcome results when the product is dependable but either lacks necessary features or possesses undesirable features. Customers

[21]Exactly how much and how soon dependability affects sales varies according to market conditions. In most cases, the return on higher dependability will not be evident in the short term, will perhaps deliver some benefits in the medium term, and will be most strongly felt in the long term. See Chap. 4 for discussion of the interactions between dependability and demand.

may choose not to buy a perfectly dependable calculator, for example, because the color offends their visual tastes; the quality function has been properly executed but the marketing department failed to accurately interpret customer preferences.

Greater dependability is not something to be sought without taking account of both the costs and benefits of the effort. At some point, further efforts to improve dependability will produce benefits worth less than the extra cost. Arriving at the ideal level of dependability is the essence of quality optimization.

Dependability is totally tangible. If the product provides the same feature set promised by the manufacturer, the customer has no right to feel misled or deprived in any manner if a more attractive combination of features subsequently becomes available from either the same or another manufacturer—unless the product ceases to provide any of the features at or above the minimum performance level promised at the time of sale. New models with extra features can be sold for higher prices not because they are necessarily more dependable than older models—indeed, the first units produced of a new model may be less reliable—but because the customer is willing to pay a premium in return for the additional functions.

Moreover, as previously stated, customers have the right to expect the product will perform its promised function(s) without interruption for the length of time promised by the manufacturer. If the product fails before that promised time is up, the average customer won't be fooled again; the next purchase may be from a competitor. Features and affordability can convince the customer on the initial purchase decision, but higher failure rates reduce the joy of ownership no matter how attractive the other attributes.

3.6 The Two Sides of Dependability Costs

For the electronics assembly company, one of the most attractive aspects to increasing product dependability is that costs decline sharply. During manufacture, most of the requirements for improving dependability—most notably (as later chapters will show) less handling of assemblies during production and evaluation—actually reduce costs when compared to producing and evaluating less dependable products. Among the cost savings will be found:

- Fewer defects to repair and rework
- Less need for postassembly evaluation such as inspection and test

- Reduced or eliminated retest
- Reduced scrap
- Less time spent by management, supervisors, engineers, other production personnel, and quality department employees discussing the acceptability (or lack thereof) of product, both at intermediate assembly stages and at final goods ready for shipping

After-sales costs are also much less for the dependable product. Fewer failures mean fewer repairs, fewer demands for field service, and fewer customer complaint calls to answer and process.

Dependable products also cost the customer less (and this can be true even if the purchase price is higher). The costs to the customer of undependability may be financial, as in the loss of productivity suffered by the customer while a critical piece of office equipment is out of service, psychic, as in the loss of enjoyment a consumer experiences while a home entertainment device cannot be used, or physical, as when a critical medical device (a heart pacemaker, for example) fails. Depending on the extent to which the customer has come to rely on the product in question, the costs of failure can greatly exceed the purchase price. Not infrequently, a product failure can threaten the very financial survival of a customer who depends on the product to conduct business.

3.7 Dependability vs. Defects

A serious misconception pervades the electronics assembly industry that quality can be measured by the first-pass defect rate. Certainly defect prevention has been the dominant theme of quality programs at least since the time of Shewhart. More recently, Crosby gained a national reputation as the spokesman for Zero Defects. In many plants even today, quality personnel devote much of their time to gathering defect data and plotting control charts. The popular concept of cost of quality includes among its most prominent concerns the losses associated with not doing things right the first time: inspecting, testing, touching up, repairing, scrap, and so on. Motorola's Six Sigma™ sets the quality level as 3.3 defects per million operations. Inside the company, in other words, the quality spotlight shines most brightly on defects.

The distinction between defects and quality (dependability), however, is profound. Traditionally, the term "defect" in the electronics assembly industry has referred specifically to soldering-related matters, primarily cosmetic in nature.[22] They are not necessarily associated with failures, whether failures that may already have occurred or those that are likely to happen in the future. Identification is typically based on subjective visual criteria, a circumstance that explains the well-known phenomenon of different inspectors producing vastly disparate "defect" results when looking at the same assembly. Many common "defects" can be prevented by techniques that compromise dependability. Poor wetting of metal surfaces during soldering is one example of how a defect can be cured while dependability is reduced. The easiest way to prevent poor wetting is the use of highly acidic fluxes that increase the probability of product failures through ionic contamination. This and related issues are discussed in Chap. 7.

The number of defects, therefore, does not necessarily measure dependability. Absence of defects may—and often does—coincide with low dependability.

Unquestionably, defect prevention generally saves impressive amounts of money for the company by avoiding the costs of doing things more than once. However, the customer is not concerned about anything other than dependability of performance. Too much emphasis on defects can obscure the only part of quality assessment that matters: determining product dependability during use by the customer.

3.8 Quality Assurance vs. Quality Control

Just as it is essential to specify the precise meaning of quality, we must also define the methodology used to manage it. The choice is between "quality control" and "quality assurance," the former normally being considered verification of conformance to specifications after production while the latter refers to techniques applied before and during production to ensure that the output will be free of defects and failures without dependence on checks after the fact.

The line between quality assurance and quality control is not always obvious. For example, [Juran 1988], p. 6.31, defines quality control as "the regulatory process through which we measure actual quality performance, compare it with quality goals and act on the differences," which on the surface appears very much like quality assurance. Note, however, the word "regulatory" in his definition. True quality does not

[22]See, for example, [Leonida 1981], pp. 504–537, and [Lea 1988], pp. 148–153 and 467–469.

emerge from regulation (though documentation and discipline must both be present); it comes from careful management using a combination of physical and social science tools.

In these pages, "quality control" will mean the postproduction policing of output to determine whether it meets specifications, while "quality assurance" means application of rigorous scientific process design and management prior to and during production. Quality control looks for defects; quality assurance prevents them. The emphasis here—as it should be in the plant itself—is on replacement of quality control with quality assurance.

3.9 Warranties

Warranties—much like primitive management views of quality—have long been perceived by manufacturers as a cost rather than an opportunity. This view is unenlightened and highly dangerous to the company's competitive position. With some aggressive electronics companies already using more comprehensive and longer warranties as marketing tools, the pressure on other companies to expand their warranties will become increasingly intense. Indeed, a chasm already exists between manufacturers' explicit warranties and the coverage expected by the customer.

3.9.1 Explicit and Implicit Warranties

What is the product's "specified" life expectancy in the definition of dependability? The manufacturer explicitly promises only that the product will continue to work for the length of the warranty. Implicitly, however, the customer is encouraged to believe that the product will last considerably longer than the warranty period. Indeed, customers typically expect at least several years of trouble-free use from an electronic product even though warranties of more than a few months are relatively rare. To the customer, the promised life expectancy that is so important in determining dependability is many times longer than the normal warranty.

3.9.2 Warranty Legalities

Whether the manufacturer agrees with the customer's interpretation of a reasonable life expectancy for the product, two points need to be recognized by the manufacturing company:

1. Customers will base their subsequent purchasing patterns on the product dependability they experience, regardless of the manufacturer's paper commitment. Products that fail out of warranty but before the customer anticipated failure may not result in legal action but will certainly bias the customer against further purchases of the manufacturer's products.

2. The customer may be so displeased with the product failure and the limitations of warranty coverage that the matter will be taken to court. If that happens, the manufacturer may well lose despite the warranty terms provided in writing at the time of sale. Courts are increasingly siding with customers, particularly with respect to what constitutes a reasonable warranty period.[23] Moreover, there is no such thing as "winning" a lawsuit; the negative publicity typically far exceeds whatever monetary sums are claimed.

3.9.3 Warranty Implications for the Quality Profession

Legal issues aside, the gap between customer expectations and manufacturer guarantees should be very troubling to the quality profession. Undoubtedly the most serious quality issue facing the electronics assembly industry today is length of warranties. Many electronic products come with warranties as short as 3 months from the time of purchase—so brief that the warranty can expire before the customer so much as takes the product out of the box.[24]

Customers normally expect that the product will continue to work much longer than the warranty period. Regardless of whether a court ruling is obtained, they are usually right—electronic products *should* last years beyond the time when they become obsolete and no longer serve any functional purpose for the customer.[25] Current warranty cov-

---

[23][Feigenbaum 1991], pp. 35, 37.

[24]We actually experienced that very phenomenon when purchasing a photocopier recently. Delays in shipping (timely delivery is also a component of dependability) meant the copier did not arrive when needed. It sat in storage at our office for some 2 months until we did have need to use it. Upon being taken out of the shipping carton, the machine was found to have a paper handling problem. The manufacturer informed us that the warranty was only for 90 days and the clock had begun ticking on the invoice date even though the invoice date was a full month before arrival at our office. The bill for repairs was more than 40 percent of the purchase price.

[25]This is not an argument in favor of "planned" obsolescence. Electronics technology naturally advances so rapidly today that there is no need for manufacturers to artificially constrain the useful lives of their products.

erage of at most 2 or 3 years is obscene and shortsighted. If the average electronic component has a minimum life expectancy of several decades—which it does—why is it necessary for the assembly company to provide warranties of just a few years? The answer—examined in greater detail in Chaps. 6 through 9—is that standard production and quality assurance practices in electronics assembly plants damage components and cause them to fail early in life. Some components are so badly damaged during assembly and postproduction evaluation that they fail almost as soon as the product reaches the customer. Not all components experience serious damage during assembly, but the failure rate is vastly higher than necessary because of abusive production and quality control procedures.

The genuinely quality-conscious electronics manufacturing company recognizes that most premature failures result from damage caused in the plant by improper production and/or quality evaluation procedures. That quality-conscious company takes whatever measures are necessary to identify those aspects of its procedures that cause damage and change their procedures to prevent failures. The company can then offer warranties corresponding to the customer's rightful expectations.

Every electronics company talks about quality being the most important consideration in its business. We wish that the words carried through to reality, but they do not. Refusal to accept responsibility for their products' performance over a realistic length of time shows how few companies are truly committed to quality as anything more than a marketing ploy.

3.9.4 More Comprehensive Warranties Need Not Increase Costs

Up to a point, some customers are willing to pay more for better dependability. The premium that customers will pay for higher dependability is steadily shrinking, however. This unwillingness to pay more for higher dependability does not mean dependability is unimportant to the customer. Rather, customers see dependability as the manufacturer's responsibility and will find a more reliable vendor at a competitive price if failure rates force them to venture back into the market.

Feigenbaum quotes an unnamed "knowledgeable purchaser" as follows: "[We] must be willing to pay for those specific costs to improve reliability; however, we must separate the traditional and basic elements of good management and engineering from those unusual and justifiable expenditures for higher reliability achievement. We must insist that the term 'reliability' is not used as camouflage for additional charges for those functions which are an intrinsic part of an effective industrial operation."[26] In this context, the reference to "reliability" can be presumed synonymous with our "dependability."

So the customer is not willing to pay extra simply because the manufacturer has trouble mastering the management and engineering fundamentals. Nor should the customer be expected to pay for the manufacturer's incompetencies. High dependability, as will be shown beginning in Chap. 6, is generally consistent with lower costs, not greater costs.

This most important aspect of the warranty equation cannot be overemphasized: In the production of electronic products, it actually costs less to employ procedures that prevent damage to components. By that, we refer only to the actual costs of product design, assembly, and quality assurance. Reduced numbers of warranty claims—even though the warranty period is much longer—and increased demand from satisfied customers (and their acquaintances) are additional benefits to the manufacturer of greater product dependability.

3.9.5 Warranties as Competitive Advantages

In coming years, as product development times shrink, competition for market share will increasingly involve warranty terms. In other words, dependability will become the vital sales tool that undefined quality was expected—but failed—to be. The importance of longer warranties can already be seen in the marketing of credit card companies; in recent years, extended warranty coverage for products charged to their cards has become an important part of the benefits packages offered by most major credit card companies. The assembly technology exists to allow warranties of a decade or more.[27]

Of course, warranty claims are not desired by the customer any more than the manufacturer. The customer's most fervent desire is that the product will not fail, whether the repairs are covered by warranty or not. At the very least, a failure requires the customer to take the product

---

[26][Feigenbaum 1991], p. 573.

[27]Some observers question the need for warranties lasting a decade or more when technological advances will render the product obsolete much earlier. Two points illustrate the shallowness of such thinking. First, some customers will be satisfied with the services provided by a product based on more primitive technology; they have the moral right to continue enjoying those services without interruption. Second, if the majority of customers will discard the product long before the warranty expires, the manufacturer's real liability decreases proportionately even though the actual warranty coverage remains intact.

out of service and return it for repairs. On easily replaced, inexpensive items, the cost and trouble of obtaining repairs may be greater than the expense of buying a replacement and discarding the original. Customers may depend on some other products to such an extent that they can't wait for even a relatively brief repair cycle. In any event, while free repair can mitigate the damage of a field failure, it will never be an adequate substitute for freedom from failures.

3.9.6 Other Advantages of Warranties

If the customer does not want to make use of warranty service, the question arises why warranty coverage is so important. Several reasons can easily be identified, including:

1. To the customer, more comprehensive warranty coverage indicates greater manufacturer confidence in the product; this can be a powerful factor in purchase decisions.

2. A warranty is a form of insurance—a hedge against disaster but not something the customer hopes to need.

3. Longer warranties make manufacturers more aware of their true product dependability. With short warranties, manufacturers develop an undeservedly high opinion of their quality. Erroneously believing their dependability to be higher than is really the case, the manufacturers become complacent. And complacent companies are vulnerable companies.

Therefore, longer warranties are advantageous to the manufacturer as well as the customer. In several respects, more comprehensive and longer warranties help the manufacturer more than the customer.

3.9.7 When Longer Warranties Are Too Expensive

Despite the many benefits to be gained from extending warranty coverage, the objection is still widely heard that longer warranties increase costs. Admittedly, this can be true, especially for mechanical parts. In general, however, any increase in warranty claims resulting from broader warranty coverage will be offset by the gains. Few electronic component failures in the first decade are attributable to component manufacturers supplying defective parts. Rather, the failures result from flaws in the assembly plant's processes. It is better for the manufacturer to receive failed products back than to have the failures sit anonymously in the field. Returned units represent more than failures: They also provide information about necessary process improvements that would otherwise be lost.

3.10 Quality Is a Constant

By historical standards, the pace of product development today is extraordinarily fast. One consequence of this ever-shorter time between release of new models is that state-of-the-art product as measured by features can become merely ordinary or even inferior in a matter of a few months (or, increasingly, weeks). More attractive combinations of features or the same features at a lower price almost certainly become available as technology and fashions change. Accordingly, today's most attractive combination of features and price can be expected to have less appeal tomorrow. Such shifts in market conditions affect demand for the product because its relative features and value have changed. However, the underlying quality of the existing model—whether it will do what the seller says it will do and keep doing it for the (reasonable) specified time—will not vary unless changes are made in processes or components used.

A surprisingly large percentage of electronics assembly companies have not yet grasped the importance of dependability to the consumer. Those companies continue to battle for customers on the basis of features and appearance the way Detroit automakers used tail fins and chrome to sell cars right into the 1970s. Detroit discovered the painful way that many customers cared more about avoiding unplanned waits for a tow truck than they did about tail fins, portholes, or hood ornaments. Now every auto manufacturer emphasizes dependability in its advertising. But electronic products are still sold by stressing the "bells and whistles" rather than life expectancy.

The computer industry illustrates the point. Computer advertising talks about the speed and accessory content. Only rarely does marketing turn the spotlight on how dependably the systems perform. But research increasingly shows that dependability is now the principal issue for computer buyers. The following item from a leading computer industry magazine[28] noted the surging customer concern about dependability:

"Above all else, reliability is the main concern of desktop PC users.

"That's the word from the J.D. Power and Associates 1994 Desktop PC Customers Satisfaction Study released [during the week of Oct. 17, 1994].

---

[28] "HP Tops Survey on PC Satisfaction While IBM Lags," *PC Week*, Oct. 24, 1994.

"The group's annual study...found that reliability has replaced previous No. 1 concerns, including ease of use and user support, in the firm's satisfaction survey."

Another prominent computer industry magazine reported in a different context that "Users consider hardware reliability and service support the most critical criteria when buying a PC."[29]

A similar story could be written for most electronics products. Even with disposable items, the average consumer anticipates that the product will work as promised over the projected (if relatively short) life span. No matter how much market research goes into the features and how pretty the package, a prematurely dead product fails to meet the customer's primary concern: dependability.

The consequences of changes in dependability—whether improvements or reductions—typically do not show up in the short term. Reputations are rarely created or destroyed overnight. Thus it is possible to sell unreliable products for a while solely on the basis of appearances and high-powered marketing. Indeed, it is possible to make a great deal of money that way—in the short run. Some unethical or extremely myopic companies actually operate field service as profit centers; this sort of short-term thinking resulted in several lawsuits being brought in the United Kingdom against photocopier companies charging grossly inflated rates for maintenance contracts.

But the buying population learns eventually, and the amount of repeat business often plummets in the longer term with even moderate increases in the percentage of field failures. By the same token, improved dependability will seldom increase short-term demand; the positive effects of greater dependability manifest themselves only over longer periods. It takes some time for a sterling reputation to tarnish—and even longer to regain lost prestige.

3.11 The Quality Department's Role in Dependability

A common theme in the quality profession today is "Quality is everybody's business." There has been much confusion concerning the exact meaning of that statement, but recognizing that quality equals depend-

[29]Vijayan, Jaikumar, "Users Want PCs They Can Rely On," *ComputerWorld*, Dec. 26, 1995, p. 61.

ability makes interpretation easier. Dependability is a function of many activities, not all of which take place in the plant itself or are influenced by opinions—and actions—of the quality department. Circuitry design, for example, can generate dependability problems long before the quality department enters the picture. The quality department can and should identify the existence of the problem and possibly determine the cause(s) and solution(s); however, unless the quality department itself takes over the design function, its influence will be restricted to collecting data, analyzing its implications, and feeding this information back to the design department along with recommendations for changes in design guidelines.

Generally, the best that the quality department can hope to accomplish is that the design department will incorporate the findings on current problems and solutions in the CAD database for future products. Even under concurrent engineering where the quality department participates in decisions from design to evaluation, ultimately the extent of the quality department's influence on design will be mitigated by the design engineers' ability and/or willingness to comprehend the issues.

The extent to which the quality department affects dependability and costs depends in no small measure on the corporate environment in which the quality function is performed. A little quiet contemplation on the part of the quality professional should lead to the realization that the all-encompassing thesis of quality advocated by the "total quality" school of thought is neither desirable nor realistic for quality practitioner or company. Assuming responsibility for the direction of the entire company is, after all, a daunting task and one which exposes the designated leader to unavoidable criticism against which there is seldom adequate defense. On the other hand, the quality department can exercise more profound influence on the company's real quality concerns—i.e., dependability and avoidable costs associated with gaining or failing to attain dependability—than has been customary. The areas in which quality personnel can best operate and interact with the rest of the company can be seen in rough terms from Fig. 3.3, a simplified flowchart of the electronics business from product concept to delivery.

3.12 Redundant Circuitry

Dependability is less common in electronics assemblies than the industry admits to the public or, for that matter, to itself. Most electronics assembly personnel live with constant denial of reality. A good example of this self-delusion can be found in so-called high reliability (or "high

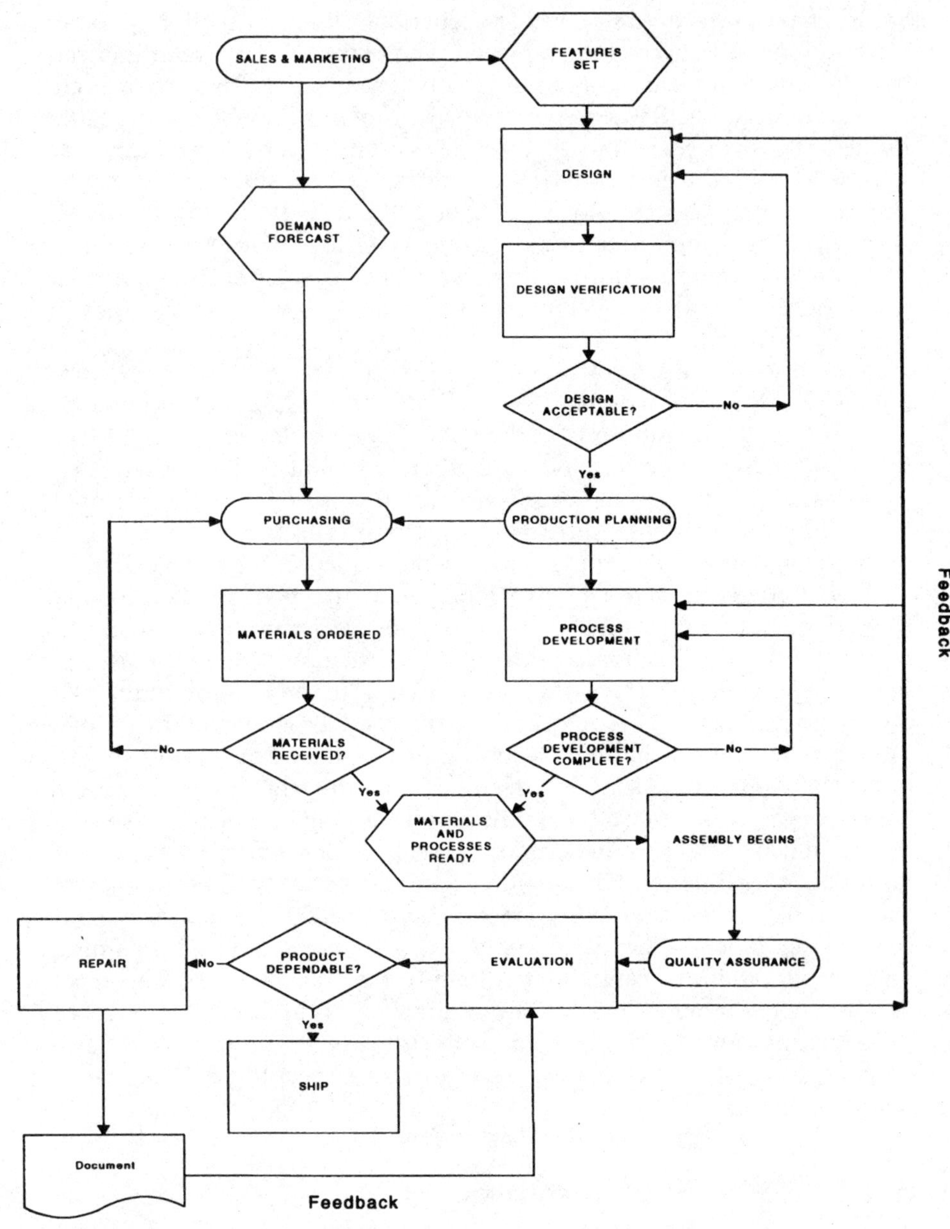

Figure 3.3. A simplified process flow from decision on features set to field performance.

rel" in the common jargon) where backup circuitry compensates for inability to prevent breakdowns in the first place.

Knowing that the failure rate of any given circuit is excessive,[30] the "high rel" manufacturer compensates by adding "redundant" circuitry that duplicates the primary circuitry. When the primary circuitry fails, the redundant circuitry takes over until the system can be taken out of service for repairs. Like touchup and burn-in, redundant circuitry is one of the prices the industry pays for the inability to get things right the first time, every time.

Exceptionally sensitive space and air guidance systems, such as inertial navigation systems, commonly employ not just duplicated but triplicated circuitry in which the central monitors rely on "two out of three voting." If one circuit goes bad, the remaining two are sufficient for continued operation until it is possible to replace the failed unit. If two circuits crash simultaneously, the system stops.

This same sort of multiple circuitry with central monitors employing either pure "two out of three voting" or some variation on that same theme is at the heart of "fly by wire" electronic aviation controls. Nuclear generating stations also depend on triplication of controls and critical circuitry.

Redundant circuitry and multiple "voting" can be found in more mundane applications than exotic avionics or nuclear power stations with their potential for widespread devastation. Ordinary electrical generating stations resort to the same redundancy in sensors—not to save lives but to eliminate costly unnecessary shutdowns. The electricity generating industry finds sensors so trouble-prone that they install multiple (usually three) identical sensors to monitor a single function such as temperature. If only one sensor indicates a system problem, it is regarded as an anomaly. The system shuts down only when at least two of the three sensors indicate a system fault. Multiplication of sensors cannot eliminate the need for greater sensor dependability in the first place, though; on occasion, two sensors have failed simultaneously and the only accurate reading comes from the one remaining instrument which majority voting systems would override.

For certain applications, the use of redundancy techniques can be the safest and most cost-effective way of providing the necessary dependability. Where the consequences of system failure would be catastrophic (financial ruin or loss of life), redundancy for the sake of safety is essential. However, redundancy must never be employed to paper over cracks left by inept design, injudicious selection of components, process

[30]"Excessive" varies from product to product depending on the consequences and frequency of failure.

management inadequacies, and damage caused by mishandling during assembly, inspection, and test. Regrettably, too many manufacturers employ redundancy techniques to obtain greater dependability rather than achieving that dependability by eliminating the causes of failures. Where failures would have catastrophic consequences, the prudent course of action is twofold: employing redundant circuits *and* tuning the production operations to prevent defects.

3.13 "High-Dependability" Components

In space systems, redundant circuitry is less feasible because every ounce added to the delivery system is an ounce less payload for delivery. Therefore, the manufacturers of electronic assemblies for use in space fall back on another questionable strategy: the use of "high-dependability" parts. The high-dependability component is produced in the same manner—and usually in the same batch—as its less illustrious siblings but is revealed through many rounds of intense screenings and measurements to possess (1) the least variability in performance and (2) a mean performance falling closest to the exact specified value. These stellar parts are sold, generally at prices 10 to 20 times higher than normal, for use in electronics assemblies destined for use in space. Some defense and communications systems also call for "high-dependability" components.

Unfortunately for all parties concerned—except the component vendors who enjoy major windfalls for delivering components that a cynic might say merely work as well as all new components should—the money spent to acquire and use these high-dependability components is pretty much wasted. The abuse to which those components are subjected in the assembly processes and subsequent test/rework/retest—all parts of the natural way of life in the "hi rel" electronics world—destroys the tight tolerances that made the parts special to begin with. For that matter, the degradation of those parts begins with the very testing used to identify the "best" parts.

3.14 Calculating the Company's Contribution to Lack of Dependability

Most electronics assembly companies do not know to what extent failures are caused by the companies' own processes. In our experience, component failures are universally attributed to component manufacturers. However, most component failures at test and in the field are the direct consequences of stresses endured during assembly and quality evaluation (inspection, tests, burn-in, etc.). Altering production and quality processes to eliminate those stresses would do away with the majority of such failures. We have seen failure rates reduced by more than 85 percent in plants where such modifications were made to processes and quality evaluation procedures.[31]

The first step toward greater dependability requires measuring the percent of failures caused by the company's own actions. The company's contribution to failures is found by comparing the actual failure rate to the failure rate expected from component reliability when those components arrive at the plant. This can be determined in a rough-and-ready but serviceable manner (through statistical sampling and simple component testing) as follows:

1. For every type of component, use statistical sampling to measure the percentage that are not dead when they arrive at the receiving dock. The percentage of component type T that are good on arrival is GOA_T. The GOA will vary from one type of component to another. Factors such as differing suppliers and different production batches can cause component reliability to vary from shipment to shipment. For accurate analysis, those factors should be taken into account but are ignored here.

2. If an assembly consists of N different component types, the predicted failure rate caused by components that were DOA is $FR_P = 1 - [(GOA_1)^{U1} \times (GOA_2)^{U2} \cdots \times (GOA_N)^{UN}]$ where ^{UT} is the number of component type T used in the assembly. This failure rate is directly attributable to errors on the part of the component manufacturers and/or distributors.

For example, assume that the assembly consists only of 5 units of component type A, 3 units of component type B, and the circuit board itself, which we will designate C. If 99 out of every 100 units of component A are good on arrival, the probability that a unit of A *will not* fail in the assembly is $(0.99)^5 = 0.951$. Similarly, if 997 out of every 1000 units of component are good on arrival, the probability that a unit of B *will not* fail in the assembly is $(0.995)^3 = 0.985$. And if 983 out of every 1000 circuit boards C are good on arrival, the probability that a unit of C will not fail is 0.983. Conversely, the probability that one or more units of component A *will* fail is $1 - 0.951 = 0.049$ (i.e., 4.9 percent); that one or more units of component B will fail is $1 - 0.985 = 0.015$; and that the circuit board C will fail is $1 - 0.983 = 0.017$. The probability that the assembly

[31]Some of the more important process changes are explained in Chaps. 7 and 9.

will fail based on the defective rate on arrival is $1 - [(0.951) \times (0.985) \times (0.983)] = 1 - 0.921 = 0.079$. In other words, the assembly has a 7.9 percent probability of failing because of a component that was dead when it arrived at the plant.

For an assembly consisting of N component types, the predicted failure rate from components that were DOA is $FR_P = 1 - [(GOA_1)^{U1} \times (GOA_2)^{U2} \cdots \times (GOA_N)^{UN}]$ where ^{UY} represents the number of component y used in the assembly. This failure rate is directly attributable to errors on the part of the component manufacturers and/or distributors.

3. Measure the real failure rate—FR_R—from plant and warranty data. Invariably, FR_R will be greater—often magnitudes greater—than FR_P.

4. Subtract FR_P from FR_R. The result approximately equals the damage caused by assembly, test, and other plant activities. In other words, $FR_P - FR_R = FR_I$, where FR_I is the induced failure rate from plant activities.

3.14.1 Complicating Factors

Although the basic methodology for determining the assembly company's contribution to failures is fairly simple, several factors complicate the exercise and challenge the abilities of the quality statistician.

3.14.1.1 Stacked Tolerances. The company's contribution to failures will be determined with greater accuracy when failures caused by "stacked tolerances" are taken into account as well. Most components as received operate at a value other than "perfect"; that, of course, is the reason component values are specified with upper and lower acceptance limits. Stacked tolerances are the total system variation resulting from accumulated variations in performance values of components as received. While each individual component may perform within the tolerances specified, the accumulation of variations can result in a product that fails to function within the required limits. Ignoring the effects of stacked tolerances in calculating probable failure rates results in understatement of probable failures.

3.14.1.2 Failure to Compile Dependability Data. Very few companies attempt to identify field failures, compile the failure rates, and analyze the root causes. In most cases, staff to collect and analyze actual field performance does not exist so the company does not own any meaningful dependability data. The lack of staff has two primary causes: budget constraints and lack of awareness by the quality department that field dependability is a concern. Our experience strongly suggests that the data on failures goes uncollected and unanalyzed because there is little comprehension that the need exists. Of course, in the absence of data showing that a problem with after-sales dependability exists, a manufacturer can easily acquire a false sense of accomplishment.

3.14.1.3 Failures Not Reported. An accurate record of failure rates cannot be established if some failures are not reported. Customers may not report failures even when the units are covered by warranty. This can happen if:

1. The cost and inconvenience to the customer of returning the failed product exceed the value of the product to the customer.
2. The product has such great value to the customer that waiting for service is not an option. The customer buys a new unit rather than sending the failed unit back for service.
3. The customer has lost the proof of purchase.
4. The customer has moved to another region (usually another country) where the warranty coverage is not available.

3.14.1.4 The Difficulty of Separating Real Failures from False Failures. Despite the company's best efforts to collect information about failure rates, the data collected may be inaccurate. Forces exist that both inflate and understate the field failure rate, including:

1. *Excessively generous customer return policies of powerful retailers.* Many retailers accept returns from customers even when it is clear there is no defect in the product. One well-known example concerns video cameras sold on Fridays and returned Monday; most of the customers involved "buy" the camera to record a special occasion such as a wedding with the full intent of returning the item after the event. When "failures" are concocted by the consumer to return a perfectly good product, identifying the true failure rate for the product becomes vastly more complicated. Such an environment also creates large numbers of "no fault found" failures. The data collected will be biased in favor of a measured failure rate greater than the actual rate.

2. *Inept design and documentation.* Some customers return otherwise dependable products only because they can't understand how to make them work. Such returns should be classified as dependability failures. Design dependability would make the operation obvious to even the technical novice. Documentation dependability would fill in what is not obvious from the design. Some consumers will never open even the

most lucid manual, but others do try, only to find that shoddy writing makes the instructions incomprehensible. It is the manufacturer's responsibility—and self-interest—to ensure that the customers receive easily understood and inviting instruction sets. But dependability actually begins with a user-friendly design; dependable design eliminates the need for most documentation, since operation of the product is readily apparent from the controls.

3. *"No fault found" failures.* The bane of every electronics assembly company is the field failure that cannot be reproduced in the company's repair department. Many companies label such returns "false failures" and believe that the only problem is in the customer's mind. Often, however, the failures are very real—if erratic—and the fault lies in the manufacturer's diagnostic abilities rather than the customer's use. Numerous conditions cause intermittent failures that cannot be duplicated on the repair bench. The most common is ionic contamination in which a layer of invisible ionic particles (generally left by fluxes, though circuit board fabrication and even airborne salts from nearby oceans are common occurrences) reduces the insulation between conductive surfaces to the point where current leakage occurs. Current leakage increases with humidity, which explains why failures happen more frequently in locations such as Florida than in Arizona. Intermittent failures induced by ionic contamination compromise dependability every bit as much as "hard" component failures.

3.15 Summary

- Quality programs can produce tangible benefits only if "quality" itself is defined in objective, measurable terms.
- Though many existing definitions of quality contain excellent concepts, they also involve subjective elements that prevent measurement of results and/or they define quality in terms that are not relevant to electronic products.
- Quality must not be defined as something determined by the customer.
- In the electronics assembly industry, quality means "dependability," that is, the degree to which the product will provide all the features promised by the manufacturer without interruption for the specified period of time under specified conditions of use.

- Dependability results from factors beyond the control of the quality department as well as forces which are the responsibility of the quality department.
- Most product failures are caused by the manufacturer's own processes, procedures, and standards. The extent to which the manufacturer creates failures can and should be measured, although the measurements will not be precisely accurate.
- Warranties are unreasonably short. A "reasonable" period of time for an electronic product to work should normally be a matter of at least a decade.
- Longer and more comprehensive warranties will become increasingly important sales and marketing tools.
- Manufacturers can measure the extent to which their own actions—including production and evaluation procedures—increase the failure rate.

4 Theoretical Benefits of Improving Quality

4.1 Chapter Objectives

This chapter examines reasons why quality improvement might be desirable for an electronics assembly company, including the propositions that:

- Quality improvement may cause greater demand, higher revenues, and added profits.
- Quality improvements may reduce product costs.

It will be shown that the benefits of quality improvement should be considered from two perspectives:

- The demand side (i.e., effects on customer purchasing decisions if quality changes)
- The supply side (i.e., reductions in costs)

Of these, the supply side benefits will normally be most significant.

4.2 Why Quality Improvement Might Be Desirable

Chapter 3 showed that quality in the context of electronics products should be defined as dependability. In other words, the quality of an electronic product is the degree to which the product will provide all the features promised by the manufacturer without interruption for a specified period under specified conditions of use.[1] The longer the product continues to provide its promised features without failure, the greater the dependability and thus the higher the quality.

Although the meaning of quality is now clear, another fundamental quality improvement question arises: "Why would an electronics manufacturer want more quality?"

All quality professionals—ourselves included—believe that quality improvements lead to a healthier manufacturing company. Disagreement enters the picture only when assessing reasons why improved quality should help the company.

Entire libraries have been published and vast numbers of speeches delivered about the ways in which better quality could benefit the company. Essentially, however, all the theoretical benefits of quality improvement can be grouped into two categories, which we will designate demand side benefits (i.e., effect on sales) and supply side benefits (i.e., effect on costs of production and customer service).

4.3 Two Sides of the Benefits Equation

The demand side benefits capture the changes in customer purchasing patterns caused by changes in quality. Generally speaking, it is believed that quality improvement will cause demand for the product to grow and produce three primary demand side benefits:

1. Sales of the product rise.
2. Revenues for the company increase.
3. Profits go up.

Similarly, if the quality level goes down, all three financial indicators are expected to fall.

---

[1]See Chap. 3.

Supply side benefits consist of the savings that the company may enjoy as a consequence of the product's working more dependably. Some of the anticipated supply side benefits from quality improvement include:

- Reductions in repair and retest because there are fewer failures—in the plant or in use by the customer—to find and correct
- Less scrap because of fewer defective parts and assemblies
- Easier inventory control, since it will not be necessary to plan for as many replacement parts
- Fewer warranty claims, translating into reduced demand for field service, repairs, and replacements
- Faster cycle time (i.e., time from start of production to shipment of product) because the number of process steps falls

In analyzing the effects of quality improvement, two important points must be kept in mind:

1. Demand side quality improvement is a relative concept. If we improve our quality but the competition matches our gains, our relative quality has not changed. If a competitor's quality improves while ours stays the same, we have effectively suffered reduced quality. Only when our quality increases relative to the rest of the industry can we be said to have experienced demand side quality improvement. Therefore, a company's quality level must be assessed in a market context. While the manufacturer typically can tell at best only if its own absolute level of quality is rising or falling and has a very cloudy picture of how the competition's quality is changing, the quality manager must be continually aware that competitors are unlikely to be resting on their laurels.

2. Both types of benefits—particularly on the demand side—are primarily theoretical. They are gains that companies embarking on quality improvement projects are told to expect. However, there are reasons to believe that at least some of the promised benefits are unlikely to be realized by many or any companies embarked on quality improvement programs.

The remainder of this chapter examines (1) the extent to which these claims of benefits from quality improvement programs are reasonable and (2) reasons why not every quality improvement initiative brings about a positive net return.

4.4 Demand Side Benefits of Quality Improvement

For many quality experts, demand side benefits have long been seen as the most pressing reason to pursue higher quality. Deming's insistence that the most important benefits of quality improvement are "unknown and unknowable," for example, originates in the closely related beliefs that (1) better quality always leads to greater demand, and (2) the magnitude of the greater demand will always exceed the costs and efforts of achieving the higher quality. Similarly, the theory of customer-driven quality depends on a strongly positive correlation between the quality level and demand. If it turns out that the links between quality and demand are weaker than has been assumed, much of the conventional wisdom of quality must be revised.

As it happens, the relationship between quality and demand may be substantially less meaningful than mainstream quality theory suggests. Although considerable effort has been expended on studying the effects of increased quality on demand, the evidence to show that demand increases appreciably with quality improvement is not persuasive. The theory is attractive but the reality remains to be proved.

4.4.1 The Relationship between Quality and Demand

The conventional wisdom of demand side quality—as in Emerson's better mousetrap advice—contends that quality improvement must lead to more demand. Deming agreed with Emerson, arguing that "The happy customer that comes back for more is worth 10 prospects. He comes without advertising or persuasion, and he may even bring a friend."[2] Juran also looked at what happens if relative quality deteriorates and concluded: "In recent years, many key industries (automobiles, color television sets, computer chips, and so on) have lost more than 25 percent of their sales to foreign competitors. A leading cause has been quality of product."[3] A study by the Technical Assistance Resource Project for the U.S. Office of Consumer Affairs shows that the cost of acquiring a new customer is five times the cost of keeping an existing customer.[4] Countless other quality theorists and practitioners have reached similar conclusions.

---

[2][Deming 1986], p. 122.

[3][Juran 1988], p. 1.

[4]Stewart, Thomas A. "After All You've Done for Your Customers, Why Are They Still Not Happy?" *Fortune*, Dec. 11, 1995, pp. 178–182.

On the surface, logic suggests that demand should vary directly with quality. Among other factors, the value equation derived in Sec. 3.5 tells us value = performance ÷ price; since value would rise with greater dependability, it would seem inevitable that demand must also increase. However, the logical outcome may not be the real outcome.

Clearly, there are many instances when demand functions according to theoretical expectations. Large numbers of real world cases can easily be identified where demand has gone up with better quality and down when quality has fallen.[5] On the other hand, real world examples also exist where demand went up without any noticeable improvement—occasionally even with a marked decrease—in the product's dependability (either absolute or relative).[6] The apparently perverse increase in demand although the quality did not improve can be explained by again referring to the value equation; although dependability may have decreased, other aspect(s) of the customer's needs or desires did change so that the value of nonquality features increased.

The question is not whether demand *can* increase with quality but whether it *must.* Unfortunately, the evidence that demand varies directly with quality is largely anecdotal or even fictitious. We can say with confidence that increased quality may lead to greater demand, but we cannot say with confidence that it will. In other words, we cannot state categorically that quality improvement is either necessary or sufficient for increased demand.

Of course, actions of many departments other than quality affect demand, as Fig. 4.1 shows. The design and production departments obviously affect dependability. Features are determined by the sales and marketing departments. And the actions of many departments affect selling price. In turn, dependability, features, and price control demand.

Our ability to predict how changes in quality will affect demand will be enhanced by recognizing the existence of two important market forces that can confuse the customer: (1) that competing products are

---

[5]The introduction of quartz watches shattered the traditional precision Swiss mechanical watch industry between 1973 and 1975. The relatively cheap quartz watches were more reliable than Swiss mechanical watches even though the precision of Swiss watches had not fallen. Ford lost substantial market share when it became apparent that its cars rusted more rapidly than other makes.

[6]Responding to the quartz watch technology, manufacturers of mechanical Swiss watches refocused on watches as jewelry. The market repositioning strategy was enormously successful, even though the Swiss watch precision remained the same as it had been before the discovery of quartz mechanisms. Similarly, demand for smaller (and therefore more fuel-efficient) cars increased sharply during the OPEC-created oil shortages of the 1970s, even though none had improved in dependability.

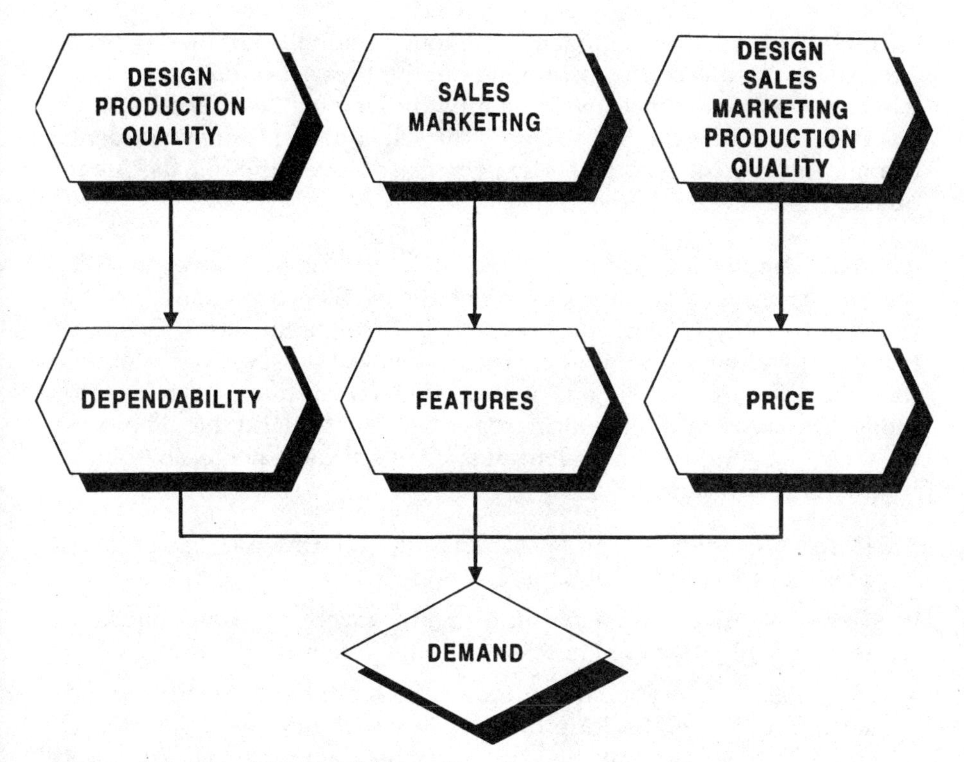

Figure 4.1. Determinants of demand.

rarely identical in all major aspects of features and prices, and (2) that the customer does not have complete knowledge of all product features including relative dependabilities.

4.4.1.1 All Other Things Being Equal. Quality improvement will most likely lead to increased demand when all other things are equal. In other words, the feature sets (excluding dependability) and prices of competing products must be the same so that the only substantive difference is in the level of quality. This condition often does not apply. Features—which include nonperformance properties such as appearance—typically vary widely among competing products; a calculator may be purchased solely because the customer prefers the shape or color of its case. Prices are often determined by subjective customer preferences for certain features over others; pink plastic flamingos may be preferred to other colors, but attempts to sell computers and television sets in nontraditional case colors (including pink) have failed.

The lack of genuine homogeneity among nominally competing products causes serious confusion among quality theorists who try to reconcile customer preferences with quality. By Deming's standard, a pink computer has low quality if it fails to sell. But if fashions suddenly change in favor of pink, sales will increase—even though the innate quality remains exactly the same.

4.4.1.2 Imperfect Information. Attaining greater dependability will not improve demand if the potential customers are unaware of the improved quality. Quality must not only be better, it must be *known* to be better. Depending on the numbers and diversity of prospective customers, the difficulties in spreading awareness of increased quality can be enormous. The barriers to convincing customers that quality has improved must not be overlooked or underestimated. A few more notable barriers to information flow include:

- *Money.* Techniques for disseminating information range from advertising to public relations, but none is free.
- *Access.* We live in an age of information overload. Many messages are physically delivered but relatively few are mentally received.
- *Obfuscation.* Every company today claims to have "quality" products and can often back up those claims with one set of statistics or another. Depending on the statistics chosen, the message conveyed can be misleading rather than truthful. As this was being written, for example, Chrysler came under attack from a major shareholder who claimed that the company's quality was faltering as evidenced by rising warranty costs; Chrysler's management responded with two points: (1) that length and coverage of the company's warranties had increased and (2) that the number of warranty failures was decreasing but the greater complexity of components increased the cost of each repair.
- *Ingrained buying patterns.* Many consumers repeatedly purchase the same brand—even the same model—because they are familiar with its characteristics and not sufficiently disturbed by its performance failings that they will search for alternatives. Other products may offer better dependability, but the brand-loyal customer is not willing to take a chance on finding out. Brand names traditionally command higher prices than identical nonbranded products made by the same manufacturer.
- *Consumer cynicism.* The days when companies could easily attract attention by advertising that they strive for quality are over. In an era

where consumers are inundated with advertising of everything from "new and improved" soaps to "higher-reliability" cars, customers have become cynical about company claims to have higher quality. Advertising copy that falls back on quality is no longer daring or even effective; it is simply clichéd, and no industry has relied more on "quality" advertising slogans than electronics products. The extent to which this industry depends on quality themes for its advertising messages is readily apparent by leafing through any of the trade publications. Different full-page advertisements in a single issue of *Electronic Business Today*,[7] for example, contained the following phrases:

- "We take customer satisfaction seriously."
- "Experience the service that wins the awards."
- "We don't just make it to market—we continually make it better."
- "Delivered on time with ZERO defects."
- "Design, quality and lasting performance."
- "It seems to be almost a daily occurrence around here to win a major quality award."[8]
- "Customer satisfaction isn't just our commitment . . . it's our way of life."
- "Total Quality Total Service."

Accordingly, word of any change—notably including quality improvement—does not spread quickly or freely. That is why advertising and other forms of promotion exist. Just as it was naive for Emerson to suggest that customers would seek out better goods in the forest, it is unrealistic to expect the customers to recognize better quality in the short term, particularly if the quality information is not spread by a powerful marketing campaign.

In contrast to the difficulties of getting across the message that quality has improved, information about serious quality problems spreads quickly, attracts considerable attention from customers, and is retained in the human memory for depressingly long periods. Ford Motor Company, for example, did not completely regain the confidence of auto buyers until more than a decade after its well-known problems with premature rusting had been solved in the 1970s. Long after Japanese

---

[7]October 1995.

[8]A sad reminder, it seems to us, of the rampant proliferation of quality awards. The number of quality awards is now so vast that no award today really carries any prestige.

companies had turned the radio industry upside down with solid state circuits, "Made in Japan" remained synonymous with inferior products. A strong argument can therefore be presented that preventing any drop-off in dependability should be of greater concern to the quality department than constantly pushing for additional incremental quality improvements—although neither aspect of quality management can safely be ignored.

The more significant the role played by the product in the customer's life, the more concerned the customer will be about a manufacturer's dependability record. Significance takes two main forms: (1) price in relation to disposable income and (2) the consequences if the product fails to work. It is reasonable to assume that patients would want to know the dependability record of whatever brand of heart pacemaker they receive because the consequences of failure would be catastrophic. Dependability records of cars and other durable goods attract considerable customer interest because failures typically involve high monetary and convenience costs. But many consumers buy low-priced products such as electric shavers or portable radios fully expecting that the products will be short-lived. (Ironically, the dependability of many mass-produced inexpensive products is at least as good as that of high-priced products which the customer cannot do without.)

4.4.2 The Relationship between Quality and Profits

The more meaningful issue, however, is not whether higher quality increases demand but whether it increases the company's long-term earnings. As Goldratt explained in *The Goal,*[9] a business activity is desirable if and only if it moves the company closer to maximizing long-term profits without reducing short-term cash flow below the level needed for survival into the long term. Conversely, any activity that does not lead to increased long-term profits or that reduces short-term cash flow below survival levels is undesirable. If higher quality is desirable for the company, it must therefore be because greater dependability produces higher long-term profits.

The link between happy customers and greater demand is intuitively apparent, though it would be comforting to possess more statistical data on whether the connection really exists. If quality improvement leads to greater demand with the same or greater profit per unit, then the invest-

[9][Goldratt 1992].

ment in improved quality would seem to meet the requirement of greater long-term profits (unless, of course, the gross extra profit generated by the additional demand is outweighed by the cost of the better quality). But, while instinct (and a certain amount of professional self-interest) tells the quality specialist that dependability improvement is consistent with the goal of higher long-term profits, our instincts may be wrong. Indeed, it is entirely possible (and not uncommon) to go broke keeping customers happy. As an example, perhaps the most direct way to satisfy customers is by lowering prices (thereby increasing the product's value to the customer). The happiest customers will be those who pay less than the manufacturer's cost—a condition that can result in extremely unhappy (and insolvent) manufacturers.[10]

It is at least theoretically possible for demand to fall but profits to rise if greater quality allows the company to increase the profit per unit by a large amount. As price goes up, the affordability factor in the demand equation forces some potential customers out of the market. However, not all customers will be equally sensitive to price increases. Price hikes therefore normally drive out some but not all customers. If the profit per unit goes up by a greater percentage than the quantity sold falls, earnings will be greater than at the previous lower price and higher sales volume. This can be seen from the equation

$$P_T = Q_T \times P_U$$

where P_T = total profit
Q_T = total quantity sold
P_U = profit per unit sold

So, does quality improvement lead to greater profitability? Unfortunately, there is absolutely no statistically valid evidence to show a direct correlation exists between the levels of quality and profits. Many studies that claim to have found a direct cause-and-effect relationship between quality programs and profits[11] suffer fatal flaws, including:

1. *Faulty definitions of quality.* As Chap. 3 explained, no accurate operational definition for quality—which we now identify as dependability—has existed. The previously proposed definitions of quality

[10]The old vaudeville chestnut "We lose money on every unit but we make it up in volume" applies here.

[11][Hiam 1993] alone lists 20 such studies, including a large-scale analysis by the U.S. General Accounting Office in 1991.

(such as "happy customers") involve considerable subjectivity. No valid measurements can be made of a subjective element.

Because the various concepts of quality on which these studies have been based are nebulous, the studies often mistake productivity improvements for quality improvements. While it is true that quality improvements often result in productivity increases (if failure rates go down, the amount of repair and retesting declines as well), productivity is quite different from quality. Similarly, the literature contains innumerable references to scrap reductions, shorter cycle time, lower inventory, fewer defects, and so on—all desirable changes to be sure but none of them specifically definable as quality improvements. There are many ways to bring about changes such as scrap reductions, shorter cycle time, and the other benefits normally credited to improvements in quality, but not all or necessarily even most of those ways relate to greater dependability.

2. *Use of short-term earnings.* The company's goal must be long-term profit maximization. In the short term, the only proper financial concern (aside from ensuring that the company is making progress toward increased long-term profits) is whether cash flow enables the company to survive. The frequently voiced criticism of American management that it emphasizes short-term results at the expense of long-term success has considerable validity. Cuts in research and development, use of inferior parts, and truncated warranty periods are all consistent with increasing short-term earnings but not with prosperity over the longer term. Or, to repeat Handy's words from Sec. 1.4, "Short-term profit at the expense of quality will lead to short-term lives."[12] While all authorities seem to agree that short-term financial performance has little significance, the financial figures used as the basis for studying the demand side benefits of quality improvement have all been for relatively brief periods. Today's sales, revenues, and profits are not necessarily good indicators of whether the company is enhancing its long-term earnings capability.

3. *Qualification methodology.* Since the studies are not based on a single objective definition of quality, the study supervisors have no way to determine whether the companies selected are actually practicing quality improvement or only believe they are. Most studies base their selection criteria on the Malcolm Baldrige National Quality Award model; however, as Chap. 16 explains, the Baldrige standards are highly subjective and therefore defy quantifiable analysis. Just as the gap between

---

[12][Handy 1989], p. 114.

how a company perceives its quality and the true dependability of its products can be enormous, many companies that consider themselves to be well advanced along the quality improvement road have not even started their engines.

4. *Lack of industry specialization.* It is entirely possible that actions which produce significantly positive results in one type of industry may have no effect or negative effects in other industries. Aside from the necessity of earning profits, just how many characteristics would an electronics assembly company share with a retail clothing store or law firm? Experience tells us there is little commonality across such broad industry borders. Therefore, when manufacturing companies are combined with service companies in the same sample population, the findings are unlikely to be accurate for either. Despite the distortions created by mixing data from different industries, however, the studies have all included a variety of industries. Attempts to establish homogeneity in the sample population have been limited to broad industrial classifications such as manufacturing.

5. *Absence of rigorous statistical controls.* A properly structured statistical study would compare results of the sample population to results of a control group lacking the relevant characteristics of the sample population. Few of the studies have involved control groups, however, which means no statistically valid conclusions can be drawn from their findings.

All told, distressingly little statistically valid evidence exists to prove that quality improvement must lead to greater profitability. On the other hand, neither is there any proof that quality improvement will *not* improve profitability. The only truly valid statement at this time is that the linkage between quality and profitability has yet to be established in rigorous scientific fashion.

4.5 Supply Side Benefits of Quality Improvement

Although the case for demand side benefits remains unproved, it *can* be proved that supply side benefits of quality improvement in the electronics assembly industry are real, tangible, and—within limits—inevitable by-products of any competently designed quality improvement program. This happy blend of higher dependability and lower costs occurs because the steps required to improve quality (dependability) in the electronics assembly industry generally involve eliminating assembly and quality assurance activities. Lower costs mean the company has greater flexibility in setting prices. If competitors reduce prices to win market share, the company that is experiencing supply side quality benefits will be more able to reduce its own prices without losing money. Where there are no competitive pressures on prices, the company can maintain its existing prices and watch its profits increase by the full amount of the supply side quality benefits.

When Crosby said, "Quality is not only free, it is an honest-to-everything profit maker,"[13] it was supply side quality benefits to which he was referring. This can be seen from his description of how quality becomes a profit maker: "Every penny you don't spend on doing things wrong, over, or instead becomes half a penny right on the bottom line."[14] Similarly, Deming was clearly thinking of supply side benefits when he wrote, "Improvement of quality transfers waste of manhours and of machine time into the manufacture of good product and better service. The result is a chain reaction—lower costs, better competitive position, happier people on the job, jobs and more jobs."[15]

4.5.1 Quality and Costs

The emphasis on demand side benefits of quality improvement is a relatively recent phenomenon. Historically, most quality specialists stressed the supply side effects of quality improvements; that is, the gains from quality improvement were expected to be more noticeable as cost reductions than through increased demand.

Juran, for example, referred to the cost reducing effects of quality improvement as "gold in the mine."[16] To make his point, Juran refers to the experience of the "ABC electronics company" with annual sales of $1 billion. In Juran's account, the company determined that a one-time expenditure of $20 million would reduce costs by $100 million per year—that is, an annual increase in pre-tax earnings of $100 million or a 500 percent annual return on investment. The same company calculated that it

---

[13][Crosby 1979], p. 1.

[14]The reason why a penny saved loses half its value when added to the bottom line has confused many of Crosby's readers. As was noted in Chap. 2, the unspoken assumptions in this statement are that the company is profitable and paying income taxes at the rate of 50 percent on every additional dollar earned. These assumptions do not apply in every case; many profitable companies pay much less than a 50 percent tax on earnings. Therefore, Crosby's message becomes clearer and more forceful if the implicit assumptions are set aside and the statement changed to "every penny saved is a penny added to pre-tax earnings."

[15][Deming 1986], p. 2.

[16][Juran and Gryna 1988], pp. 22.13–22.16.

would need to spend $500 million to increase sales sufficiently to achieve a demand-driven annual profit increment of the same $100 million.

While Juran's story—based on a "universal sample of one"—has no more empirical or statistical value than Deming's claim that one happy customer is worth ten prospects, it does illustrate the greater importance of cost reductions compared to demand increases in the minds of the quality pioneers. It also drives home the noteworthy point that an extra dollar of sales likely adds only between a few pennies and perhaps 40 cents to the well-known bottom line while a dollar of cost reductions equals a dollar of extra profit. That is, the multiplier effect of cost reductions is considerably greater than the effect of sales increases. So the supply side benefits of quality improvement should be more significant than demand side benefits.

Lower costs have wide-ranging implications in addition to contributing directly to higher profits. A manufacturer with lower costs is better positioned to respond if competitors lower prices. Alternatively, the company can boost demand by reducing prices in line with cost reductions. The extent to which price reductions affect demand depends on the sensitivity of customers to price changes; economists refer to this as the "elasticity of demand." If the customer base is characterized by tight budgets, the affordability factor plays a major role and demand should rise by a greater percentage than the price reduction. The disciplined company always establishes price at the point where total profits are maximized. Since total profits equal the profit per unit multiplied by the number of units sold, the company's interests may be best served by holding the line on prices even though costs have decreased.[17]

4.5.2 Advantages to the Quality Department of Supply Side Benefits

For the quality specialist, the most attractive feature of supply side benefits from quality improvement should be that they are easily identified and readily quantifiable. Armed with hard data about cost reductions attributable to quality improvements, the quality department is better positioned to convince nonquality executives that past activities to increased quality were wise investments and/or that additional quality improvement projects are justifiable on the basis of standard business criteria.

---

[17]Unfortunately, few things in social sciences are as easy as they first appear. If holding prices up provides an umbrella to shelter new competitors, a better long-term strategy will be to set prices lower and discourage new entrants in the market.

Demand side benefits do not offer such opportunities. The intangible nature of demand side quality benefits has long troubled nonquality specialists by emphasizing the importance of gains that are, in Deming's words, "unknown and unknowable." Executives outside the quality department do not like acting on faith; they are infinitely more comfortable acting on results that can be identified and measured (i.e., supply side benefits). Quality personnel should be equally reluctant to base the company's welfare and their own careers on the hopes and hunches that characterize most demand side quality benefits.

4.5.3 Means of Improving Dependability

Dependability can be improved in a variety of ways that produce supply side benefits (i.e., reduce costs), the four most notable being:

1. Changes in design (making the product more "robust," in currently favored jargon)
2. Use of more dependable components
3. Eliminating handling steps (including reducing the number of quality verification steps that product experiences before shipping)
4. Process modifications to reduce the stresses to which components and assemblies are subjected (e.g., use of less aggressive fluxes, scientific heat control during assembly)

It must therefore be recognized that the quality department does not operate in isolation from the rest of the company. The quality assurance function does affect dependability; so, however, do activities in the design, purchasing, and production departments. The linked effects of these departments on dependability are shown in Fig. 4.2.

4.6 Implications of Quality Improvement in the Electronics Assembly Industry

The demand side consequences for the electronics industry from increased quality should be no different from the consequences in other industries with one possible exception. In many industries, dependability is less significant in the value equation than is the case in electronic

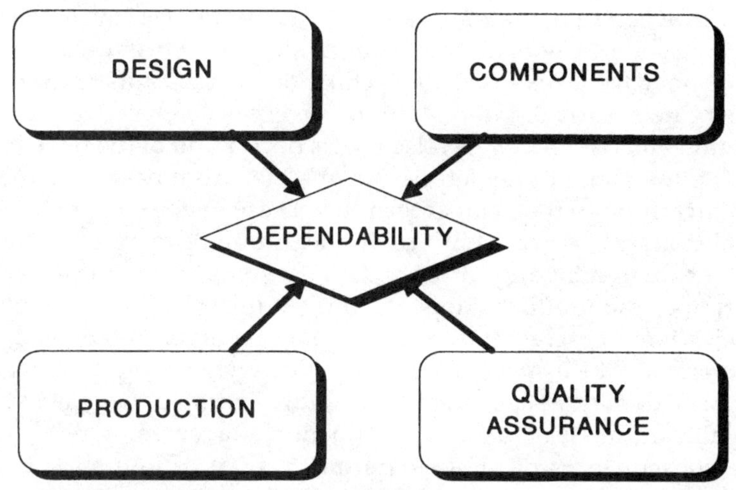

Figure 4.2. Determinants of dependability.

products. The dependability and operating features of most electronic products can be more easily measured than in less tangible businesses such as clothing or food. The results can also be expressed in the same numerical terms for all products in the category. An oscillator's features, for example, can be expressed in nominal frequency and range of variability while customers will discover the dependability even without input from the manufacturer.

Although demand side considerations cannot be overlooked in electronics assembly, supply side returns should always dominate the industry's benefits equation. The only meaningful exception to this rule occurs if a manufacturer's quality falls so noticeably relative to its competitors that customers eliminate the company as a supplier. Having lower production costs provides little comfort to—and cannot save—a company whose products cannot be sold.

Quality analysts have long recognized the supply side savings associated with quality improvement. However, the analysts' focus has normally been on defect prevention. Defect scores can be terribly misleading, however. While real defects certainly represent a component of quality, the vast majority of "defects" in electronics assembly operations are visual and subjective. It is not unusual to find that 90 percent or more of all visual rejects are actually perfectly reliable. Those false defects are the consequences of improper specifications and interpretation. Being unrelated to product design, production management, or any other aspect of operations, false defects cannot be prevented by any company action except changes in the evaluation procedures.

While not a trivial exercise, the inclination of inspectors to reject good product can be turned around with relatively little investment in documentation and education. As false rejects decline, all of the rework associated with making those defects "acceptable" disappears and the flow of product through the plant accelerates. These results are often praised as quality improvements but are, in fact, nothing more startling than proper evaluation of output; since the false defects never threatened to compromise the product's dependability, it is difficult to see how ending rejection of acceptable product can be interpreted as quality improvement. The inspection personnel do, however, typically work for the quality department, so the failure of evaluation personnel to accept acceptable output is a quality department management issue.

The consequences of failures are much more serious than those of defects. Failures prior to shipping require—at a minimum—troubleshooting to determine the causes of failure, replacement of the responsible components, and retesting. Failures after the product reaches the customer are still more costly, even if the possible demand side harm is discounted. Employee time is required to process customer calls, receive incoming systems for repair (or dispatch service personnel to the customer site), troubleshoot, replace defective components, conduct validation exercises such as testing, package, and return the unit to the customer. Some element of disassembly and reassembly may also be required.

In work done by the authors with respect to cost of failures at various stages in the assembly and use of avionics equipment, it was found that the cost of failures was relatively low early in the cycle but grew more than exponentially as assembly progressed and the product reached the customer. A typical sequence in the unit's production through to use would have the following sequence of stages:

Stage 1: PC board assembly

Stage 2: Subassembly

Stage 3: Line replaceable units (LRU)[18]

Stage 4: System

Stage 5: System burn-in

---

[18]"Line" refers to a maintenance level where first line = aircraft; second line = service maintenance workshop; third line = base or deep level workshop; fourth line = return to supplier.

Stage 6: System sell-off

Stage 7: In the field

Setting the cost of failure at stage 1 (PC board assembly) as 1, the cost of failures detected at stage 2 would be 3 and at stage 3 would be 9. That is, the cost of failures grew by a factor of 3 from stage 1 to stage 2 and stage 2 to stage 3. The costs of a failure at stages 4, 5, and 6 are essentially identical and equal to 10 times the cost of failure at stage 3 (i.e., 90). By the time the unit had entered field service, the relative cost had risen by another factor of 10 to 900!

The customer, meanwhile, generally pays shipping costs; expends time recognizing the existence of a problem, arranging for service, and perhaps delivering the product to a shipping company; and, of course, suffers modest to serious or catastrophic inconvenience while the system is out of service.

4.7 Summary

This chapter presented the following points:

- The benefits claimed for quality improvement result from (1) increased demand and (2) reduced operating and production costs.
- The benefits can be described as (1) demand side benefits and (2) supply side benefits.
- Demand side benefits are difficult to predict and measure; they may not even exist for many products.
- Many variables complicate the relationship between quality levels and demand; lack of customer awareness is a significant factor.
- Supply side benefits are predictable and can be measured with considerable confidence.
- Data demonstrating supply side benefits from quality improvement programs are extremely valuable to the quality department's influence within the company.

5 Quality in the Marketplace: Lessons from Japan

5.1 Chapter Objectives

Japan's successes in international trade are often cited as "proof" that American quality has been inferior and must improve. But just how important to Japan's industrial success has quality been? This chapter provides some answers, including:

- Many factors helped Japanese industry become internationally competitive. Higher quality was just one factor—and not necessarily among the most important.
- The assertion that Japanese goods provide superior quality to goods from American factories cannot be substantiated from any existing evidence.
- For decades, Japanese products offered better value than comparable goods made in America. Those who believe Japanese products have higher quality confuse value with quality.

- American government agencies provided the financial, economic, and political assistance that enabled Japan to become an international industrial power—often at the expense of American companies and their employees.
- Japan's Ministry of International Trade and Industry (MITI) has been less important to the country's industrial successes than is generally believed.
- Japan's Ministry of Finance (MoF) has controlled the country's industrial environment for the past decade.
- Deming and Juran were less important than Joseph Dodge in Japan's development.

5.2 The Search for Proof of Demand Side Benefits

As we saw in Chap. 4, making a strong theoretical case for the value of quality in a company's fortunes is not at all difficult. Unfortunately for the quality profession, however, proving the importance of demand side benefits—that demand varies directly with the level of quality—has not been easy. We know that the correlation exists, but exactly how to quantify that correlation has caused quality professionals many sleepless nights. Out of that frustration have emerged many inadequate "proofs," including attempts by business professors to show higher sales and profit margins for companies that profess to practice some form of total quality.

The most elaborate undertaking is the American Consumer Satisfaction Index (ACSI), a joint undertaking by the University of Michigan school of business and the American Society for Quality Control. Begun in 1994, the ACSI surveys some 30,000 Americans a year to obtain a numerical score for the companies and public agencies providing a total of about 3900 products and services. The sample population provides about 250 individual opinions for each company of the "service, quality, value, how well the product or service lived up to expectations, how it compared to an ideal, and how willing people were to pay more for it."[1] The maximum possible score is 100; the minimum possible is 0. The range of products and services included in the sur-

[1]Fierman, Jaclyn, "Americans Can't Get No Satisfaction," *Fortune*, Dec. 11, 1995, pp. 186–194.

vey—delivery services, food packaging companies, television networks, cigarette manufacturers, the IRS, and hotel chains are just a few of the various industries scored—is mind-boggling. This is a survey that not only tries to compare apples and oranges but throws cabbages, kumquats, wild rice, and dog food into the mix. Despite the obvious dependence on emotions rather than science, the index operators—in one of those all-too-common triumphs of academic method over logic—claim the scores are "statistically significant and representative of national trends.... The statistical margin of error is 0.2% for the national average, 1.3% for industries and government agencies, and 1.6% for individual companies."[2] Some businesspeople who should know better actually take this seriously.

The numerically literate individual can readily see the folly in bizarre "research" like the ACSI. For several decades, however, a more subtle and insidious trap has tricked countless knowledgeable executives with the illusion of proof that demand side quality benefits exist. The trap involves superficial analysis of Japanese exporters' successes in the last 30 years.

5.3 Quality and the Japanese "Miracle"

When quality advocates make their case that quality improvement must be the foremost strategy of any company, they normally bring up the Japanese "miracle." The conventional wisdom about quality's part in Japanese companies' successes in international markets was summed up well by journalists Lloyd Dobyns and Clare Crawford-Mason: "Quality costs less, which helps explain why the Japanese could sell a better car for less money in the 1980's, why a Japanese stereo system in the 1970's could produce better sound for a longer time with less trouble than a more expensive American model."[3] In a similar vein, *Business Week*[4] noted: "Just four decades after digging out from under the rubble of World War II, Japan has transformed itself from a maker of knickknacks into an economic powerhouse through its single-minded focus on quality."

While most of the world agrees with those statements, they gravely distort the historical reality. In particular, the popular belief that quality

[2]p. 188.

[3][Dobyns and Crawford-Mason 1991], pp. 2–3.

[4][*Business Week* Editors and Green 1994], p. 1.

improvement was the key to Japanese success fails to recognize that economic and political factors were even more important in the development of Japan's industrial power. Accordingly, any analysis of how and why to pursue higher quality in the electronics assembly industry (or any other business sector, for that matter) must include thorough investigation of the factors other than quality that enabled Japan to make such dramatic advances in world markets.

An accurate look at the realities of Japanese industrial, political, and trade practices is hard to find, primarily because Japanese companies and the Japanese government have operated remarkably effective lobbying and public relations programs. Among other tactics, Japanese institutions have:

- Contributed to American political campaigns at every level from Congress to civic elections.
- Funded "independent" departments of Japanese studies at leading American universities. Many prominent professors augment their university salaries with income from make-work projects sponsored by Japanese companies or the Japanese government.
- Bought the influence of myriad highly placed civil servants—primarily in the Department of Commerce, which oversees international trade—in some cases apparently while those civil servants were still on American government payrolls and shaping official policies.
- Cultivated business and general reporters in all branches of the news media.

5.3.1 "Japan Bashing"

The extent to which Japan has coopted American institutions is distressing. For example, "Japan bashing"—a term used aggressively and effectively by Japan's supporters against any person, government agency, or company that tried to cast light on the less savory Japanese competitive practices—was coined by American political scientist Robert Angel during the 1980s. At the time, Angel was president of Washington-based Japan Economic Institute, a Japanese-funded lobbying group. He was subsequently hired as a professor at the University of South Carolina,[5] an appointment that should cause some concern about the influences to which students at that school are subjected.

[5]*USA Today*, Mar. 5, 1992.

In 1988, American professor George Packard, dean of the Paul H. Nitze School at Johns Hopkins University in Maryland, was paid by the Japanese Chamber of Commerce to write an advertising supplement on Japanese business for *U.S. News & World Report.* Sections of the supplement bore titles such as "Toyota U.S.A.: An American Phenomenon," "Mitsubishi Forges Strong American Ties," "Suzuki: Partners in Prosperity with the American People," and "A Call for Autonomy at Ricoh."[6] The following year, Packard wrote an article published by *The Washington Post* attacking four of the most respected Western authors of books and articles criticizing Japanese industrial, economic, and social policies.[7] Packard's "The Japan-Bashers Are Poisoning Foreign Policy" accused Chalmers Johnson (author of *MITI and the Japanese Miracle*), Clyde V. Prestowitz (*Trading Places*), Karel van Wolferen (*The Enigma of Japanese Power*), and *The Atlantic*'s James Fallows of "Japan bashing" and McCarthyism.

In 1994–1995, a total of 45,276 Japanese students—the greatest number of students from any foreign country—were enrolled at all levels in U.S. colleges and universities. In contrast, only 2200 Americans attend Japanese schools.[8] Those students have become such an important source of tuition revenue for some well-known universities that their admissions officers travel to Tokyo specifically to recruit Japanese candidates for MBA programs. *The Wall Street Journal* in 1991 identified the University of Chicago, New York University, and Cornell as three universities recruiting MBA students from Japan.[9] That year, 700 Japanese students were enrolled in MBA programs at American universities, providing more than $5 million in tuition fees alone. The numbers have evidently increased each year since 1991. The possible conflict-of-interest implications for universities that derive substantial graduate school income from Japanese students are troubling, inasmuch as the same faculties shape the attitudes of America's business and political leaders. Equally troubling is the fact that the university administrators themselves do not recognize the ethical questions.

The ease with which Japanese interests have been able to sway America loyalties contrasts sharply with behavior inside Japan. Japanese car buyers never considered a foreign (particularly American)

---

[6][Eberts and Eberts 1995], p. 310.

[7]Packard, George N. "The Japan-Bashers Are Poisoning Foreign Policy," *The Washington Post,* Oct. 8, 1988, p. C4.

[8]Henry, Tamara. "Japan Tops in Foreign Students," *USA Today,* Nov. 8, 1995, p. 1D.

[9]*The Wall Street Journal,* June 10, 1991.

make because their neighbors would shun them. Japanese trading groups would not distribute American goods inside the country. And, until the realities of depression hit Japan in 1993, no Japanese national would trade a career in a Japanese company for a position with a foreign company. With glacial slowness (even though the purchasing power of the yen has increased by more than 100 percent in this decade), the Japanese are shifting their behavior away from blind allegiance to domestic goods. At this point, however, it is hard to imagine prominent retired Japanese bureaucrats selling their souls to American interests.

In such an environment, separating the truth about Japanese development from the fiction has been difficult. Much of the real story may never be known. However, the following chapters of Japan's postwar history are well documented.

5.4 Six Critical Factors in Japan's Success

Six critical factors unrelated to quality gave Japan an overwhelming advantage in competing for international markets. The factors consisted of:

1. U.S. government financial support for Japanese industry
2. Deliberately undervalued currency
3. Manipulation of land and stock values by Japan's Ministry of Finance
4. Predatory pricing in foreign markets
5. Exclusion of foreign goods from the Japanese domestic market
6. Sophisticated lobbying and public relations

The effects of those six critical factors were reinforced by:

1. The Pentagon's determination to prop up Japan as a bulwark against the spread of Russian (and, by the early 1950s, Chinese) communism in Asia.
2. Naive support—based on antiquated beliefs in the superiority of American industry—by American government and business for free trade.
3. Japan's unwavering determination to sustain enormous trade and current account surpluses even though the inevitable outcome of

this extraordinarily aggressive trade policy was global recession beginning in 1991 and, subsequently, the most serious economic crisis in Japan since World War II.

5.4.1 American Aid

Few, if any, conquering nations have ever treated a defeated enemy as generously as the United States treated Japan after World War II. Often, Japanese companies were helped at the expense of American industry. Among the benefits provided by the American taxpayer to Japan and its business sector were:

- Instruction on advanced manufacturing and quality assurance techniques by some of America's leading authorities (including Deming and Juran)
- Access to U.S. patents at remarkably low royalty rates
- Favored-nation admission to the U.S. domestic market for Japanese products
- Subsidies in the form of vast assembly subcontracts from American corporations (often at the request of the U.S. government)
- Substantial purchases by the U.S. military during the Korean and Vietnam wars
- Direct financial aid from the American treasury

The amounts involved were astoundingly large. The aid bill alone between September 1945 and December 1951 amounted to $2.118 billion,[10] an average of almost $1 million a day. Additionally, the Department of Defense bought $4 billion of Japanese goods including uniforms and electronic products, providing an extraordinary windfall for fledgling Japanese companies that otherwise would have faced a bleak future. In the decade following the end of the Korean War, U.S. military purchases from Japanese suppliers amounted to another $3.2 billion. Then, between 1965 and 1970, American defense procurement of Japanese goods for use in the Vietnam war approached another $3 billion. All told, the American military spent almost $10 billion on Japanese goods between 1950 and 1970, which translates into an average annual military subsidy to Japanese industry of approximately $500

[10][Nester 1990], p. 106. Interestingly, this book by a lecturer from the University of London was found in a bookstore in central Tokyo.

million at a time when an ounce of gold traded for $35.[11] Japan might well have become an industrial success without American aid but almost certainly not so powerful and likely not in this century.

5.4.2 The MacArthur Occupation

The conventional wisdom of history believes that Japan was completely restructured—socially, politically, and industrially—by General Douglas MacArthur, Supreme Commander for the Allied Powers from the Japanese surrender on Sept. 2, 1945, until his transfer as Commander of NATO forces in South Korea in June 1950.[12] More recent studies, however, tell a very different story. MacArthur was interested solely in winning favorable notice from American newspapers and magazines that would pave his way to the U.S. Presidency. Concerned about his image, MacArthur was determined that Japan be seen to shed its militaristic past and adopt democratic trappings. To that end, he quickly made the Japanese government and civil service aware that he was concerned only with appearances, not substance. The dynamics between the occupying forces under MacArthur and the Japanese civil service—where the Japanese were judged on the basis of public statements rather than behind-the-scenes actions—undoubtedly established the precedent for the Japanese government's standard operating strategy under which officials make promises to the American government that are never kept. This strategy includes the actions of every Japanese prime minister for more than four decades.

At one point, it was reported back to Washington by an American observer (who was not part of the occupying force) that MacArthur aides were telling Japanese officials that a law to break up the large family-held conglomerates known as *zaibatsu* "had to be passed so as not to embarrass the Supreme Commander who is expected to run for president." Should passage of the law be delayed, the warning continued, such action would "prejudice the future of Japan when the Supreme Commander became President."[13]

By 1949, under MacArthur's careless watch, Japan was in disarray. Chronically short of money—its deficit for 1948 was 160 billion yen—

---

[11][Schaller 1985], p. 296.

[12]The Allied (though in fact totally American) occupation of Japan officially ended in April 1952, almost 2 years after MacArthur was reassigned to command NATO forces in Korea.

[13][Schaller 1985], p. 47. The author quotes from a memorandum of Dec. 14, 1947, found in Records of the Office of the Secretary of the Army.

the Japanese government paid its debts by running the mint's printing presses overtime. More than 12 times as much money was in circulation in 1949 as in 1945. Hyperinflation inevitably accompanied the increased money supply. Using 1945 prices for a base index of 100, by 1949 the wholesale price index was 5965 and the retail price index 6867.

5.4.3 The Joseph Dodge Legacy

Late in 1948, having been consistently prevented by MacArthur from implementing the economic changes it wanted, the U.S. State Department sent a delegation headed by Joseph Dodge to establish economic stability in Japan. The key elements of Dodge's mandate were "to order the Japanese government to balance its budget, establish a single exchange rate (for the yen against the dollar), and create new institutions to guide economic development."[14]

Dodge fulfilled his mandate totally. The 160 billion yen deficit of 1948 became a surplus of 260 billion yen in 1949. Government's share of Japan's GNP fell from 51 percent in 1949 to 34 percent 2 years later, reflecting the combined effects of Dodge-imposed government austerity and sustained real growth in the private sector.

Of greatest consequence for the future of American industry, Dodge set the value of the yen at 380 against the dollar, a low value that would promote Japanese exports to the United States. The real value of the yen even then was higher than the value set by Dodge. Helping Japan enter the American market was compensation for a U.S.-imposed embargo on Japanese trade with mainland China, where Mao had finally gained unquestioned control. At the time of the embargo, China had been Japan's major trading partner.[15] The undervalued yen would become the most important negative influence faced by American manufacturing in the twentieth century.

In addition to setting the new low exchange rate for the yen, Dodge replaced the existing Ministry of Commerce and Industry with a new Japanese government agency: the Ministry of International Trade and Industry. Two laws were then passed to enhance MITI's powers:

1. The Foreign Trade and Exchange Control Law of 1949 gave control over all foreign exchange to MITI—together with the right to allocate that foreign exchange to industries targeted by MITI for development.

---

[14]pp. 107–108.

[15]p. 108.

2. The Foreign Investment Law of 1950 prevented most foreigners from investing in Japan unless they were willing to forgo repatriation of any profits.

The effects of these laws were profoundly harmful to foreign manufacturers, including American companies. Faced with the virtually insurmountable barriers to investing or trading in Japan erected by the new laws, many foreign companies ultimately sold their technologies to Japanese firms, generally for a fraction of the technologies' real value. More than 10,000 foreign technologies were licensed by Japanese firms between 1950 and 1968 at a total cost of only $1.8 billion. Those cheap licenses allowed the Japanese licensees to avoid expensive research and development work. The money saved on R&D was available for investment in new plants and equipment. Japan was thus able to develop major industries such as electronics, automobiles, computers, electrical machinery, petrochemicals, or synthetic textiles in a fraction of the time and cost that would have been required under normal circumstances.

Dodge returned to Japan upon America's formal acceptance of Japan as a sovereign nation in 1951. Having already established monetary and legal conditions that would prove disastrous to American industry a quarter of a century later, Dodge's assignment on his return was to create financial institutions dedicated to promoting Japanese industries. With 5 billion yen (slightly less than $41.7 million) from the U.S. government, Dodge set up the Export Bank. Another 10 billion yen ($27.8 million) of American money was used to create the Japan Development Bank. Dodge also directed the founding of the Fiscal Investment and Loan Plan (FILP) funded by post office savings accounts. All personal savings had to be deposited at FILP in return for a meager interest rate.

The Export Bank, the Japan Development Bank, and FILP were placed under MITI's control. Overnight, MITI became the most powerful engine of Japanese industrial expansion. Control over these financial institutions allowed MITI to distribute more than 169 billion yen ($469 million) to Japanese industry in 1956 alone even though the ministry's official budget for the year was only 8.2 billion yen ($22.8 million). From 1951 to 1961, the Export Bank, Japan Development Bank, and FILP together provided between 19 and 38 percent of the total capital raised by Japanese industry every year.[16]

In other words, through the combined efforts of the Pentagon, the State Department, Douglas MacArthur, and Joseph Dodge, the stage

---

[16]pp. 108–109.

was set for Japan's conquest of the American market. Although the conscious intent does not appear to have been to make Japan powerful at America's expense, that was the outcome.

5.4.4 The MITI Factor

Although Dodge provided MITI with remarkable tools and power to carry out an aggressive program of managed industrial development and trade, the ministry on many occasions managed to prove that "government help" is often an oxymoron.[17] MITI's more memorable missteps over the years include:

1. *Retarding development of the domestic auto industry.* During the 1950s, MITI leaders tried to force Japan's auto industry to consolidate into a single company. They failed but, in 1961, were still trying to restrict the number of auto companies to, at most, two or three companies. Honda, for example, was actively discouraged from entering the auto industry or, later, from exporting to America. Japan now has nine domestic auto companies (although admittedly not all are paragons of strength).

2. *Supporting the wrong industries.* MITI in the 1950s believed electronics was a poor use of Japanese capital. In 1953, when Sony needed permission to license the transistor from Bell Laboratories, the ministry refused the request. Sony succeeded only because its president, Akio Morita, persistently lobbied other ministries until he was given permission, still against MITI's wishes.

MITI believed steel, aluminum, and nonferrous metal smelting were strategically important industries. During the 1960s and 1970s, the Ministry force-fed a huge buildup of steelmaking capacity. The results were disastrous; countless billions of dollars were wasted and tens of thousands of steelworkers lost their jobs when international and domestic demand for steel collapsed in the 1980s. Upon discovering they had grossly misjudged market needs, Japan's bureaucrats resorted to global

[17]MITI is not the only Japanese ministry whose interference in industrial policy has cost the country dearly. The Transport Ministry has also made several expensive errors in judgment. One of the ministry's most expensive blunders—the "bullet trains"—became widely admired in the United States, where it was not widely known that the Japanese rail system cost the government $20 million a day in subsidies before the system was privatized in 1987. In order to make the railways attractive to private investors, the government had to pay off accumulated debts of $280 billion. The Transport Ministry was also responsible for Japan's becoming—to its regret—the world's largest shipbuilding nation. Japanese ships, developed in the 1950s, were unsuited to cargo needs in the 1970s. Once again, the result was massive losses.

dumping—a common practice of Japanese business—disrupting the metal-producing industries throughout the world. The turmoil was so severe that President Clinton applied antidumping duties to foreign steel as his first major trade initiative after taking office.

Under MITI prodding, Japan continues to make its own aircraft, although its national competency in the design and production of aircraft is nowhere near the standards of American and European companies. The products are inferior and the prices high, so the international demand for Japanese planes is essentially nonexistent.[18]

3. *Biotechnology and energy blunders.* Attempts to build a meaningful biotechnology industry failed. So did entries into alternative energy. In traditional fossil fuels, MITI's national oil program is generally regarded as the world's most inefficient. The result: All users of petroleum, from industry to consumers, pay significant penalties. Under the circumstances, it is no wonder that Japan's petrochemical industry has floundered.

4. *Supporting failing industries.* Although MITI tried to restrict the number of Japanese auto makers in the 1960s by discouraging Honda from entering the auto industry, in recent decades it has acted in the belief that the country really needs nine major car and truck makers. When smaller auto or truck companies ran into trouble, MITI pressured healthy companies to buy parts from—or stakes in—the troubled ones to sustain them until the economy improved. In September 1995, for example, Toyota was persuaded to double its interest in Daihatsu to 33.4 percent. There was little reason for Toyota to invest voluntarily since Daihatsu was puny in comparison to Toyota and very unprofitable. For the fiscal year ending March 1995, Daihatsu had sales of $9.4 billion but earned only $14 million while Toyota's sales were $91.2 billion and earnings $1.5 billion for just the first nine months of its corresponding fiscal year.[19]

---

[18]In the long term, aerospace could become yet another field of Japanese dominance despite the country's current shortage of technical and production knowledge. MITI has been encouraging foreign aircraft manufacturers to license their technology to Japanese companies, the strategy that worked so well for Japanese industries like television production in the 1950s and 1960s. Although history provides ample warning of the dangers inherent in providing knowledge to the Japanese, both American and European aircraft companies have already embarked on a few joint ventures with Japanese companies. If the greatest demand for new planes in coming years turns out—as industry analysts predict—to be the Pacific rim, and Japan can entice some gullible Western aircraft companies into joint ventures, it is not inconceivable that Japan could become a force in this industry.

[19]Healey, James R. "Japanese Automakers Make Comeback," *USA Today*, Nov. 5, 1995, p. 1B.

5.4.5 The Finance Ministry Saved the Day

Most Japanese success stories—such as consumer and office electronics, watches, cameras, NC machine tools, precision equipment, and cameras—developed despite rather than because of MITI. The government department responsible for the most important commercial breakthroughs tended to be the Finance Ministry, which worked separately from (and often in competition with) MITI. The Finance Ministry provided the crucial funding and monetary policies that enabled manufacturing industries to build factories cheaply and sell abroad at a loss.

5.4.6 The Undervalued Yen

Twenty years after Dodge returned to America, the yen still traded at 360 to the dollar. During that time, Japan had become the world's second-largest economy and was running surpluses in its trade with the United States. In 1971, American negotiators acting for then-President Nixon forced Japan's Ministry of Finance to accept a 16.9 percent increase in the value of the yen, from 360 to the dollar to 308.[20] In February 1973, the world's industrialized nations (the "Group of Seven") agreed despite Japanese opposition to eliminate the fixed exchange rates originally established decades earlier under the treaty of Bretton Woods. With abolition of fixed exchange rates, the yen nominally "floated" against the dollar, but the float was almost pure illusion. Constant covert buying and selling of yen on the international spot market by Japan's Ministry of Finance ensured the yen never rose beyond 175 to the dollar, a level it reached in October 1978 before spiraling downward again. In 1974 and again in 1978 when the yen appreciated rapidly against the dollar, the Ministry of Finance imposed temporary restrictions on capital leaving the country. Between 1983 and 1985, the yen never rose above 240 to the dollar or fell below 260.[21] By every reasonable standard, the yen was grossly undervalued, providing the best possible environment for Japan's export ambitions.

Over all these years, with the sole exception of 1971, Japan's desire for a weak currency coincided with the desire of U.S. politicians to sustain a strong dollar. Rightly or wrongly, a strong dollar is often interpreted by the public as synonymous with a strong nation. A strong dollar is good for consumers of imported goods and Americans traveling

---

[20][Burstein 1988], p. 143.

[21][Emmott 1989], pp. 55–56.

abroad. It reduces inflationary pressures by reducing the prices of imports. And interest rates can typically be kept to low levels when there is heavy international demand for the nation's currency. However, all these "benefits" carry a heavy price. In particular, since labor and other inputs cost them more than their competitors located in low-wage countries, domestic manufacturers may be unable to match or undercut the prices of imports in the domestic market. Often, meeting the prices of foreign rivals means accepting minimal profits or even incurring losses.

Historically, American producers coped with higher labor costs by utilizing more advanced machinery and other manufacturing technologies. That is, they increased productivity. This ability to devise and carry out technological breakthroughs led to the term "good old American know-how." But the importance of American knowledge to the nation's successes during the twentieth century was often confused with good luck. Particularly important in the years immediately following both world wars was America's enviable status as the only major industrialized country whose factories had not been destroyed during battles or air strikes. American factories were not only intact but also equipped with state-of-the-art production equipment paid for by the federal government as part of the war mobilization effort.

For more than two decades, the war inheritance provided a comfortable cushion for American manufacturing. And the consequences were tragic. The business community became arrogant, wrongly believing that the prosperity enjoyed in the first three postwar decades was just another example of American expertise. But by the 1970s foreign competitors had rebuilt their manufacturing capacity—often and not just in Japan with American financial assistance. Their plant and equipment, being newer, tended to be more efficient. Suddenly, for the first time in this century, American manufacturers faced foreign competitors who had comparable or better production capability *and* lower labor costs.

Domestic producers simply cannot compete in free markets against foreign manufacturers when those foreign competitors are equally well equipped with plant and equipment and also pay less for labor. Many of the most prestigious names in American manufacturing responded by sending production offshore to low-wage nations, in effect mutating from manufacturers into distributors of imported goods. But this strategy entails significant hidden costs. For example, exporting manufacturing jobs creates unemployment in the domestic market. Since unemployed workers cannot consume, the domestic market—which normally accounts for the greatest and most profitable demand for domestic manufacturers—shrinks. A downward spiral sets in; declining domestic consumption capability forces employers to lay off more workers, thus further reducing domestic demand and leading to additional reductions in the workforce.[22]

By 1985, Japanese companies had so successfully penetrated the traditional markets for American-produced goods at home and abroad that serious doubts arose about whether the U.S. manufacturing base might ever recover. The concern about the future of American industry was so great that President Reagan, a profoundly dedicated free trader, took unprecedentedly strong action against Japan. A meeting of the five leading economic nations was convened by the White House and agreement was reached—despite almost frantic Japanese opposition—to act in concert to reduce the value of the dollar. The most important consequence of that agreement, known as the Plaza Accord, was a rapid appreciation of the yen from 250 to the dollar in September 1985 to 140 a year later.

According to all principles of Western-style economic theory, Japan's exports should have crumbled in the wake of such a dramatic increase in the value of its currency. But Japan is no ordinary nation and the government—specifically the Ministry of Finance, which early in the 1980s had surpassed MITI as the country's most influential industrial development agency—approached this crisis in a familiar Japanese bureaucratic manner: from a wartime perspective.

5.4.7 Tokyo's Inflationary Cure

As it did during the years of postwar budget deficits, the Bank of Japan (BoJ) responded by running the printing presses overtime and holding interest rates far below what other countries were charging. The inevitable—and desired—result was rapid inflation of property. At the market's peak in 1989, the land surrounding the emperor's palace in Tokyo had a paper value equal to the entire GDP of Great Britain; 2 years later, after the market for land turned from extraordinarily bullish to distinctly bearish, the grounds were worth only about the total value of the whole state of Florida. Owners of modest homes in urban centers became multimillionaires in the blink of an eye.

[22]This very downward spiral occurred in Canada when the central bank brought about a substantially overvalued currency in comparison to the U.S. dollar through a relentless campaign of high interest rates beginning in 1988. This artificially high currency value coincided with the introduction of free trade with the United States. U.S. manufacturing companies simply closed the doors of their Canadian subsidiaries and began supplying the Canadian market out of the American plants. At the end of 1994, the national unemployment rate in Canada exceeded 11 percent even though the Canadian dollar had fallen from 88 cents U.S. in 1988 to below 73 cents U.S.

The artificially inflated land values enhanced the abilities of Japanese corporations to raise capital for expansion or to ride out the anticipated drop in foreign demand due to the higher dollar prices of Japanese exports. This easy money came in two forms. First, individuals and companies, having gained easy access to substantial loans from banks using real estate holdings as collateral, went on a speculative buying spree in the Japanese stock market; the demand for stock was so great that Japanese corporations raised enormous amounts of capital by issuing warrants that paid no interest but were convertible into stock 5 years later. Second, corporations with land holdings—which included in essence every meaningful Japanese corporation—could borrow against that property for capital expansion. Kenneth Courtis, an economist with the Deutsch Bank in Tokyo, estimated that, because of these central bank actions, Japanese corporations between 1988 and 1990 raised *$630 billion* at an effective interest rate of zero.[23]

During the same period, Western companies were being battered by exceptionally high real interest rates[24] artificially induced by their central banks. The high real interest rates further undermined the ability of Western companies to compete against Japanese products. As it turned out, the Western central bankers' preoccupation with inflationary pressures was unfounded and their interest policies drove the world into the worst recession in six decades.

5.5 A Japanese Financier Speaks Out

Akio Mikuni, president of the only independent credit-rating agency in Japan, provided a rare peek behind the silk curtain of Japanese bureaucratic and financial behavior in a *New York Times* article early in 1993.[25] Unlike the customary shallow looks at Japan, from Japanese or Western sources, Mikuni's article pulled no punches. It may be one of the most courageous reporting exercises by any postwar native Japanese still living in his homeland.

---

[23]*Business Week,* Aug. 26, 1991.

[24]The "real" interest rate equals the interest rate charged minus the rate of inflation. If the interest rate charged is 10 percent but inflation is 8 percent, the real interest rate is only 2 percent. On the other hand, if the interest rate charged is 6 percent and inflation only 2 percent, the real interest rate is 4 percent. The real interest rate determines the true cost to the borrower.

[25]Mikuni, Akio. "Behind Japan's Economic Crisis," *The New York Times,* Feb. 1, 1993, p. A13.

"The Japanese economic crisis, reflected in a 60 percent decline of the stock market over the past two and a half years and by declining growth, should not be comforting to foreign competitors," Mikuni warned. "Japan's manufacturers will be concentrating afresh on the world market to keep themselves going. They have no choice."

The key to understanding Japanese industrial performance, Mikuni argues, is to "recognize the extraordinary powers of the Finance Ministry, which, for all practical purposes, has been running a wartime economy from almost the moment U.S. occupation forces departed.

"The ministry can shape and control the economy through the informal right to intervene in any and all transactions. It can control which corporations receive bank credit and which do not. It determines interest rates and controls the prices of stocks, real estate and other investments.

"It need not explain its actions to the Diet, the Prime Minister or anyone else. It controls the financial information necessary for an open discussion on economic policy and even for an accurate analysis of individual corporations. It is expected to be infallible.[26]

"Japan's 'bubble economy' of the second half of the 1980s was implicitly encouraged by the ministry. This provided the opportunity for large manufacturing companies to continue to expand capacity despite the reduction of their export income by almost half after the value of the yen shot up from 240 yen to the dollar in August 1985 to 50 yen to the dollar one year later.... The same financial authorities decided to let air out of the bubble in 1990 when they concluded that too much money was in too many hands, threatening their ultimate control of the economy."

5.5.1 Deflating the Bubble Economy

Still, Mikuni notes, not even the mighty Japanese bureaucrat has a clear crystal ball to tell the future. So, when the Ministry of Finance punctured the bubble economy, it set the nation up for more economic anguish than it imagined. In Mikuni's words, "The deflation of the bubble economy has resulted in Japan's worst banking crisis since the 1920's."

[26]Mikuni's argument suggests that it is unrealistic to believe Western observers will ever be able to accurately predict movements in the Japanese economy. If only a few well-placed members of the Ministry of Finance know the true state of any Japanese economic matter, Western predictions of Japan's economic performance can be no more precise than reading tea leaves.

But that is not all. "To make matters worse, the ministry demands that financial institutions stick to policies that are ultimately unsustainable.

"To see this in perspective," Mikuni argues, "financial institutions, principally the banks and insurance companies, have been subsidizing Japan's all-important manufacturing sector since the mid-1970s. Because the value of the yen rose from 360 to the dollar in 1971 to 120 at present, manufacturers' operating profits as a percentage of sales declined from an average of 9 per cent to about 3 per cent.[27] Returns on plant and equipment investments have been correspondingly dismal.[28]

"Against this background, the banks were forced to plunder their own earnings by lending at unprofitably low rates and by buying up shares from clients even though dividends were marginal and prices (of shares) were vastly inflated. Gradually, the banks began to assume risks for which they were not prepared; they began shoveling credit at the much less controlled second tier of small and medium-sized companies. These companies, which have always provided the essential subcontracted cheap labor, went on a borrowing binge, bought stocks and land and added to their production capacity faster than economic growth justified.

"The banks overlooked the fact that they were no longer dealing with protected large-scale entities enjoying the benefits of lending costs arranged by the Finance Ministry. The returns were always generous when measured against those costs, because the returns were also arranged by Japan's famous Ministry of International Trade and Industry. With the bubble leaking air, thousands of companies are now unable to service their debts. Banks are repossessing collateral that has lost much of its value and dealing with the worst losses in their history.

"Inflated stock and real estate prices created the illusion that Japanese banks rested on a sizable capital cushion. Thus, while the banks were barely making money from their main customers (the large companies that they were expected to subsidize), they reported profits in the second half of the 1980s from the rising values of the stocks and real estate in their possession. And now, after the managed 'collapse' of the bubble economy, the ministry must try to maintain some of that illusionary cushion to prevent a collapse of the entire economic system.

---

[27]The period of time in which this decline in return on sales took place is not entirely clear. Presumably Mikuni means the decline occurred between 1975 and 1992.

[28]Mikuni's concern about the decline in return on sales may reflect his Japanese heritage. Profit as a percentage of sales is measured by Western companies but is considered less significant than return on equity (i.e., return on investment) and trends in absolute profits. If return on sales is reduced by two-thirds but sales increase by more than threefold, absolute profit increases; so, potentially, does return on investment.

"Officials cannot allow stock and real estate prices to fall to a level that would attract private investors. In fact, officials are heavily sabotaging market forces in an all-out attempt to keep the stock markets at an absurdly high average price-earnings ratio of 50, while keeping dividend levels at less than 1 per cent and propping up a real estate market that yields an average income of not more than 2 per cent. Banks and insurance companies are being forced not only to retain their unprofitable stocks and real estate but also to buy more. Large corporations that own stock are pressured to follow this example. With borrowing costs now well above 4 per cent, a loss is guaranteed."

So the economic bureaucrats faced a new dilemma. If stocks and real estate continued to be valued at unrealistically high levels, the financial and manufacturing sectors would go on deteriorating. Japan's overpriced land depressed the construction market. Overpriced stocks drove off individual investors, so trading volume plummeted. Why, then, when it became abundantly clear that the inflationary cure had become an inflationary curse, did the Finance Ministry continue to pursue this debilitating practice? Because, Mikuni concluded, "if the Finance Ministry allowed the prices of land and stocks to be determined by market forces, a genuine capital market would emerge, ending the ministry's power to allocate funds without regard to profitability. Obviously, Japan's economic mandarins would never relinquish this power voluntarily."

5.5.2 Global Effects

Unfortunately for the rest of the world, Japan's domestic economic conditions now have global ramifications. When Japan sneezes, Western economies and their corporations catch pneumonia. "So once again Japanese exports are being used to help close the gap in the current downturn," Mikuni noted. "Tokyo's trade surplus is now three times higher than the 1984 surplus, which the U.S. said was politically intolerable and which led directly to the agreement to increase the value of the yen.

"Profitability has not been the guiding principle in this campaign, as in the Western economies. Only now, Japan's greatly expanded production capacity has made the global market even more important if Japan's corporations are to earn a slim profit.[29]

[29]Mikuni may regard excess capacity as a uniquely Japanese phenomenon. IBM, GM, Wang, Digital Equipment, and countless other U.S. companies could teach him that excess capacity exists outside Japan.

"At the same time, there is less reason to expect the world to cling to the illusion that Japan runs a market economy—an illusion necessary for maintaining global hospitality to Japanese products, especially when the sales push is being intensified.

"The situation reminds me of the time, four months before the end of the Pacific war, when the Japan's [sic] 64,000-ton Yamato, the largest battleship ever built, sailed for Okinawa. It never got there, never had a chance, given American air supremacy.

"Without considering the requirements or tolerance of the global market, Japanese manufacturers built excessive production capacity with the same blindness that went into building the Yamato. In desperation, they are now sending out their unprofitable exports. This will prompt more adverse reactions throughout the world, foreshadowing huge trade-induced political problems. Ultimately, the ability of Japan's economic mandarins to close the circle of the war economy at home depends on how long the rest of the world, especially the United States, will accept their methods."

5.6 Predatory Pricing

The trade practices described by Mikuni add up to a classic case of predatory pricing, which economists define as selling below cost in some markets. The effect of predatory pricing is to destroy or critically weaken competitors—existing and potential—in those markets.

Influential Americans, including not a few reporters and columnists with important business publications, have consistently ridiculed the idea that Japanese companies would indulge in predatory pricing or dumping. One good example of this attitude appeared in *Fortune* magazine's "Keeping Up" section in mid-1992. Daniel Seligman, the *Fortune* contributing editor who writes the "Keeping Up" items, argued in the issue of June 29, 1992, that "very few economists take seriously the alleged menace of dumping. The model of the problem retained in many noneconomists' noggins is one in which the foreign predator diabolically accepts huge losses so as to drive his American competitors out of business—and then, once they are gone, raises his prices to extortionate levels. Oddly enough, it is more or less impossible to cite real-world examples of this popular scenario. Steven Plaut's (1980) article in *Fortune* explained why: however predatory, the rational foreign producer would know that new market entrants would surface as soon as prices rose, so he would have absorbed all those losses for nothing."

But Seligman's arguments were riddled with serious mistakes, among them:

- It is by no means certain that most economists do not take seriously the alleged menace of dumping.
- Examples of dumping and predatory pricing are easily found, starting with Japan's destruction of the U.S.-owned television receiver industry, spreading throughout consumer electronics, and more recently involving computer memory chips.
- The predator need not presume that new competitors would emerge as soon as prices rose above predatory levels. Once the predator owns the market, channels of distribution are controlled. The predator has a larger market over which to amortize costs such as research and development, advertising, and distribution. Then there is the issue of name recognition, which in itself is good for a substantial competitive advantage.
- Seligman obviously does not understand the structure of predators, particularly the Japanese variety that typically belong to enormous diversified conglomerates (known in Japan as *keiretsu*). The conglomerates practice predatory pricing (and thereby sustain losses) only in some industries at any given time; the remaining member companies of the conglomerate subsidize the loss leaders.
- Japan, in the words of *The New York Times*,[30] is populated with "long suffering consumers accustomed to the world's highest prices in a closely protected economy." The high prices paid by Japanese consumers offset losses on products dumped abroad. Non-Japanese competitors are not protected in their home markets and are effectively excluded from the Japanese market where they could earn higher profits. Predatory behavior in a global context makes sense when profit margins can be sustained domestically.

5.7 The Television Takeover

The most graphic example of predatory pricing and collusive trade practices by Japanese companies involves the American electronics assembly industry. Though the fact is largely unknown today, in the 1950s U.S. companies dominated the world's consumer electronics markets. No product was more effectively dominated by American compa-

---

[30]Sterngold, James. "In Japan, the Clamor for Change Runs Headlong into the Old Groove," *The New York Times*, Jan. 3, 1995, p. 1.

nies than television receivers. The decline and fall of the domestic television industry is the most important chapter in the story of America's consumer electronics collapse.

Revisionist history now says that Japanese companies won control of the international consumer electronics industry by providing higher quality. The truth is quite different: America's consumer electronics companies, led by marketing and finance graduates of U.S. business schools, essentially gave the market to the Japanese, beginning in the 1950s. At the time, Japan specialized in low-cost manual assembly, so American companies began subcontracting with Japanese manufacturers for assembly of simple products. This progressed to Japan making the components as well as the basic subassemblies. From there, it was a short hop to producing the entire product. With no real manufacturing activity at home, the American companies found it expedient to transfer research and development to Japanese firms as well. Having lost the lead in research and development and being utterly dependent on foreign manufacturers for supply, the American firms lost any reason to exist aside from serving as sales outlets for foreign interests. Once the Japanese firms decided to open their own North American marketing offices, there was no longer any function for the American company to perform.

Not all American consumer electronics managers willingly conceded their markets. Zenith, the last remaining American name with any domestic manufacturing, fought on until last fall when it transferred its remaining manufacturing operations to Mexico. Others, now deceased, struggled valiantly but unsuccessfully against unfair Japanese trade practices. The extent to which the contest was one-sided is breathtaking and includes the following Japanese trade practices and American responses:

1. In 1956, the largest Japanese manufacturers who together controlled 90 percent of the Japanese market, formed a cartel called the House Electric Appliance Market Stabilization Council. The cartel's immediate goals were (*a*) to monopolize the Japanese market for consumer electronics including television sets and radios, (*b*) to exclude foreign competition, and (*c*) to use the domestic market as a financial base from which to attack the American market.

The cartel established minimum domestic prices, imposed profit margins for retailers and wholesalers (22 and 8 percent, respectively), and boycotted nonmembers. Equally important, the cartel closed domestic distribution channels to foreign companies and successfully lobbied the Japanese government to impose an impenetrable wall of tariffs and nontariff barriers to imports.

Under Japanese law imposed by the American occupational government after World War II, this cartel was illegal in all respects. Indeed, the Japanese Fair Trade Commission (JFTC) twice (in 1956 and 1966) filed suit against all the major Japanese consumer electronics companies except Sony. Both times, the accused companies pleaded no contest and walked away without being prosecuted; the cartel hung together and became stronger as time went on.

2. The JFTC was unable to act effectively against the cartel because the Japanese Ministry of International Trade and Industry (MITI) took the side of the producers.

3. To offset losses that would result from dumping the same sets in America below cost, domestic prices were set at more than twice the price that would be charged abroad. To sustain such an operation required cartel cooperation on domestic prices and production volumes. A three-tiered council of working committees (with the so-called Okura Group, consisting of the CEOs of the cartel companies, having the final word on disputes) was set up to mediate issues of quotas, prices, and customers.

4. A key element of the cartel's plan was to gain access to America's television technology, which at the time was the world's best. American television manufacturers had demonstrated competence in global trade, even building factories in Canada and Europe to bypass tariff barriers. But the Japanese market was completely closed through measures designed by the U.S. Secretary of State to compensate Japan for exclusion of the Chinese market. On the theory that some income was better than none, the American companies decided to license their technology to Japanese firms. RCA, GE, and Westinghouse, the leading names in the U.S. industry, licensed their black-and-white technology to members of the cartel. In 1962, RCA, in what would prove to be the most disastrous decision, licensed its leading-edge color technology as well.

5. To avoid American charges of dumping, the cartel members told U.S. Customs officials that they were receiving much higher prices per set than was actually the case. They then rebated amounts to their American distributors through practices that included free goods and Swiss bank accounts. The free goods were imported under the guise of "market research" merchandise.

6. In 1968, America's Electronics Industry Association petitioned the U.S. Treasury to investigate the dumping of Japanese sets. In June 1968, the Treasury Department sent questionnaires to five major television manufacturers—Hitachi, Matsushita, Sharp, Sony, and Toshiba—inquiring

about their prices, costs, and sales volumes. Requests for information about import conditions were sent to Sears, J.C. Penney, and Singer. None of the questionnaires were answered in the first year. When the Treasury Department pressed for responses, the Japanese Embassy and the U.S. importers asked for more time. In fact, the questionnaires were never completed. Finally, in 1970, Treasury decided that even the artificially inflated prices on the import forms were low enough to constitute dumping. Japanese exporters were liable for damages equaling the difference between the selling price in America and the price for the same goods in Japan. But the damages would only relate to activities after the finding of dumping; in other words, there was no penalty for the more than a decade that Japan had conducted predatory pricing prior to Treasury's decision in September 1970. Meanwhile, employment in the U.S. television industry fell by more than 50 percent between 1966 and 1970. The number of jobs would decline by another 30 percent between 1971 and 1975 and 25 percent more between 1975 and 1981.

Although it had determined that the Japanese were dumping in American markets, Treasury was then obliged to calculate the amount of the damages. As it turned out, Treasury never imposed any penalties on Japan. Moreover, it effectively quashed private actions brought to Treasury by the National Union Electric Corporation (NUE) in 1970 and Zenith in 1974. NUE was better known by its brand names, Emerson and Dupont. The two suits were rather different, NUE asking for general damages while Zenith challenged the sale of Motorola's entire Consumer Products Division to Matsushita. Reflecting the desires of Richard Nixon and, subsequently, Gerald Ford, the Treasury Department refused to act on these suits as well. Finally, Congress itself realized that the nation's trade policy was in total disarray and gave Treasury one year to rule on the cases. Exactly one year to the day later, Treasury rejected the cases. Zenith was forced to take its case to court.

7. Between the first request for action by American television manufacturers in 1968 and the time the Treasury Department shut the door in 1976, 28 of America's 34 television manufacturers went out of business. But the 6 survivors were feisty. And there was a new administration headed by Jimmy Carter. GTE/Sylvania brought the International Trade Commission into the fray. However, when the ITC asked to see Treasury's files, Treasury refused. Then details of the kickbacks to distributors and customers were unearthed, causing the Justice Department to take up the cause. Uncertain what to expect with this new government agency in charge, the main Japanese television manufacturers hired one Harold Malmgren in 1977 to represent their interests. Malmgren was

not an innocent bystander; under both Nixon and Ford, he had been employed as America's top trade negotiator with Japan.

Only in America could a former senior civil servant play both sides of the street in such a flagrant manner. In any event, he produced results quickly. According to *The Japan Economic Journal,* "In three short months, Mr. Malmgren was able to talk to all sides in the dispute, and work out a compromise that was later called the orderly marketing agreement for Japanese color TV exports to the U.S."

Under the Orderly Marketing Agreement (OMA), Japan would limit its color television exports to 1.75 million units a year for the first 3 years. However, no restrictions were placed on the number of sets that the Japanese could supply from their newly acquired American production facilities. With amazing short-sightedness, RCA and GE supported the OMA, calling it "a good compromise." Of course, by this time most RCA and GE sets were made in Japan anyway. The remaining manufacturers, led by Zenith, strongly opposed this one-sided arrangement without success.

Meanwhile, President Carter appointed as Special Trade Representative Robert Strauss, former chairman of the Democratic Party and a top Democratic fund raiser. Describing his credentials for the job, Strauss told *Newsweek:* "I know something more than absolutely nothing, but less than a little." Strauss then proceeded to prove the accuracy of his self-assessment by signing a secret letter with the Japanese in which the United States promised to:

- Limit the ITC investigation into the Japanese cartel's predatory pricing.
- Greatly reduce or eliminate antidumping duties as quickly as possible.
- Quickly inform the Japanese government of any important findings arising out of the various U.S. probes into Japan's television industry *and* discuss these findings informally with the Japanese.
- Appeal a ruling that Zenith had finally won from the Customs Court requiring the payment by the Japanese cartel of damages.

8. As a final note, while the ITC was still reviewing the GTE/Sylvania case (the investigation was dropped shortly thereafter in accordance with Strauss' promises to the Japanese), its chairman was in Tokyo briefing the vice minister of MITI and members of the Japanese Diet. In surely the most flagrant of all the sell-outs and double-crosses of the American television industry by politicians and civil servants, the chairman of the ITC negotiated a contract to lobby for the Japanese once he retired. He was still ITC chairman when the contract was finalized.

5.8 Effects on the Auto Trade

The auto industry—one of the largest sectors of the global electronics industry—provides a particularly good example of the way Japanese fiscal and monetary policies were used to keep prices of Japanese products down in the face of strong upward pressures. In a perfect market, the 1985 Plaza Accord, which forced rapid appreciation in the yen, would have caused prices of Japanese goods in America to increase by 78 percent in the following year. However, prices for Japanese products did not rise along with the yen. Thanks to the Finance Ministry's actions providing manufacturing companies with virtually unlimited access to what then seemed like free capital, Japanese exporters were able to hold prices unreasonably low. Thus Japanese cars continued to be praised for high quality when what they really offered was better (albeit artificially induced) value.

If the price increases had materialized, a Japanese car priced at $15,000 in 1985 would have increased in price to $26,000 the following year. At $15,000, the same car would have been evaluated in the consumer's mind against a medium-priced Ford or General Motors car. Had the price increased in step with the revalued yen, the same car would have had a sticker price of $24,000 a year later and been compared to a Lincoln or Cadillac that, in terms of price, the car should have more closely approximated.

5.8.1 Japanese Auto Production Moves West

By 1995, the Japanese had run out of economic miracles. With the yen valued at around 100 to the dollar, Japanese auto companies began moving increasingly large amounts of production out of Japan to the United States. Nissan already owns one of the largest auto plants in America in Smyrna, Tenn. Toyota, which had largely forgone North America production, was planning a factory in Indiana to make 100,000 pickup trucks a year and another factory to produce 350,000 engines a year. Honda announced that it was permanently reducing yearly production at its plant in Suzuka, Japan, from 1.43 million vehicles to 1 million while increasing production in the United States to 750,000 from 630,000. Other companies were making similar plans, and the Japanese auto parts suppliers—including the very important electronics auto components makers—were also preparing for expansion in America.[31]

Such massive reductions in domestic production in favor of manufacturing in America would have been impossible and unthinkable as recently as 1992. The shifts violate Japan's long-standing commitment to lifetime employment and constant domestic industrial expansion. More importantly to American industry, perhaps, the necessity for Japanese companies to break the traditional social contract shows the importance of exchange rates in the marketplace.

5.9 Value vs. Quality

The significance of the economic and political conditions prevailing in Japan during the decades after the end of World War II can be seen clearly by reference to the definition of value presented in Chap. 4:

$$\text{Value} = \text{performance} \div \text{price}$$

where price is the cost to the customer and performance is a combination of features and dependability. In other words, since dependability and quality are identical, quality is one of several variables determining value. On the other hand, the influence of price on value is proportionally inverse. Therefore, factors that cause variations in price will affect value and thus demand far greater than can changes in dependability.

Was the quality of Japanese products better than that of the American products they displaced in the marketplace? It is difficult, if not impossible, to know. How could the "quality" of a Japanese television set be compared with the "quality" of an American set when the price and other market conditions differed so greatly between the two countries? How could the "quality" of an artificially low-priced Japanese car be assessed against the "quality" of higher-priced American cars? The answer, for both television sets and automobiles, was that Japanese quality could not be compared against the American quality because the two sides were not competing in the same markets.

While the question of whether Japanese products offered higher quality cannot be answered, it can be said unequivocally that, for the customer, the Japanese television came to represent a better value than a set from an American manufacturer. Similarly, a Japanese car possessed better value than an equivalent American-made car.

---

[31]Healey, James R. "Japanese Automakers Make Comeback," *USA Today*, Nov. 5, 1995, pp. 1B–2B.

With the increase in value of the yen during the 1990s, price distortions between Japanese- and American-made goods sold in the United States have been reduced or eliminated. Now that Japan's ability to undercut American prices has been restricted, indications are appearing that Japanese quality is not as good as its reputation. A few random selections to show that Japanese quality may be overrated include:

- An estimate by the Electronic Industries Association in the United States that the return rate for consumer electronics products—an industry dominated by Japan—in 1992 was between 5 and 25 percent. The EIA blamed the returns on functional failures and difficulty of use.[32]
- A 1993 J.D. Power and Associates survey for *Automotive News*[33] suggests that Japanese cars are no longer better than American makes. After surveying more than 6000 owners of cars registered in the United States between May 1991 and October 1992 to determine the probability that those owners would buy another car from the same manufacturer, the Power study found that owners of Cadillacs, Lincolns, Fords, and Saturns were all more likely to buy another car from the same manufacturer than would owners of Toyotas—and Toyota was by far the best-liked Japanese brand. Chrysler, Buick, Dodge, Chevrolet, and Pontiac rounded out the top 10 brands that current owners would buy again while Honda, Nissan, and Mitsubishi were—in descending order—the least likely of the makes included in the survey to enjoy repeat business. Another study released in late 1995 by Chrysler Corporation based on figures from Maritz Research, Consumer Attitude Research, and Strategic Vision showed that between 1988 and 1995 when owners of American cars bought replacement cars they were more likely to buy from the same auto company than Japanese car owners who were buying new vehicles.[34]
- Eberts[35] studied the probability of manufacturer's recall for various American and Japanese cars sold in the first half of 1994. The conclusions were that Chrysler products were only 40 percent as likely as General Motors or Ford cars to be recalled. Nissan models were 70 percent more likely than GM or Ford models to be recalled. Subarus

---

[32]Coates, James. "It's High Noon in Showdown over Electronics Simplicity," *The Chicago Tribune*, Jan. 24, 1993, p. 4-1.

[33]Diem, William R. "Bond Stronger with Age," *Automotive News*, Mar. 28, 1994, p. S6.

[34]"Loyal to the Brand," *The New York Times*, Nov. 3, 1995, p. C2.

[35][Eberts and Eberts 1995], p. 104.

were more than twice as likely to be recalled as the worst American makes, Mazdas 4.7 times more likely, Suzukis 11.2 times more likely, and Hondas 12.9 times more likely.

- Not of any statistical value but profoundly important to the authors, of more than 20 high-end Japanese VCRs we have owned over the years, every one has failed within the first year. Repair times of more than a year have been experienced—and some could not be repaired because the parts were not available.

Other studies can doubtless be found that show diametrically opposing results to those listed above. Moreover, the statistical validity of the studies listed—most notably the survey of car owner inclinations to purchase another car of the same brand—are all questionable.

Of course, the relative quality of Japanese goods vis-à-vis goods produced in the rest of the world has been largely irrelevant because the forces that enabled Japanese manufacturers to sell at artificially low prices overwhelmed any quality issues. Thanks first to the grossly undervalued yen and subsequently to the inflationary policies of the Bank of Japan, Japanese quality did not need to be better than comparably priced American products.

Because artificial price considerations generated such profound statistical distortion, those who argue that Japanese products offered higher quality cannot prove their positions. All that can be said for sure is that Japanese products long offered better value to American customers. And—provided the price passes the affordability test—better value is always the ultimate competitive advantage.

5.10 In Defense of the Japanese

If the account of Japan's postwar industrial development presented in this chapter seems critical of the country's actions, that is not our intent. Some acts by Japanese government agencies and companies—most notably the deliberate inflation of land and stock prices—were certainly shortsighted and harmful to both the global and Japanese economies. Irresponsible would not be too harsh a term to describe a number of Japan's international trade practices. However, over the decades, Japanese companies have shown determination and initiative not always found in their American counterparts. The American weaknesses exploited so ruthlessly by Japan should not have existed in the first place.

American government has shown itself willing to sacrifice domestic businesses in exchange for a military presence near China and Russia. That the international trade war between Japan and the West could have more serious consequences than the Cold War apparently escaped the government's attention. Japanese money has improperly induced America's trade negotiators and policymakers to switch allegiances; however, the option of passing more stringent conflict-of-interest legislation to control the activities of retired government employees was always available but never exercised. The recent willingness of U.S. corporations to joint venture with Japanese partners—despite decades of evidence that such cooperation results in a one-way flow of technology and competitive strengths from west to east—cannot be blamed on underhanded Japanese tactics.

Similarly, if America chooses to believe that Japan's success in international trade is attributable primarily to higher quality, the Japanese cannot be criticized for capitalizing on that misconception.

The Japanese experiences offer valuable lessons for American manufacturing companies about what it takes to succeed. Those lessons extend far beyond how government intervention in the economy affects a company's ability to compete internationally. At the level of the individual company and its plants, the more important lessons to be learned from Japan are:

- *Manufacturing cannot be a sideline.* Too many American companies want to shift responsibility for manufacturing to outside contractors—even if the contractors are owned by the very Japanese companies against which the American companies will be competing. Japanese companies, in stark contrast, believe that manufacturing competence must be their primary concern. A production culture must be nourished in order to sustain other vital business skills such as product concept, development, and even marketing. Put another way, success requires the will to succeed. Japanese companies have that will; few Western companies can make the same claim.
- *A services-based economy does not sustain prosperity.* The world offers countless examples of how industrial strength turns to fat when the manufacturing sector atrophies. Glasgow, Liverpool, and Manchester were among the world's wealthiest cities before their manufacturing disappeared; only through enormous effort to attract new manufacturing in the past decade have they begun to regain jobs. This country's national preoccupation with services rather than manufacturing will leave America a second-rate industrial nation.

- *Copy good ideas shamelessly.* A large percentage of Japan's most successful products and management practices were "borrowed" from other countries—primarily from the United States. The transistor, perhaps the most important single item in the rise of an electronics industry in Japan, was of course created at Bell Laboratories. The radio—which, when combined with the transistor, became Japan's first globally dominant product—was largely American as well. The television technology that proved Japan's success with the transistor radio was not accidental was licensed from American companies. The VCR was an American invention. As Chap. 12 explains, the greatest strength of Japan's manufacturing companies is the apprenticeship system; Japanese manufacturers aggressively pursued apprenticeship programs at the same time as American companies were cutting back on development of manufacturing skills. The list of foreign ideas adopted by Japanese companies is almost endless.[36]

  Japan's ability to see opportunities in the ideas of other countries has been poorly understood in this country. The many Americans who regard the Japanese as weak innovators prospering primarily by stealing the best ideas of others miss two critical lessons: (*a*) Japanese companies succeed in large measure because they have not been cursed with the NIH (not invented here) syndrome; and (*b*) when foreign ideas are adopted in Japan, it is generally in a modified form designed to better meet Japanese conditions. American companies, in contrast, greatly distrust any procedures not created by their own people—except for "Japanese management" techniques that have been blindly adopted in countless U.S. companies without modifications to compensate for unique strengths or weaknesses of the company or its workforce. Thus the contradiction prevails in American companies of rampant NIH syndrome coexisting with an almost naive willingness to accept practices that are said to have worked in Japan.

- *Capitalize on our technology.* Why were the Japanese able to turn languishing American technologies (e.g., the transistor, the VCR, the fax machine) into enormously profitable products? The answer seems to be that Japanese companies allocate more of their R&D investment to applied technology—often finding practical applications for foreign research results—while American research has been more theoretical. The lesson: If American companies fail to bridge the gap between the lab and production, other countries will gain the benefits.

---

[36]Copying ideas from other countries is a time-honored Japanese practice. The kimono is only the most prominent example of foreign ideas imported by Japan over the centuries.

Japanese companies have little to teach America about quality. To the extent that quality improvements result from attitude adjustments, however, there are valuable lessons for any manufacturer to be found in Japan.

5.11 An Unanswered Question

Japanese manufacturers have proved themselves willing to accept losses for what, by Western business standards, is a very long term. No examples come to mind of American companies selling below cost for decades at a time. Indeed, considering the increasingly strident demands by large shareholder groups of American companies for greater short-term earnings, it is unlikely that any company based in this country could take such a long-term outlook even if it possessed sufficient capital to operate at a loss for several consecutive years. Competitive pressures may prevent American companies from charging prices that deliver a profit but, if prolonged losses seem inevitable, the American company will generally leave the industry. A leading model of American unwillingness to endure losses is the General Electric Company. For the past decade, GE has closed or sold off most divisions that did not rank first or second globally in their industries. This triage approach to business (which has seen GE transformed from a huge manufacturing company to a modest manufacturing company with massive service operations) is widely considered to be a paragon for modern pragmatic business planning.

To this point, the Japanese willingness to sacrifice earnings for market share has worked well. In many segments of the electronics industry—especially in consumer electronics—Japan has exterminated all meaningful U.S. competition. However, these gains in market share have been obtained only at a heavy price:

- By late 1995, stock prices had fallen 52 percent from their 1989 peak.[37]
- Five Japanese financial institutions failed during 1995 and many others were technically insolvent.
- The official government estimate of the outstanding bad debts held by Japanese banks in the fall of 1995 was $370 billion, of which at least half was said to be unrecoverable; other estimates place the total bad debt at $735 billion.

---

[37]Montague, Bill. "Daiwa Debacle Erodes Trust in Japan's Banks," *USA Today*, Nov. 7, 1995, pp. B1–B2.

- The official bad debt amounted to 8 percent of Japan's gross domestic product—compared to the worst level in American banking history of 2.5 percent in 1991.
- The return on assets for the top 50 Japanese banks in 1995 was on average only 0.002 percent compared with 1.3 percent for a typical American bank—despite the fact that the Bank of Japan's discount rate (the rate it charges commercial banks to borrow money from it) was only 0.5%.[38]

By the end of 1995, many members of the international financial community openly wondered whether the entire Japanese banking system might be on the verge of collapse. The unanswered question is whether the Japanese economy can endure the financial strain much longer. As one Japanese automaker after another shifts production from Japan to America, it is clear that at least one industry that was commonly believed to base its competitiveness strictly on quality has been badly damaged by changing financial conditions.

5.12 Summary

This chapter contained the following key points:

- Japanese goods were sold at artificially low prices outside Japan.
- Because prices of Japanese goods were unnaturally low, the Japanese goods offered better value to American consumers.
- Although value and quality are different concepts, many Americans wrongly believed that Japanese products possessed higher quality.
- No evidence can be found to substantiate the claim that Japanese products are more dependable.
- American companies have been wrong to copy Japanese management techniques without modifying them to better meet the special characteristics of American companies and their employees.
- Japan can teach American industry many important lessons. However, those lessons have very little to do with quality and much to do with manipulation of exchange rates, capital availability, and competitive trade practices.

[38]WuDunn, Sheryl, "The Wounded Giants of Japan," *The New York Times*, Nov. 24, 1995, pp. C1, C3.

6
The Science of Quality Management

6.1 Chapter Objectives

Items to be examined in this chapter include that quality management:

- Has only one purpose: to oversee changes in the dependability of product.
- Can be based on science or cosmetics—but only science has meaning.
- Requires working knowledge which combines experience and academic knowledge.
- Requires identification of quantifiable parameters by which progress can be measured.
- Must put aside its traditional reliance on visual assessments of "quality."

6.2 Factors Influencing Dependability

Personnel at most electronics assembly companies erroneously believe that they are engaged in rigorous quality management through scientific process control. In reality, all plants practice—to varying degrees—both science and art. While some facilities are less reliant on art than others, the presence of *any* art in the design and management of processes is usually disastrous. Artistry in the design or operation of a process destroys the company's ability to manufacture dependable products on a consistent, predictable basis and is therefore the enemy of genuine quality assurance.

Scientific process control is simply an alternative term for rigorous quality management. Just as process control and process "artistry" are incompatible, rigorous quality management can exist only in companies operating entirely on the basis of scientific process control.

The first requirement of scientific process control is acceptance that the causes of most failures and defects are identifiable, predictable, and preventable. Of course, as statisticians for the past seven decades have noted,[1] some causes will be random and beyond the company's control. In electronics assembly, however, the percentage of dependability problems that cannot be prevented by scientific process control is much lower than commonly accepted.

Controllable factors determining dependability of electronics products can be grouped into four primary categories:

1. Design
2. Component selection
3. Handling
4. Processes

6.2.1 Working Knowledge

Two facts about the controllable factors of dependability deserve special attention:

1. Each of the four groups of controllable factors can be harnessed for improvement of dependability and reduction of cost.
2. The enormous opportunities represented by controllable factors can be turned into realities only with thorough working knowledge of physical sciences that include physics, chemistry, metallurgy, and materials properties. In the absence of this working knowledge, the forces will run out of control and make consistent output of dependable product impossible.

---

[1]See, for example, the discussion of Shewhart's steady and intermittent defect causes in Sec. 2.3.

Working knowledge differs greatly from academic knowledge. In particular, working knowledge does not require advanced degrees in the sciences concerned. Indeed, although the products that result from electronics assembly often represent spectacular technological advances, the assembly processes themselves are rather mundane—to the point where advanced study of the sciences affecting the processes yields few benefits. Diligent application of some lessons from high school science works wonders where advanced scientific theory often serves no purpose but to complicate otherwise straightforward activities. It is fair to say that the industry uses low-technology processes to create high-technology products.

Stated another way, humans belong to the species *Homo sapiens*, literally "wise men." That, however, does not mean we are born with wisdom. Rather, life can only be learned by living; it cannot be learned from books. By the same token, operations can only be learned by operating. Books and schools can provide explanations and historical perspectives, but that academic instruction has little practical value until combined with hands-on experience. Operating experience helps make us aware of the questions that need to be answered; formal education provides "answers" without the context of determining the necessary questions.

6.2.2 Requirements of Dependable Production

To produce consistently dependable products, companies must comply with each of the following three requirements in the exact sequence that they are presented here:

1. Master the underlying sciences that affect production.
2. Use that understanding to develop and manage their processes.
3. Root out all elements of "artistry" in production and quality verification.

Companies that meet these three requirements can properly be described as process-driven. The process-driven electronics assembly company reallocates its quality verification resources from post-production detection of failures to pre-production process validation. Preproduction process validation involves four broad groups of actions that must be carried out in the following order:

1. Determining the causes of failures and defects
2. Discovering how those causes can be prevented

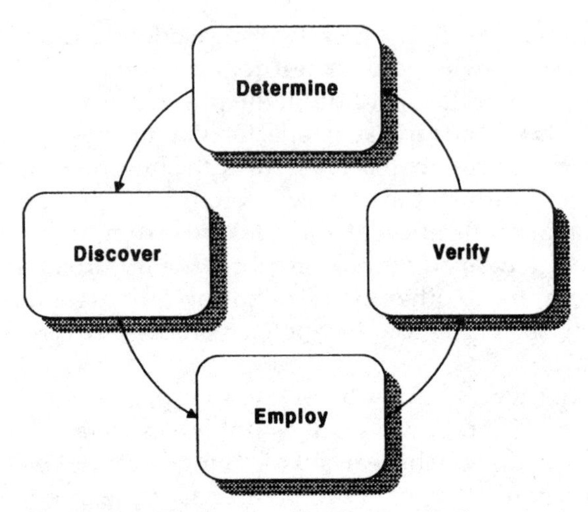

Figure 6.1 The Process Verification Cycle.

3. Employing the preventive measures throughout the company
4. Verifying the results

The series of actions 1 through 4 have characteristics similar to those of a Shewhart Cycle. That is, the series begins with step 1 and progresses through step 4, which in turn leads back to the first step, and the series begins anew. Accordingly, the repeating sequence can be termed the Process Verification Cycle, as shown in Fig. 6.1.

The Process Verification Cycle which characterizes the process-driven firm ensures that the output contains negligible numbers of defects. Inasmuch as there will be inconsequential numbers of defects to correct, the familiar post-production inspection and much test can be eliminated. The process can be accurately monitored by small-lot sample testing.

6.3 Cosmetic vs. Scientific Quality

Cosmetically driven "quality"—epitomized by but not limited to the familiar MIL specifications—contends that dependability problems can be recognized by visual inspection and corrected through appearance-altering procedures such as touchup.

Process-driven quality, on the other hand, contends that dependability results primarily from the actions taken by the company, from the earliest stages of design through to the final handling as the product ships to the customer.

The concept that electronics assembly somehow involves "skill" and "art" is solidly embedded in the beliefs of many industry veterans. All total quality systems also implicitly endorse reliance on appearance of output by emphasizing the need for feedback loops between inspection and assemblers. The theory of the feedback loops is that the assembly personnel will know immediately when their workmanship is unsatisfactory and take the necessary remedial actions to prevent recurrence. Visual inspection plays a prominent role in cosmetic quality.

Scientific quality assurance looks at the issue from a profoundly different perspective, starting by recognizing the limited capacity of the assembler to improve output. Assemblers can never be better than their tools—including processes specifications and materials—allow. Workers cannot deliver intrinsically dependable product if the processes or materials with which they must work are flawed.

As Chap. 7 will show, rework and touchup can alter the appearance of connections so they *appear* perfect, but looks in such cases are deceiving. Dependability cannot be added retroactively. The "art" possessed by the experienced assembler is that of camouflaging rather than curing potential failures. If the wrong materials, chemicals, or other ingredients in the processes are incorrectly specified, the output of those processes will be undependable. Since the process parameters are beyond the control of assemblers, sending unattractive work back to assemblers only makes them paranoid; this paranoia will then manifest itself in considerably greater amounts of informal touchup and rework occurring before the assemblers permit any work to be sent for initial inspection. If any inspection occurs under a system of scientific quality management, it consists of statistical sampling to find indicators of process deficiencies.

Cosmetic quality is completely illusory; indeed, the cover-up that takes place in the name of cosmetics is responsible for the majority of product failures in the average electronic assembly plant. Scientific quality is definitive, leaving no room for doubt about dependability. A further attractive feature of scientific quality concerns cost: Scientific quality management is always much cheaper (not to mention more effective) than cosmetic quality.

We can therefore distinguish between cosmetic and scientific quality. However, two important questions remain unanswered:

1. How can the two forms of quality be distinguished?
2. What knowledge is vital to achieving scientific quality?

6.4 Basic Requirements of Effective Quality Management

For the quality department to help rather than hinder the company, the following conditions must be met:

1. The quality department must have clear knowledge of its purpose.
2. Complete knowledge of the relevant process sciences must exist and guide all decisions.
3. Adequate tools for managing and improving quality must be available.
4. The responsible personnel must know how to use the tools.
5. The quantifiable attributes that determine quality must be identified.
6. Efficient systems must be devised and implemented for collecting and analyzing the attributes data.
7. The company must have sufficient scientific knowledge to determine the process modifications dictated by the attributes data.
8. Systems must be in place for implementing the necessary corrections to the processes as determined from the attributes data.

These requirements are *fundamental* and *essential.* Each of them is necessary but none is sufficient in itself to allow scientific management of quality. Despite the unconditional need to fulfill each and every requirement, however, few electronics assembly companies at this time can honestly state that their quality operations meet more than a few of the requirements. We have yet to find a single company that meets all seven requirements. Moreover, the quality profession as a whole seems to be losing ground rather than making progress toward fulfillment of all the requirements.

6.4.1 Knowledge of Purpose

The first of Deming's 14 Points for Management (see Sec. 2.4.2.1) states, "Create constancy of purpose toward improvement of product and service, with the aim to become competitive and to stay in business, and to provide jobs." This advice is defective in several critical aspects, not the least of which is the reliance on "improvement" without explicitly stating what constitutes improvement.

Despite Deming's argument, the company's purpose is not to provide jobs. Although the desire to create more employment is unquestionably humane, it is also an ethical issue that is unrelated to the quality function of advancing the company's prosperity.[2] And a company that lacks financial strength also lacks the ability to provide lasting, rewarding jobs—not just for potential new hires but for the existing workforce as well.

The company, through its sales and marketing department, is responsible for setting the specifications according to its interpretation of what will motivate customers to buy the product at a price that provides a satisfactory profit to the company. The customer's subsequent purchasing behavior shows whether the sales strategy was valid but does not alter the product's quality in the least. The quality department exists to ensure that the product meets the specifications *in a cost-effective manner.* Whether customers buy the product must never be a factor in evaluating the quality department's performance (although it is highly pertinent in assessing the competency of the sales and marketing personnel).

Quality practitioners who evaluate their actions—existing and contemplated—by any standard other than (1) increased product dependability at the same cost or (2) lower costs while maintaining the same dependability are not in fact engaged in the pursuit of quality.

6.4.2 Possession of Scientific Knowledge

Quality assurance is a system under which the dependability of output is determined by controlling the steps in the product's creation—from design to final assembly—in a rigorous fashion so that the characteristics of manufactured goods are consistent from one unit to the next. "Consistency" refers to the condition of the product without any additional steps for rework or repair. It might best be considered the first-pass performance.[3]

Consistency is necessary for operation of statistical controls, for both production and quality. All quality practitioners know the importance of consistency in output, though universal comprehension has not yet been reached that "consistency" achieved through rework is merely consistently undependable. Less well recognized is the fact that consis-

[2]Of course, the profitable company will be able to provide jobs that the failing firm cannot. For this reason, two otherwise competing groups—shareholders and employees—have a common interest in seeing the company prosper. Jobs are only a consequence of the company maximizing long-term profits; jobs are not the objective.

[3]Consistency of output is necessary for statistical process control. This must not be interpreted as support for the statement (see Sec. 3.3) that "quality is the absence of variation."

tency can mean consistent failures as well as consistent perfection. In other words, consistency allows the state of the process(es) to be measured with greater confidence but certainly does not mean there is no further need for improvement.

6.4.2.1 Processes vs. Natural Laws. In assessing any manufacturing operation, we must always keep in mind that three operational possibilities exist:

1. The real process—what is actually done
2. The specified process—what the process engineers decreed should be done
3. The natural process—what scientific forces require to be done

Dependable output will result only when all three operational possibilities are identical. In other words, the real process is identical to the formal process, which in turn conforms to the laws of nature. The effectiveness of manufacturing systems depends on interacting natural scientific forces. Systems that work with rather than against those forces result in products of high dependability and low cost. Systems that oppose—whether wittingly or not—the natural forces cause low dependability and high cost of products. A properly functioning quality department helps bring about the condition of convergence.

6.4.2.2 Dependability Requires Knowledge of Science. High dependability and low cost are attainable if and only if the system designers and managers—notably including the quality department—work with thorough knowledge of the ways natural forces (i.e., science) affect performance of the processes and the products. Gaps in knowledge about the underlying science in a single key process can easily cause high ongoing product failure rates and excessive operating costs.

Virtually every electronics assembly company suffers from serious gaps in its collective knowledge of process sciences. Consequently, companies operate with:

- Designs that violate critical rules of manufacturability
- Processes that defy natural forces
- Faulty workmanship (i.e., quality evaluation) standards
- Needless excess handling that increases costs but reduces dependability

Knowledge typically requires a combination of experience and formal education. Neither experience nor education alone equals true knowledge. Experience in the absence of education, for example, causes intellectual inertia; many of the industry's worst practices persist because they are the methods with which the veteran personnel are familiar. The situation is particularly serious with respect to the core process of soldering where processes and quality evaluation procedures have not changed in decades, although the components subjected to those procedures have become smaller and more fragile than anyone could imagine as recently as two decades ago.

6.4.2.3 Limitations of Education without Experience. Just as experience without education equals failure, education without experience has little immediate usefulness. Over the years, many companies have recognized that electronics assembly dependability depends less on electrical knowledge (at least outside the circuit design department) than on proficiency in sciences such as chemistry, metallurgy, mechanics, and thermodynamics. The usual response of such companies is to hire inexperienced graduates of university science programs and assign them to process design and management. Since the proper application of the sciences also depends on familiarity with the activities in the production environment—familiarity that new graduates cannot possess—the expectations that companies hold for these science specialists are unrealistically high.[4]

Without understanding the context in which the science will be applied, it is impossible to recognize the problems that must be solved or see that a range of possible solutions exists. A chemist may, for instance, be able to analyze the compounds left behind after fluxing but not necessarily realize that capillary forces render post-assembly cleaning and cleanliness testing ineffective.

Quality improvement requires more than desire to improve. As one of the popular slogans from the early 1980s put it, work should be performed "smarter" rather than "harder." Unfortunately, we have confused our priorities and work harder rather than smarter. The case can even be made that the workloads placed on remaining personnel by the downsizing mania of recent years have left no time for scientific learning.

---

[4]A similar observation can made with respect to the large numbers of electrical engineers whose first assignment is management of solder reflow activities. Plant management seems incapable of seeing the obvious: that an electrical engineer learns as much in university about such processes as a fine arts major—and is equally well prepared for life as a process engineer.

6.4.3 Availability of Adequate Quality Management Tools

In the quality world, "tools" refer to intellectual systems rather than physical instruments and equipment. A manual of workmanship standards is a quality management tool; a more precise component placement machine is not, even though the greater accuracy of the new placement equipment may reduce the incidence of failures. The properly selected and employed quality tool will lead to decisions such as whether the merits of purchasing new equipment outweigh the liabilities.

Tools for managing and improving quality are readily available and well known. They range from analytical techniques such as statistical methodologies to humanistic matters such as employee involvement. The most valuable total quality management tools are presented in Chaps. 12 to 14.

6.4.3.1 Tools vs. Programs. Tools differ from programs, which are blends of tools directed at accomplishing specific objectives. A total quality program, for example, typically includes statistical devices, teams, and training as a minimum set of tools. The program can never be any stronger than the effectiveness of the selected tools.

There is no shortage of tools. Indeed, so many tools exist that the sheer number of choices can be overwhelming to the quality program designer/manager. Moreover, while tools may be ubiquitous, few will actually bring about the desired effect no matter how energetically they are employed. Other tools may be capable of achieving some desired changes but require excessive effort on the part of managers and the broader workforce. Statistical tools, for example, require considerable insight into the characteristics of key parameters: What they are, how they should be measured, and where in the sequence of processes they should be measured are just three very challenging questions that must be answered accurately before the statistics have any practical value. Other readily apparent questions include:

- Are control charts necessary?
- Is Pareto analysis useful for our purposes?
- Can inspection data be considered statistically relevant?
- What is the best way to display statistical findings—bar charts, pie charts, scatter diagrams, some combination of these, or are there others we overlooked? (For that matter, do data maps serve any useful purpose?)

The challenges presented by statistical tools are not unusual; almost every tool involves difficult questions. Employee teams, very fashionable at the moment, are every bit as troublesome as statistics. Only the forms of the problems differ between the two types of tools. Simply forming teams and "empowering" workers will not help the company prosper; teams require considerable support from management in the form of supervision, education, and encouragement.

6.4.3.2 Subsets of Tools. No magic formula exists to identify which tools are best for the needs of any given operations. Compounding the difficulty of selecting the right tool is that a subset of the tool may not be suitable in certain situations.

A screwdriver, for example, is a tool for mechanical purposes. However, the screwdriver will have limited effectiveness if it is the wrong configuration to mate with the head of the screw; it is necessary to know not just that a screwdriver is needed but what kind of screwdriver meets the specific conditions of the job at hand. Similarly, some form of team management may be the proper tool for resolving a certain problem or problems. The fact that a team approach will be helpful in some circumstances does not in itself imply that wholesale adoption of employee teams throughout the company is desirable. Nor is it obvious that all teams in a plant or company should have the same mandates, composition, or supporting resources. The best choice may be a team or small group of teams with explicit responsibilities in a particular department; alternatively, a corrective action team drawn from cross-functional departments expressly to correct a single known problem can be impressively productive.

6.4.3.3 Avoiding Obsolescence. Choosing the proper tools in the first place poses serious difficulties. Even more troublesome can be ensuring that the selected tools do not linger on long after they have outlived their usefulness or become less effective than alternatives. If a quality procedure such as test reveals no failures, it is probably time to reconsider whether that procedure is still appropriate to the operation's needs. Returning to the example of teams, they may serve best in a temporary capacity; when the issue has been resolved or the team is clearly stymied, it is often time to look for alternative tools.

Selection of tools is second on our list of requirements but only for purposes of explanation. In practice, the tools should not be chosen until many of the other requirements have been met. At that point, the best tools for the job will usually be self-evident. Selecting the tools before possessing the complete scientific knowledge about what problems the company faces will most likely result in poor choices.

6.4.4 Competence in Using the Tools

Deming, a passionate believer in statistical methodology, misled his constituency when he said "Engineers and scientists need rudiments of experimental design."[5] Actually, anyone possessing only the "rudiments" of experimental design will quickly prove the validity of the adage "a little knowledge is a dangerous thing." Productive use of statistical tools becomes possible only when the user has thorough knowledge of the experimental requirements *and* detailed familiarity with the operation to be studied. Many tools are deceptively simple, and competence in their use is not all that easily acquired.

6.4.4.1 The Need for Competence in Use. The effectiveness of any tool—even when properly chosen—depends on the user's expertise in use of that tool. A chain saw, for example, may be a superb tool for an experienced lumberjack but lethal in the hands of a novice. In the same vein, we have seen more than a few operations badly injured by flawed attempts to use over-powered tools such as partial factorial experiments. Decentralized decision making in such forms as employee teams will also be regretted by those who undertake such radical restructuring of operations without foreknowledge of the many ways even the most talented teams can go astray.

Competency entails much more than preventing the tool from running amok. Speed of operation is another factor to be considered; the competent user accomplishes much more in any given time span. Too often, companies feel real progress has been made when they have actually lost ground to the competition. The company that needs 5 years to accomplish what competitors manage in 6 months has demonstrated incompetence and fallen behind, not progressed.

6.4.4.2 Contributions of Experience and Education to Competence. Familiarity with the tool is necessary but not sufficient to ensure a happy outcome in its use. Competence is a combination of understanding the tool and familiarity with the special requirements of the industry in which it will be applied. For that reason, the academic statistician (like any other academic lacking deep hands-on experience in the operations to which a tool will be applied) without thorough grounding in the industrial processes which are to be optimized cannot be said to possess competence. The academic has only a portion of the necessary proficiency to design a satisfactory statistical quality management program for use in electronics assembly.

---

[5][Deming 1982], p. 47.

Academic consultants have created enormous grief in the quality field by applying their theories to operations in which they have no real experience. Indeed, it is typically easier to learn the mechanics of operating the tool than to become thoroughly acquainted with the unique aspects of the industry. The company wishing to gain mastery of both quality and process management tools is better advised to educate its veteran operations personnel in the techniques of the tool than to try teaching plant operations to a generalist practitioner of any tool.

6.4.5 Identification of Quality's Quantifiable Attributes

Satisfaction with the results of any endeavor comes more easily when there is no requirement that the results be quantified. As Chap. 3 emphasized, *effective* quality management depends on identifying quantifiable parameters that accurately reflect what the quality program is trying to achieve—i.e., in the case of electronics assembly, dependability—and following through with measurement of changes in those parameters as the program progresses.

The typical electronics assembly company has no idea what parameters need to be measured in determining quality effectiveness. Since the parameters are unknown, it is impossible to quantify the company's quality performance or to determine when it changes. The failure to recognize the real measures of quality has long been the most serious obstacle to achieving genuine quality and avoiding unnecessary expenses in quality verification.

6.4.5.1 Defects vs. Quality. Except for blatant hard failures such as missing components, reversed components, breaks in circuit tracks, solder shorts, or solder opens, the dependability of an electronic assembly cannot be determined visually. Even these few "hard" defects are often hard or impossible to find by inspection. Moreover, most "defects" are fully reliable and cannot be improved by any means—including replacing the component. A defect in electronics assembly jargon, it must be emphasized, is different from a failure. Defects are determined by inspectors on the basis of appearance rather than function; a detected "defect" may or may not be an indicator of dependability, although it often provides useful information about the state of the process. For this reason, measuring "quality" according to the defect rate is quite misleading.

The validity of data depends on the point at which it is collected. Data collected after touchup or rework has little meaning. But touchup and rework are more prevalent in electronics plants than is generally recognized, and much of it takes place prior to inspection. Whenever an assembly passes through the hands of an operator equipped with a soldering iron, for example, the tendency is for the operator to conduct a visual check and "improve" some solder joints by touching them up. In hand soldering operations, it is normal for an operator to rework most connections several times until becoming satisfied with the appearance. Little to none of this activity is captured in statistical data.

6.4.5.2 Invisible Damage. The most common failure modes for electronic assemblies cannot be seen and often cannot be measured. Static damage occurs inside components, as do the more insidious forms of heat damage. Reductions in resistance between current-carrying services is invisible. With modern multilayer PCBs, the most serious failure modes are located in internal layers or through-hole plating where they cannot be seen.

Even more troublesome, a substantial percentage of damaged components do not cause failures before being put into use by the customer. Heat damage during touchup and rework (see Chap. 7) shortens the life of components but not by predictable amounts. The damaged component may fail before final test, during the early days of use by the customer—or 5 years after assembly. Ionic residues between conductive paths affect performance only under certain conditions of high humidity, temperature, and bias voltage; failures from ionic contamination may never be seen in an arid climate like Arizona's but be regular events in coastal environments.

Burn-in and thermal cycling during test accelerate the aging of all components, not only those that would fail during infancy. Causing those components to fail by applying stresses during test comforts some quality practitioners, but the more enlightened understand that the life expectancy of every component in the assembly has been shortened to find the weakest parts. (The always controversial issues of burn-in and stress testing are examined in Chap. 11.)

Of course, quality departments in electronics assembly plants do reject a substantial number of assemblies for visual reasons. If the first-pass yield is calculated on the basis of those assemblies that are subjected to no touchup or rework, the first-pass yield in many plants is effectively zero. That is, *every* unit is deemed to contain one or more defects. However, the reasons for rejection are usually unfounded. The defect standards in use today have existed virtually unchanged in large part since before the first printed circuit boards appeared; others originated with the introduction of single-sided PCBs (see Chap. 9).

The only meaningful quality attributes are found at test and in field use. Those meaningful attributes are the numbers and manners of product failures. All other information is relevant only to process management. Quality departments that agonize over the acceptability of product may be working hard but they are not working well.

6.4.6 Collection and Analysis of Data

Scientific management of quality requires the collection and careful analysis of dependability data. Most electronics plants believe this is what is already happening in their operations. Certainly it is true that numbers are being dutifully recorded and compiled, often even graphed and displayed for public consumption. Unfortunately, the "data" concern inappropriate attributes, are gathered at the wrong points in the product's creation and use, or are incorrectly interpreted. Under any of these circumstances, data are liabilities rather than assets.

6.4.6.1 Meaningful and Irrelevant Variables. The first rule of useful data collection is that only the meaningful variables be tracked. The effectiveness of data is not determined by weight or volume but by the insights it provides. Amassing great quantities of information about irrelevant variables only increases the difficulty of finding the vital facts about the product's dependability and adequacy of processes.

Data must be collected at several points in the creation and use of the product. The first point of measurement should be at incoming materials to determine the failure rate of components; this information is necessary before the expected failure rate of the completed product can be computed (see Sec. 3.14). Inspection should be conducted—if at all—only with respect to the few failure modes that can be detected visually; as will be explained in subsequent chapters, appearance is seldom indicative of dependability. The frequency and mode of failures at any test point is clearly relevant; indeed, any test failure that is not fully documented must be considered a fatal breakdown in the quality management system. Finally, the dependability of the product in use must be tracked and compared to the predicted failure rate.

6.4.6.2 Data Collection Problems. As Sec. 3.14 noted, gathering meaningful data about field performance is not a trivial undertaking. A few of the problems inherent in collecting accurate field dependability data include:

- The item was needed and used for only a brief period, after which the customer used misrepresentation to return it for refund.
- The customer is unable to ascertain how to properly operate the product (a failure in design for use and documentation).

- The repair technician wrongly diagnoses the nature of the failure.
- The customer does not report the failure.
- The failure happens out of warranty.

Data collection efforts should be scaled to the magnitude of the quality problems observed. For example, if the failure rate at an interim test point prior to full functional test is negligible, that test operation should be eliminated. Other tests conducted only for verification of general product dependability should be conducted on a strict sampling basis rather than universal test being employed. After all, quality cannot be inserted retroactively; except for those devices which must be tuned, every test reflects lack of confidence in the processes. (For the defense electronics company, of course, every test may merely be confirmation that the customer has no knowledge of how to manage quality. More than a few DoD inspectors have demanded that every conceivable test or inspection at every stage in the operation must be carried out—completely and always!)

6.4.7 Ability to Act on Data

Competently compiled data will reveal dependability crises. But there is little to be gained from knowing about failures if the technical and scientific resources do not exist to determine (1) the cause(s) of the crisis and (2) the changes in design, materials, or processes that will prevent recurrence. Since so many failures result from internal damage to components that cannot be seen, the usual "corrective action" in the event of failure is to blame the component manufacturer rather than investigating what aspects of the assembler's own operations could be contributing to the component failures.[6]

Ultimately, the issue returns to whether the company possesses the relevant scientific knowledge and the organizational talent to gainfully employ that knowledge. As was seen in Sec. 6.4.2.2, the desired scientific knowledge probably does not exist within the company. Depending on the size of the company, it may be most cost-effective to retain the

[6]The value of comparing predicted failure rates to actual failure rates should be apparent at this point. A real failure rate significantly greater than the anticipated rate is reason to presume that the assembler's processes are to blame. Resources currently being directed into inspection and touchup would provide a higher return on investment if allocated to developing the statistically probable failure rates, measuring the actual failure rates, comparing the two, and taking corrective action.

services of independent failure analysis specialists or qualified freelance troubleshooters.

The temptation to call upon the "expertise" of suppliers in troubleshooting problems is strong. This road may lead to misfortune, however, because suppliers are not necessarily scientifically knowledgeable themselves. Lack of true expertise is particularly likely in sales personnel. And distributors are still less likely to have competence in resolving a process issue.

Fortunately, the underlying causes of most component-related failures involve stresses to which those components are subjected during assembly and subsequent quality verification. Those causes and methods of prevention are found in the discussion of soldering in Chap. 7.

6.4.8 Implementation of Corrective Action

While the quality department collects and compiles the data, the responsibility for changing operations so that failures do not recur normally falls to other functional departments such as process engineering, design, purchasing, and production. If those departments external to quality cannot be induced to act on the information revealed by the data, there is no point in maintaining a quality department. The frequency with which the quality department simply records ongoing failure modes is distressingly high.

Many platitudes are uttered about the need for support of quality from the company's most senior executive ranks. This is too often interpreted to mean that upper management will decentralize some decision-making authority and make funding available for "quality improvement" activities. Far more important but much less common is insistence by the senior executives that dependability problems identified by the quality department be corrected by the responsible departments. Tragically, senior executives most commonly come from the ranks of sales or financial personnel and are either intimidated by or indifferent to production matters. Outside the quality department, the vital quality issues are decided not in terms of dependability but by effects on short-term earnings or other expediencies.

The realized worth of even the most brilliant quality plan is zero (or, more accurately after taking account of the preparation costs, negative). All the returns depend on the effectiveness with which the plans are implemented. In the quality business, actions not only speak louder than words—words do not even whisper.

6.5 Summary

- Dependability in electronics assemblies can be achieved only by scientific control of processes and evaluation.
- True quality depends on accurate and complete knowledge of physical sciences.
- Knowledge has value only when it determines all actions from design of product and processes to evaluation of results.
- Few assembly plants have eliminated formal visual assessment of output; among those that have, hidden rework often remains.

Many factors from design to test methodology affect ultimate dependability. Ultimately, however, dependability is most affected by the soldering process sciences. Soldering is the core process of electronics assembly; the majority of components are involved in soldering operations either through direct application of solder or through interaction with components whose performance is affected by soldering. Furthermore, no process better epitomizes the nature and benefits of scientific process management than soldering. For those reasons, soldering is the subject of the next chapter.

7 The Soldering Process

7.1 Chapter Objectives

This chapter examines the core activity of all electronics assembly: soldering. By reference to specific facts about the soldering process, it will be shown that:

- The normal concept of the soldering process is too restricted.
- Soldering—like all processes—is a science.
- Most of what is called "soldering" in electronics assembly is actually welding.
- The key to analyzing any soldering result involves knowledge of four interacting natural forces that determine "wetting."
- The dependability of any connection is a function of thermal and chemical forces.
- The outcome of every soldering event is perfectly predictable and controllable.
- Improper design and application of the soldering process causes product failures.
- The most important quality aspect of soldering—absence of failure modes—cannot be seen or measured.
- The most common forms of damage caused by soldering involve ionic residues and excess heat.

- Scientific process management will prevent that damage.
- The scientific approach to soldering differs from the traditional techniques for setting up, managing, and evaluating the process.

7.2 Importance of the Soldering Process

Chapter 6 submitted the propositions that (1) true quality in electronics assembly can be achieved only through rigorous scientific control of processes, (2) scientific process control is possible only when the underlying sciences are known and understood, and (3) true working knowledge requires both hands-on experience and academic knowledge. It should therefore be apparent that meaningful quality assurance requires substantial effort to learn and apply subjects that traditionally have not been part of mainstream electronics assembly quality practices.

The need to master so many new disciplines can be intimidating to prudent individuals or wildly exhilarating to the overeager. This chapter is designed to encourage the faint of heart and rein in the reckless. This will be achieved by applying the lessons of previous chapters to design and management of the soldering process.

7.2.1 Goldratt's Problem-Solving Formula

The choice of soldering for illustrative purposes often surprises electronics assembly professionals who see soldering as a dirty low-technology aberration in an industry constantly challenged by issues at the very limits of technology. The reason for focusing on soldering—probably not surprisingly—is found in Goldratt. In *The Theory of Constraints,*[1] an important follow-up to *The Goal,* Goldratt provides a highly effective three-step formula for problem solving:

1. *Find what to change.* "Pinpoint the *core* problems," Goldratt argues. "Those problems, that once corrected, will have a major impact, rather than drifting from one small problem to another."

2. *Clarify what to change.* Before rushing to make changes, it is necessary to know what the desired outcome should be.

---

[1][Goldratt 1990], pp. 7–8.

3. *Determine how to cause the change.* Construct simple, practical solutions rather than complex approaches that are less likely to work. This will require use of psychology as well as technology. Indeed, Goldratt regards this step as the most difficult because it depends on convincing others to buy into the solution ("take ownership" in the language of contemporary quality).

The immediate issue facing us is the first step: pinpointing the core problems. Later chapters will address the remaining two steps.

7.2.2 Soldering Is the Core Problem

Soldering turns out to be the core process and the source of the core problems. Most other electronics assembly processes are either preparations for soldering or attempts to evaluate the outcome of soldering. In a very real sense, soldering is the very heart of electronics assembly. As Goldratt's problem-solving formula predicts, curing problems associated with the soldering process makes identification of all other issues—from design to materials to field performance—vastly less complex.

7.2.3 Soldering in Daily Plant Life

No book on quality in electronics assembly can have much practical value without explaining key scientific principles of the soldering process. Soldering, after all, is a daily topic of conversation among most electronics assembly personnel. The discussions about soldering generally occur in the context of attempting to determine why certain cosmetic soldering conditions exist, whether they are cause for concern, and how they can be corrected.

Soldering outcomes have long been the most common means by which electronics assembly companies measure their "quality" and the effectiveness of their quality departments. Within those companies, a substantial portion of the quality manager's life is spent wrestling with soldering concerns. Assembly companies generally evaluate their products according to standards maintained and managed by the quality department. Product may function perfectly well but cannot be shipped until quality personnel verify that the units conform to the visual standards. As will be demonstrated, those standards are dominated by soldering evaluation criteria—and most of those criteria are wrong. Consequently, quality personnel devote much of their time to futile screening of solder joints and enforcement of archaic regulations.

The importance of the soldering process is poorly recognized by anyone employed in electronics assembly who has never been personally involved in soldering activities. Admittedly, soldering is not a glamorous activity and holds little appeal for the industry's technical elite. At the same time, soldering is the most pervasive operation in most assembly plants as well as the source of most quality debates. Years of experience show that the greatest opportunities for fast, substantive improvement in any electronics company normally result from overhauling the prevailing soldering processes. Moreover, of all the opportunities that new approaches to soldering offer the company, the greatest return on investment is invariably obtained by adopting more scientific methods of evaluating the quality of output. In other words, the sciences of soldering directly and profoundly affect the optimal operation of the quality department.

7.2.4 The Soldering Process Defined

We have always found soldering to be one of the industry's most fascinating processes. Our attraction to a topic that others find drab probably can be attributed to the way we define soldering. To us, the soldering process encompasses much more than simply applying liquid metal to another metal surface. The soldering process actually includes:

- Design layouts conforming to the natural forces governing whichever technique will be used to create solder connections (wave soldering, reflow tunnels, hand soldering, vapor phase, hydrogen ovens, and so on)
- Solderability management—i.e., the science of selecting and storing parts that can be deoxidized by the chosen soldering method
- Flux chemistry
- Thermodynamics
- Application of the filler metal (solder)
- Post-soldering cleanliness
- Dependability verification

When reference is made to "soldering" alone, the reader may presume that the activity is restricted to actual formation of the solder connection. "Soldering process," on the other hand, indicates the entire range of issues, concerns, and responsibilities from design through purchasing and production to customer use.

7.3 The Science of Soldering

Of all the activities in electronics assembly, probably none epitomizes the scientific nature of process management better than soldering. If the soldering operations are designed and managed in conformance with natural forces, the output will be consistently free of defects. Conversely, each defect shows that the process has broken a natural law. Scientific process management eliminates the need to search for defects after production.

The soldering process causes more grief than any other electronics assembly process—quite possibly more than all other processes combined. While scientific management of the soldering process makes the outcome predictable and painless, few electronics companies have mastered the science of soldering.

Why is soldering so much harder to understand than other processes? After all, the scientific forces are not especially difficult, particularly when compared to the remarkable scientific accomplishments seen every day in component technologies; most aspects of soldering are based on simple high school chemistry and physics. Since the actual science is fairly simple, the fundamental obstacle to mastering soldering seems to be that significant determinants in the process are not readily apparent. Moreover, no educational facilities exist where the critical aspects of designing, operating, and evaluating a soldering process can be learned. Curriculums of engineering schools rarely even mention soldering.

The vital attributes of other links in the assembly chain tend to be reasonably obvious and their measurements straightforward. For example:

- The need for bare circuit boards to be flat during assembly will become apparent to all process engineering personnel in very short order. Once that need for flatness is recognized, coplanarity specifications and techniques for evaluating bare circuit boards can be developed fairly readily and accurately.
- The switching characteristics of circuits, though too complex to be grasped by the average person, present few problems to the electrical

engineer; the circuit properties can be computer simulated or measured in real life.

- The accuracy of component placement equipment, durability of various circuit board laminates, attributes of conformal coatings—these, too, pose few conceptual or practical challenges to engineers and scientists.

In all such cases, the appropriate parameters can be found without much difficulty because we know exactly what outcome is required, can see the results, and therefore instinctively comprehend the most significant attributes to monitor.

In contrast to other processes, little about the soldering process is instinctive or obvious. The chemistry and metallurgy, though reasonably straightforward, continue to fascinate scientists and baffle engineers. The changes induced by physics are subtle and mysterious to the point where few electronics assembly professionals are even aware of their existence. The number of interacting factors is so high that worthwhile experiments to prove cause and effect require impractically large numbers of test runs. Above all, the legacy of misinformation created by decades of myths and unqualified consultants discourages questions about the validity of industry practices.

7.4 The Purpose of Soldering

Soldering has been practiced for several thousand years. The use of soldering for electronics assembly, on the other hand, began only a few decades ago. Not surprisingly, therefore, many attitudes about soldering in electronics assembly come from earlier uses of solder for mechanical purposes such as constructing jewelry or connecting water pipes. Only recently has the possibility been recognized that electronic components should not be processed in the same manner as rings, pipes, or stained glass.

To properly appreciate what role solder *should* play in electronics assembly, it is essential to first understand what it cannot do. Solder[2]

[2]In a global technical sense, solder can be any metal or alloy that melts at a lower temperature than the base metal surfaces it is joining. In an electronics context, however, the term "solder" typically refers to alloys of tin and lead of which tin typically represents 60 to 63 percent of the alloy by weight, with lead making up the balance. It must be kept in mind that solders are not exclusively combinations of tin and lead. For example, solder used in assemblies built to MIL Specs must contain small amounts of antimony (a requirement based on a scientific misunderstanding by early standards setters that has never been corrected).

should not be employed for mechanical purposes; the solders used in electronics production possess little tensile strength, though they are rather ductile and thus quite successful at withstanding levels of expansion and contraction that would shatter most other metals. If components will experience meaningful physical stresses—such as vibration or mechanical shock—some reinforcement type of mounting is the only certain way to guard against joint failure. Happily, the typical electronics assembly never experiences forces greater than the modest thermal expansion and contraction associated with temperature cycling between ambient and operational levels.

Though most solder joints will maintain their integrity under surprisingly intense stresses, the risk of failure increases significantly if solder alone is expected to sustain the mechanical integrity of joints subjected to physical stresses. For through-hole components, simple crimping of leads provides more than enough mechanical strength. Surface-mounted components present a greater risk of joint failure because crimping is not possible; gluing heavier components to the circuit board is a good precaution if the assembly will experience severe physical trauma. Some failure analysts regard such reinforcements as the electronic equivalent of wearing suspenders *and* a belt, but experience has proved that lack of reinforcement equates to oversized pants worn without either suspenders or belt.

The function of solder is to provide electrical continuity between current-carrying surfaces.[3] Therefore, the acceptability of solder connections should be determined entirely by whether that electrical continuity exists and will remain intact for the specified life of the product. With the exceptions of solder shorts and opens, visual inspection will not provide any information about electrical continuity—and, as component densities increase while sizes shrink, the ability of the human eye to find opens or shorts is considerably more limited than the emphasis on inspection implies.

7.5 Soldering vs. Welding

The electrical continuity resulting from soldering is created by a bridge of metal fill (the solder) between current-carrying metal surfaces (generally referred to hereafter as "the surfaces") such as component leads

[3]Alternatives to solder exist, among them conductive epoxies, but soldering is generally more convenient and less expensive.

or circuit pads. The solder is melted and flowed over the surfaces to be connected. But the surfaces themselves do not melt. Only if the melting temperature of the solder is lower than the melting temperature of the surfaces will it be possible to melt the solder without also melting the surfaces.

The state of the surfaces—whether they remain solid or melt—is the critical trait for determining the type of operation. If the surfaces melt, the operation can no longer properly be termed soldering; rather, a new operation known as welding has come into the picture. It is widely believed that the vital distinction between soldering and welding is the absolute temperature at which the operation takes place. Welding, of course, is presumed to require higher temperatures than soldering. But that is incorrect. Soldering can involve much higher temperatures than welding. The crucial distinction between soldering and welding concerns the state of the surfaces being connected: with soldering, the surfaces remain solid while welding melts the surfaces.

It is essential to know whether the plant is engaged in soldering or welding. For reasons that will become apparent in this chapter:

1. Soldering is easier than welding.
2. Electronics assembly typically involves welding, even though the process is usually called soldering.
3. Problems arise most frequently when the surfaces fail to melt.

Reference is frequently made to an alleged third process called "brazing." In both brazing and soldering, a melted filler metal is applied to metal surfaces that do not melt. Brazing is distinguished from soldering only by the melting temperature of the filler metal, with the demarcation point arbitrarily established, typically at approximately 700°C (1300°F). Since pure metals and alloys with melting points between 350 and 700°C are almost unknown (zinc and magnesium are the two most common exceptions), the border between soldering and brazing effectively falls in the range 400 to 500°C (roughly 900°F).

Although the properties of metals with higher melting temperatures differ from those commonly used in soldering, soldering and brazing both create intermetallic bonds between the filler metal and the surfaces being joined. Welding, on the other hand, fuses the filler metal and surface metal together; if the filler metal is a different element from the surface metal, the product is an alloy.[4]

---

[4]Normally, the filler metal used in welding is the same as the metal of the surface being bonded. No new compound—whether intermetallic or alloy—is created, but the atoms of filler metal mingle freely with atoms of the surface metal.

In electronics assembly, the primary concern involves the interaction between the filler metal and the surface being bonded. In other words, our interest lies in the surface chemistry. Since the surface chemistry of soldering and brazing are identical, it is helpful to eliminate brazing from the terms of reference. In this industry, we need consider only two processes by which metal surfaces are joined using a melted filler metal: soldering and welding.

7.6 Wetting Forces

The outcome of any soldering operation can be predicted from knowledge of the four natural forces that control the flow—generally known as the "wetting" action—of solder. Those forces are:

1. *Gravity.* Gravity can encourage or oppose solder flow, depending on whether the solder is being applied from above or below the PCB. When solder is applied from above the circuit board, as in hand or paste reflow soldering, gravity pulls the solder toward the circuit board. The resulting solder fillet will be flatter and broader. When solder is applied from below the circuit board, as in wave soldering, gravity pulls the solder away from the circuit board. The solder fillet will be longer and narrower compared to the hand or paste reflow fillet. The effects of gravity from above and below are represented in Fig. 7.1*a*.

The influence of gravitational force on perceptions of "perfect" solder joints has been profound and costly. Although hand-soldered fillets are flatter and broader than wave-soldered joints, the visual standards found in most electronic assembly plants do not take this difference into account. The pictures and diagrams of those standards are based on hand soldering only. Consequently, anyone schooled in the traditional visual standards tends to regard wave-soldered fillets as inferior to their hand-soldered equivalents.

2. *Surface tension of solder.* Reference to the traditional Bohr atomic structure of elements shows that the atom consists of a core around which electrons move. The electrons are arranged in layers known as "shells" which may each contain more than one electron; if an atom possesses more than one electron shell, the inner shells will always have an even number of electrons but the outermost shell can have any number between one and eight.

Chemical reactions normally involve only those electrons in the outer—or "valence"—shell. If the valence shell contains eight electrons, the atom is stable and the least reactive of all elements. Only a handful of elements—known as "noble gases"—have eight valence electrons.

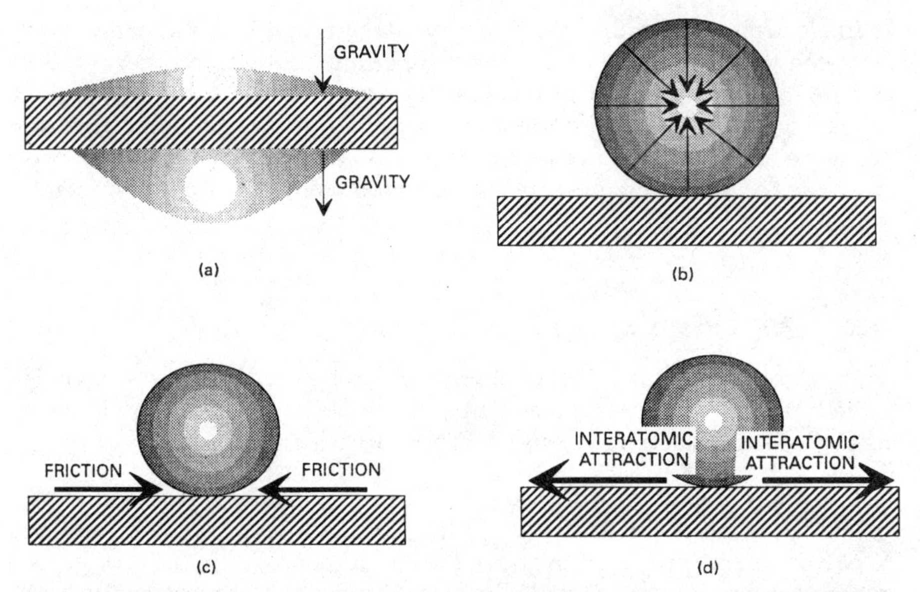

Figure 7.1 Wetting forces. (*a*) Effect of gravity on solder flow. (*b*) Effect of internal surface tension on solder flow. (*c*) Effect of friction on solder flow. (*d*) Effect of interatomic attraction on solder flow.

The valence shells of all other atoms contain fewer than eight electrons, a condition that causes the atom to be unstable and chemically reactive.

An atom with fewer than eight valence electrons can become stable by losing, gaining, or sharing electrons with other atoms through formation of a molecule. The process by which atoms combine into molecules is a chemical reaction. Two or more atoms bonded together make up a molecule, even if the atoms are all of the same element.

Atoms in a bonded state are more electrically stable than unbonded atoms (with the exception of noble gas atoms). Therefore, elements other than noble gases normally exist as parts of molecules, whether bonded to identical atoms or with atoms of other elements.

Very few metal elements naturally exist in a metallic state. One exception is gold, which most commonly forms bonds with other gold atoms to form crystals; silver does the same. On the other hand, the metals most common in the electronics industry—including copper, tin, iron, nickel-generally bond with nonmetallic elements. Breaking those chemical bonds to separate the pure metal from the nonmetallic elements requires energy. The most common form of energy used in metal refining is heat, although some metals can be separated using electricity.

The attractive force exerted by one atom toward another can be quite powerful and affect the outcome of soldering in several ways. One of the most notable forces—surface tension—influences the behavior of melted solder. When a material is in a liquid state, molecules at the surface behave differently from molecules beneath the surface. Any molecule is attracted by the force of cohesion to neighboring molecules of the same material. Beneath the surface of a liquid, molecules are evenly distributed and thus the cohesive forces acting on any molecule in the interior of a liquid are in equilibrium. Molecules on the surface of the liquid, however, are not evenly surrounded by molecules of their own kind. There will be unbalanced forces that attempt to draw the surface molecule into the interior. Surface molecules will be pulled to the interior until the minimum surface area has been achieved; at that point, no molecule could be drawn into the interior without displacing an interior molecule to the surface. At this point, the forces are equalized.

The minimum three-dimensional surface area is formed by a sphere. Every other surface geometry involves greater area. Therefore, surface tension exerts a powerful opposing force to solder flow. This is diagramed in Fig. 7.1*b*.

3. *Friction.* As the liquid solder attempts to flow over the solid surface, it encounters resistance in the form of friction. The magnitude of that negative wetting force is determined by the configuration of the solid surface and is unpredictable. Surface ridges, for example, may act as barriers to flow or capillaries to assist flow. See Fig. 7.1*c*.

4. *Interatomic Attraction.* Just as attraction exists between atoms of the same element to cause "surface tension," atoms on the surface of liquid solder are attracted to atoms on the metal surface being bonded. The interatomic (sometimes termed "intermolecular") attraction between solder and base metal strongly promotes solder flow, as shown in Fig. 7.1*d*. The product of the bonding between solder and base metal is a molecule known as an intermetallic (see Sec. 7.8).

The strength of the attractive force will vary among different combinations of metals. Lead, for example, has little affinity for other metals and provides poor wetting of most surfaces. Tin, on the other hand, has a strong affinity for the other metals commonly found in the electronics industry. The interatomic attraction between tin/lead solder and the base metal is almost entirely a function of the tin content. (Lead also has poor electrical conductivity. The only reason for employing lead in solder for electronics applications is to lower the melting temperature. Certain combinations of tin and lead melt at lower temperatures than either tin or lead alone.)

Metals need not bond with other metals to achieve stability in their outer electron shells. A powerful attraction exists between metals and oxygen to form metal oxides. Oxides are electrically stable and have no surface energy to contribute to interatomic attraction; consequently, solder—with one notable exception to be explained in Sec. 7.9.3—will not flow over an oxidized surface. Therefore, the surface must be free of oxides for wetting to occur. The liquid solder's surface is also coated with oxides, but the oxides, being lighter than pure solder, are displaced by the gravitational pull in hand soldering or the mechanical force in wave soldering.

If any material—even a seemingly insignificant layer one atom thick—coats the base metal before application of solder, interatomic attraction will be prevented. The barrier will be oxides or contamination. As Sec. 7.9 explains, oxides are not contaminants, and even a "clean" metal surface can be oxidized.

Any soldering operation therefore always has at least two forces opposing wetting: surface tension and friction. Gravity may help or hinder wetting. Therefore, solder will flow only if the positive wetting force of interatomic attraction exceeds the sum of the forces exerted by gravity, surface tension, and friction. The soldering process must be designed to ensure that the net wetting effect of all four forces combined is positive.

7.7 The Intermetallic Bond

When direct contact is established between solder and the metal surface(s) to be joined,[5] atoms of the surface metal and atoms of solder bond to form a molecule known as an intermetallic compound.[6] In the case of tin/lead solder bonding to copper, tin is more reactive than lead and the resulting intermetallic is a compound of tin and copper. The usual chemical structure of tin/copper intermetallic is Cu_6Sn_5 (i.e., six

[5]Of course, electrical continuity means that more than one metal surface will be involved in the joint. For purposes of illustration, however, only a single surface (e.g., a component lead or a pad on a circuit board) is necessary.

[6]Note that the reaction involves only those metal atoms on the surface of whatever metal is being bonded. Unless the surface is eliminated during the soldering, other materials beneath the surface have no consequence. For example, terminal clips for automotive rear window defrosters are soldered onto a metal surface bonded onto the window glass.

copper atoms bonded to five tin atoms) or, less frequently, Cu_3Sn (three copper atoms bonded to one tin atom). Other metals common in the electronics assembly world—such as nickel, silver, gold, and steel—also readily form intermetallics with tin.

Intermetallics are closely related but not identical to alloys. Alloys are formed by blending liquid metals while intermetallic formation involves at least one metal that is not melted. Tin/copper intermetallic has features similar to bronze—both consist of tin and copper, and are rather brittle and less conductive than either of the metals composing them—but it is not an alloy.

The intermetallic is a permanent condition. Once it has formed, it cannot be removed by any means short of abrading or etching. Attempts to remove solder by melting the connection only make the intermetallic thicker. One of the most costly myths about soldering, perpetuated over the years by Manko in particular,[7] is that soldering is a "reversible process." Clearly, this is not true. Even if components are replaced between desoldering and resoldering, the intermetallic layers on the tracks and through-holes of circuit boards will become thicker.

7.7.1 Requirements for Intermetallic Bonding

"Direct contact" between the atoms of surface metal and atoms of solder means nothing—not a single atom or molecule of any other substance—lies between the solder and surface atoms. This requires that:

1. The solder is liquid.[8]
2. The surface metal is absolutely clean—i.e., no foreign matter of any kind is present.

---

[7]See, for example, [Manko 1986], p. 367: "Soldering is a reversible process, and remelting the solder makes it possible to correct industrial defects."

[8]Technically, the solder need not be liquid for intermetallic bonding to occur. The only requirement is that the solder and surface atoms be in full contact. Theoretically, if solder is brought into contact with a surface shaped in the exact mirror image of the solder, intermetallic bonding would take place. Since oxides would prevent the necessary atom-to-atom contact between solder and surface, the contact must take place in a complete vacuum. While such bonding is possible, it is hardly practical. It is worth noting that early space missions experienced problems with bearings bonding to their metal shells when the bearings were not moved for long periods. This bonding occurred because friction between the bearings and the shells during movement of the bearing abraded the oxide layer established on earth; when rotation stopped, the bearing's surface atoms were in contact with the shell metal, and intermetallic bonding occurred.

3. The surface is absolutely free of oxides while the solder is being applied.

Many times, requirements 2 and 3—i.e., clean and oxide-free surfaces—are erroneously thought to be identical. Oxides occur naturally and inevitably on almost all metals exposed to oxygen. Only four metals—gold, platinum, rhodium, and iridium—do not oxidize, and of those only gold is likely to be encountered in the normal electronics assembly environment.

All soldering processes should therefore begin with the presumption that oxides will be present. In addition, the attributes of the oxides—particularly whether the oxides will be removed by the flux employed—must be known before soldering begins.

Oxides are a normal condition for metals. We expect metal surfaces to be oxidized and set up our soldering processes to ensure that the oxides will be eliminated. Knowing that oxides will be present at the commencement of soldering, we should not think of them as contaminants. On the other hand, any other materials than oxides or pure metals are present only by accident. The presence of these other materials—generally greases or oils, sulfides, silicon, dust, or similar materials—cannot be assumed. Since their presence and composition are unpredictable, these substances other than oxides and pure metal make up the category of contaminants.

When all three requirements for direct contact between solder and surfaces are satisfied, the formation of an intermetallic bond such as that shown in Fig. 7.2 cannot be avoided. On the other hand, if *any* requirement is not met, no intermetallic bonding can take place.

Figure 7.3 shows the effect on solder flow of contamination or oxide on a surface to be bonded. As previously noted, a barrier layer of contamination or oxides only one atom (or molecule) thick will prevent solder flow. In the absence of gravity, solder applied from above the PCB would assume the shape of a sphere. But gravity does exist in our manufacturing environments and pulls the solder toward the surface to create an ovoid shape such as that shown in Fig. 7.4. Of course, if solder is being applied from below the surface, as in wave soldering, gravity pulls the liquid solder away from the surface and the resulting solder will be pendulous rather than flattened.

7.7.2 Properties of the Intermetallic Bond

Though necessary for electrical continuity, intermetallics possess two undesirable properties. Intermetallics are:

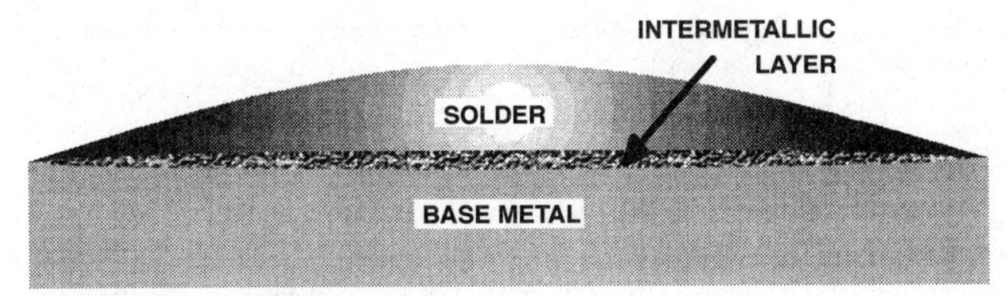

Figure 7.2 The intermetallic bond.

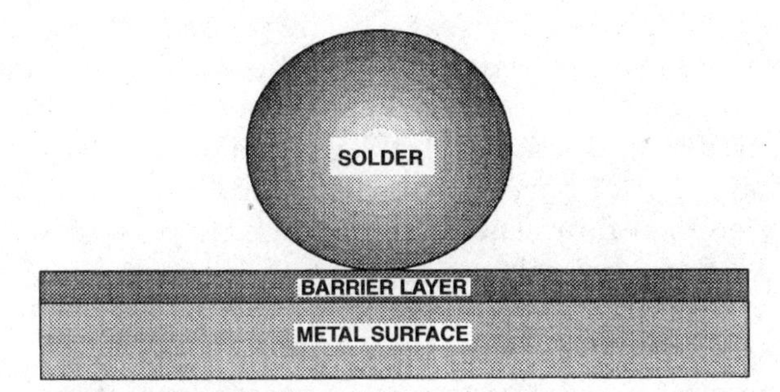

Figure 7.3 Nonwetting.

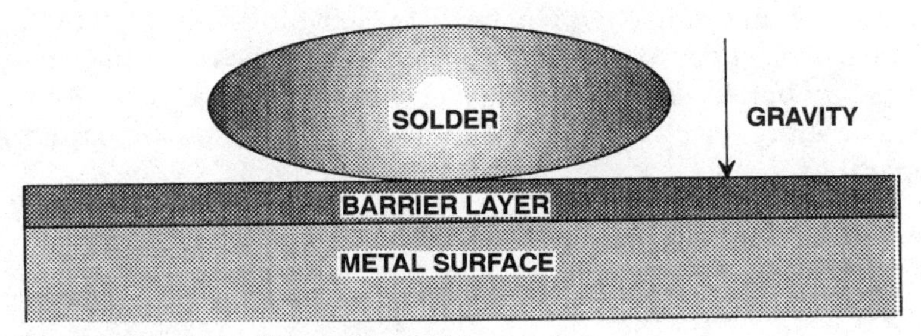

Figure 7.4 Nonwetting influenced by gravity.

1. Less conductive than either solder or the surface metals being bonded.
2. Brittle; it is not unknown for vibrations that have no effect on either the solder or base metal to fracture thicker layers of intermetallics.

The perfect intermetallic bond would therefore be only a single molecule thick. In real life, though, intermetallic layers are invariably considerably thicker than one molecule and will continue to expand over time. Immediately after soldering, the thickness of the intermetallic bond reflects the amount of heat applied during soldering: Greater amounts of heat cause thicker intermetallic bonds.

The existence of an intermetallic bond is of consequence only because formation of the bond is inevitable if direct contact is established between the solder and the metal surface. In the absence of direct contact, no intermetallic bonding occurs—but neither will electrical continuity exist between the metal surface and the solder.

More accurately, *some* electrical flow often results even when a barrier layer prevents intermetallic bonding. However, the barrier layer will impede electrical flow to some extent. The problem is that the extent to which the barrier layer impedes electrical flow is unpredictable. Oxides are semiconductors rather than true conductors, while contaminants may be partially conductive but are more likely to be insulators. Of course, the amount of barrier material in addition to its composition affects the electrical flow. Therefore, it can only be said with certainty that electrical flow through the circuit will be partially or completely disrupted when atom-to-atom contact cannot be established between the solder and the surface of the base metal.

Weaker currents may be unable to overcome insulating properties of a barrier layer that does not block a stronger current. Even if some flow takes place, however, the switching characteristics and other electrical properties of the connection will be degraded. Since the effect of the barrier on electrical continuity is uncertain but never desirable, the presence of *any* barrier between solder and metal surface cannot be tolerated. Therefore, electrical dependability and intermetallic bonding are inseparable.

7.7.3 Visual Attributes of Correctly Bonded Joints

Many electronics assembly authorities believe that solder cannot cover (or "wet") a contaminated or oxidized surface. The anticipated result of applying solder to a contaminated or oxidized surface is shown in Fig. 7.3. This leads to the expectation that, even though it is impossible to see an intermetallic bond because of the solder covering it, the *absence* of bonding is visible in the form of non-wetting. The reliance on visual inspection standards follows from this belief that any obstacle to the formation of the intermetallic bond must result in visibly poor wetting.

However, under certain conditions solder *can* be flowed over a barrier layer. Although it will not form intermetallic bonds with the base metal, at high temperatures solder can adhere to oxides or contaminants coating the metal surface. The appearance of this artificially induced solder "connection" is identical to the appearance of perfectly bonded solder. Only destructive analysis of the solder joint will reveal whether the solder has bonded with the metal or adhered to a foreign substance on the metal's surface. But adhesion and soldering are very different operations. For example, adhesion is strictly a mechanical property whereas the intermetallic bonding creates new molecules. Further, although the conductivity of the intermetallically bonded joint is predictable and stable, the presence of the barrier layer in the adhered joint makes the electrical integrity suspect.

The amount of heat required to achieve adhesion is above 320°C (600°F), well beyond the temperatures produced by wave soldering, I.R. reflow, or other forms of automated soldering. Only soldering irons generate sufficiently high temperatures to cause adhesion. Touchup transforms clearly unreliable machine-soldered joints into seemingly dependable joints by causing the solder to adhere to surfaces that it would not wet at the lower machine-soldering temperatures. The touched-up joint is not more dependable; it only appears more dependable to the inspector's eye.

What conclusions can be drawn from this information?

1. Unwetted joints certainly lack dependability.
2. Even the most attractive joint may be undependable if operators with soldering irons have been at work on the assembly.
3. Visual appearance can be changed, but the underlying dependability cannot be improved.
4. Inspection after hand soldering provides little useful information.

Ironically, the greatest emphasis on inspection of solder joints is found in MIL Specification operations where most of the soldering is performed manually.

7.8 Oxides

All metals—and all alloys—except gold, platinum, rhodium, and iridium combine with oxygen to create metal oxides. The chemical reactions that cause bonding between metals and oxygen occur for the same reason as intermetallic bonding during soldering: The valence shells of the metals and oxygen are incomplete and stability is achieved by borrowing, lending, or sharing electrons with other atoms.

7.8.1 Time and Temperature in Oxide Formation

The amount of oxide formed—like the intermetallic bond—is a function of time and temperature. The rate of oxidation rises at a greater-than-linear rate as temperature increases. Consequently, when two pieces of the same metal are compared, the greatest amount of oxide will be found on the surface which has been subjected to the greatest temperature (assuming both pieces are the same age and have been stored in identical atmospheres containing oxygen). Similarly, the greatest amount of oxide will be found on the surface that has been exposed to oxygen for the longest period (assuming the surface metals are otherwise identical and have been stored in the same oxygen-bearing environment).

7.8.2 The Effect of Surface Contamination on Oxidation

Just as the chemical reaction producing intermetallic bonds requires interatomic contact between the metals being joined, oxidation requires contact between the metal atoms and oxygen. Once contact is made between metal surface atoms and oxygen, an oxide layer at least one molecule thick develops instantly. But in the same way that a continuous layer of contamination only one atom thick prevents intermetallic bonding, a contaminated metal surface presents a barrier to oxygen, and oxides will not result.

Over the years, manufacturers of PCBs and wires have occasionally coated the surfaces with contaminants to prevent oxygen from reaching the underlying metal and building oxide layers. The results of this strategy were inconsistent, however. The barrier layers did exclude oxygen but at times also prevented flux from reaching the existing oxides during soldering. The most common example of this practice today is the coating on wires used in coil winding.

7.8.3 Oxide Structures

Oxides coat the metal surface in lattice-like formations. More accurately, the first oxide layer—one molecule thick—that forms instantly when metal meets oxygen is a lattice, while further oxidation establishes a pattern of overlapping lattices. Although the basic format of the oxide lattice is essentially the same for all oxide-forming metals, the exact pattern varies from metal to metal. The lattice of some metals is tightly "woven" while other metals form very loose oxide lattices. The size of the gaps or pores in the lattice is critical in determining the extent to which the metal surface will oxidize. Specifically:

1. If the gaps are larger than a diatomic oxygen molecule (two oxygen atoms sharing electrons, which is the most common form of free atmospheric oxygen), oxygen can penetrate the existing oxide layer to reach and bond with metal atoms at the interface between the metal surface and the oxide layer. The oxide layer will continue to grow.

2. Gaps smaller than a diatomic oxygen molecule but larger than an oxygen atom also allow oxygen to penetrate the existing oxide layer and bond with metal atoms. However, the relative scarcity of single oxygen atoms means oxidation will occur at a much slower rate. Single oxygen atoms (generally in ionic form) are relatively more common in water than in air; this explains why some metals oxidize to a greater extent in water.

3. If the gaps are smaller than an oxygen atom, no oxygen will reach the pure metal below the oxide layer, and oxidation will cease.

Aluminum and iron are examples of metals whose oxide lattices are very large relative to diatomic oxygen molecules; those metals readily develop quite thick oxide coatings. Copper and tin oxide lattices are much more tightly interwoven, with the result that they oxidize only moderately. The gaps in the lattice of stainless-steel oxide are smaller than the diameter of single oxygen atoms, with the result that a very thin coating of oxide actually acts as the agent keeping "stainless" steel stainless.

The strength of the metal-oxygen bond varies according to the type of metal. Stainless steel and aluminum bond strongly to oxygen. Copper and tin oxide bonds are much weaker.

7.8.4 Oxide Layers as Protective Barriers

Inasmuch as the lattices of oxide layers overlap rather than coinciding, the gaps become smaller as oxide buildup progresses. Eventually, the oxide becomes sufficiently thick that oxygen cannot reach the metal atoms beneath the oxide layer. At that point, the oxide layer itself becomes a protective barrier against additional oxide development. Oxidation never actually comes to a complete halt in most metals because atoms are always in motion; some oxygen atoms will continue to find their way through the oxide layer to the underlying metal and some metal atoms will reach the surface. However, while oxidation may not stop absolutely, the rate of oxidation becomes inconsequential.

7.8.5 Oxides and Wetting

As Sec. 7.6 explained, solder can only wet a metal surface where the attraction between surface metal and solder atoms is greater than the net effects of solder surface tension, gravity, and friction. However, when a metal atom bonds with oxygen, the attractive force that the metal atom itself might exert on other metal atoms is neutralized. Thus, solder cannot wet an oxidized surface. An oxide layer one molecule thick effectively neutralizes the metal surface.

7.9 Flux

Every metal surface normally encountered in the electronics assembly environment (except gold) instantly develops an oxide layer when exposed to any atmosphere containing oxygen. The oxide layer may grow thicker over time, but every contamination-free non-noble metal surface that has ever been exposed to air is fully coated with an oxide layer. In other words, no metal surface (except gold) can be wet by solder in its natural state. Before solder can be applied, the oxides must be removed and prevented from reforming. The soldering operation therefore requires a means of eliminating oxides *and* blocking any contact between the deoxidized surface and oxygen until the solder has wet the surface.

7.9.1 Defining Flux

Oxides can be removed by abrasion or through chemical reactions that separate the oxygen from the metal atoms making up the oxide. Abrasion presents several problems, including physical damage to the component. More important, unless the abrasion and subsequent soldering are carried out in an atmosphere devoid of oxygen, the surface will immediately reoxidize.[9] Accordingly, oxides are removed in electronics assembly by chemicals known as fluxes.

---

[9]Over the years, soldering operators have used everything from pencil erasers to penknives to abrade heavily oxidized surfaces. The amount of oxide removed in this fashion is considerably greater than the amount that can form if soldering takes place soon after. Since the amount of oxide to be removed chemically is much less after the abrading, milder fluxes than would otherwise have been necessary can typically be used to chemically eliminate the newly formed oxides during soldering. Plumbers, for example, customarily use emery paper to remove accumulated oxide when soldering metal pipes. The potential physical damage to components during abrading is so great, however, that removing oxides in this manner can never be sanctioned in an electronics context.

Flux must do more than break down oxides. It must also erect a barrier between the deoxidized surface and oxygen until application of solder is complete. A material that performs only one of these functions—removing oxides or preventing reoxidation—cannot be considered a flux. On the other hand, any material that does fulfill both roles is a flux. Therefore, flux is defined as any substance that eliminates oxides through chemical reaction and prevents formation of additional oxides until after soldering is complete.

The fluxes normally encountered in electronics assembly consist of acids ("activators") to break down oxides and a material to coat the surface until solder is applied. The activators may be in the form of compounds that are neutral at normal storage temperatures but break down into acids when heated.

For many years, most fluxes for electronics applications were formulated with a rosin base. Although rosin is slightly acidic at certain temperatures, its primary function in fluxes is to protect the deoxidized metal surface from reoxidation. Not being soluble in water, rosin's popularity has waned as CFC-based solvents have given way to water washes. Other chemicals have taken rosin's place in the prevention of reoxidation, but none of the replacements withstands the heat of soldering as effectively as rosin.

7.9.2 Time in Oxide Removal

Chemical reactions take time. Fluxes therefore require some interval to break down oxides. Depending on the nature and amount of oxide and the activity level of the flux, the time required may be insignificantly short or impractically long—but *some* delay is always involved. Solder must not be applied until flux has removed the oxide layer. When solder is applied before the flux has broken down all oxides, the heavy solder will displace the lighter flux on the surface; deoxidation will end and the solder will not succeed in wetting the surface.

7.9.3 Temperature in Oxide Removal

Chemical reactions either absorb or give off heat. The chemical reactions encountered during electronics assembly require application of heat. Below some minimum temperature, no reaction will occur. The minimum temperature varies among fluxes; some highly acidic fluxes can perform well at room temperature, but others are ineffective until heated beyond the boiling point of water. However, all fluxes work faster at higher temperatures, and rate of reaction has a greater than linear relationship to temperature. At some temperature ceiling, however, the flux materials that prevent reoxidation decompose (usually combining with oxygen to create a carbonized residue), and the metal surface is exposed to oxygen again. Therefore, every flux has a minimum temperature below which it is neutral and a maximum temperature beyond which it burns off.

7.9.4 Flux Strength and Dependability

Though the correlation is not exact, flux strength is best specified in terms of pH. The lower (more acidic) the pH, the stronger the flux.[10] The pH that matters is at full activation temperature rather than room temperature. So-called neutral pH fluxes become acidic when heated to their activation temperatures.

Stronger fluxes can remove more tenacious oxides and/or thicker oxides than is possible with weaker fluxes. The easiest way to be sure of obtaining cosmetically attractive solder joints is therefore to employ a highly activated flux. Indeed, this approach is employed in most electronic assembly plants today, often without awareness that consistently attractive connections are being achieved only because the flux is extremely strong.

Stronger fluxes reduce the number of visible solder defects. However, reducing soldering defects in this manner involves very high hidden costs. As a rule of thumb, increasing the flux strength reduces dependability of the resulting assembly.

The activators that make fluxes stronger are conductive. If activators remain on the assembly after soldering, the surface resistance between current-carrying surfaces such as circuit traces or pads is compromised. This condition is known as "ionic contamination." The consequences of ionic contamination range from moderate changes in an assembly's switching characteristics to permanent failures.

Perhaps the most frustrating problem associated with ionic contamination is intermittent failures that cannot be reproduced in test or repair. These intermittent failures are attributable to variations in atmospheric conditions between the plant or repair site and the product's operating environment. Higher humidity greatly increases the conductivity of

[10]Fluxes can also be alkaline (i.e., basic) rather than acidic, in which case their strength would be measured by the amount by which their pH exceeds 7.2. Alkaline fluxes are not normally found in electronics assembly.

ionic residues so that units operated in conditions of high humidity will be more likely to fail. Therefore, one way of identifying failures due to ionic contamination is by looking at the geographical patterns of failure instances. If failures are more frequent in high-humidity regions such as Florida or the northwest coast than in arid regions such as Arizona, the most likely cause is ionic contamination. Similarly, higher failure rates during summer than winter also indicate ionic contamination problems.[11]

In extreme cases, combinations of ionic contamination, humidity, and bias voltage will cause metal atoms from the surfaces of circuit traces or pads to migrate from one surface toward another. The process is analogous to electroplating and forms dendrites between the current-carrying surfaces. Dendrites are most commonly formed by tin, although copper dendrites are not unknown; they permanently short the electrical paths of the assembly.[12]

The consequences of ionic contamination are well known in most electronic assembly plants. To prevent flux residues from undermining the dependability of their assemblies, many plants still wash their units after soldering. Often, samples of the washed assemblies are checked by a cleanliness testing machine[13] to ensure that the washing process is operating properly.

Unfortunately, no cleaning system can protect an assembly from failures induced by ionic contamination. Nor can a cleanliness tester reveal the presence of potentially disastrous concentrations of ionic particles. The reason in both cases is the same: Flux enters tiny gaps (such as the space between a low-lying component and the circuit board surface) that cannot be reached by the solvents used for cleaning or cleanliness testing.

[11]Not all ionic contamination failures can be attributed to flux. Chemicals used in fabrication of the bare circuit board have the same effect if the boards are not adequately cleaned before assembly.

[12]More accurately, dendritic failures are permanent unless the dendrite breaks. Dendrites break in four ways: (1) a surge of electricity melts—i.e., fuses—the dendrite; (2) the flexing action of removing the defective PCB assembly from the product cracks the dendrite strands; (3) vibration during shipment of the assembly back to the plant breaks the dendrite; or (4) the service technician removes it. Of course, if the ionic residues remain in place, a new dendrite can take the place of the old.

[13]The assembly to be checked is immersed in a tank containing a given volume of solution (typically deionized water and alcohol), the electrical resistance of the solvent before immersion of the assembly and after. Using a formula based on the extent to which the resistance falls when the assembly is inserted and the surface area of the assembly, the cleanliness tester computes the level of ionic residue (expressed as the concentration of chlorine ions that would produce an equal increase in the solution's conductivity) on the assembly.

Through a combination of low surface tension and high capillary attraction, flux flows readily into small spaces. The surface tensions of solvents used for cleaning and cleanliness testing are much greater than the surface tension of flux; therefore, the flux residues in tight spaces are protected from the solvents.

Two assumptions with respect to ionic cleanliness of assemblies must always be kept in mind:

1. Even the most thorough possible post-solder cleaning will leave behind some ionic residues that may cause failures.
2. The real cleanliness will be less than that reported by any cleanliness tester.

The only certain means of preventing flux-induced ionic contamination failures is to avoid applying excessively strong fluxes. Any flux stronger than RA (or its non-rosin equivalent) is unacceptable in modern electronics assembly. RMA-strength flux is considerably safer and more desirable.

Using rosin-based fluxes offers one very important advantage over non-rosin fluxes: greater certainty about the actual strength of the flux. In North America, the activity levels of rosin-based fluxes are spelled out by various DoD and telecommunications reliability specifications. Manufacturers test their fluxes against those standards and label the containers accordingly. However, no such standards exist for fluxes that are not rosin-based. The lack of standards places the onus for determining flux strength on the user. Bans on the use of CFC-based solvents have caused mass migration to water-soluble fluxes from the rosin-based fluxes that are not soluble in water. Since few users know how to determine the strength of fluxes, the typical electronics assembly plant now unwittingly employs extremely acidic water-soluble fluxes. The seriousness of this dependability problem cannot be overstated.

Post-solder cleaning of assemblies is discussed in Sec. 10.3.5.

7.9.5 Fluxes in Soldering and Welding

In Sec. 7.5, the distinction between soldering and welding was presented. The meaningful difference between the two processes is that soldering does not melt surfaces being bonded whereas welding does. In that earlier section, it was stressed that the distinction has more than academic significance.

We have seen that component leads and other surfaces to be soldered are often reflow plated or presoldered to maintain solderability. Tin and the tin/lead alloys used for soldering by electronics assembly companies melt at below 233°C (450°F), but soldering temperatures range upward from 233°C. (The temperature of the liquid solder applied during wave solder, for example, is often as high as 260°C, or 500°F.) Therefore, tin and tin/lead surfaces melt during "soldering" and the filler metal bonds with the intermetallic lying beneath that surface. In other words, "soldering" of tin and tin/lead surfaces is in fact a welding operation. Further, since the vast preponderance of surfaces bonded during electronics assembly are either tin or tin/lead, very little soldering occurs in electronics assembly. Most of the bonding activity is welding.

The pervasiveness of welding in electronics assembly cannot be overstated because welding is a substantially more forgiving process than soldering. During welding, oxides and contaminants are resting on a liquid metal surface whereas in soldering the surface remains solid. When the liquid solder (which should more accurately be termed "bonding metal") is applied, it pushes aside the much lighter oxides and contaminants to easily and unfailingly make contact with the (oxide- and contaminant-free) underlying surfaces. Any oxides or contaminants that may have been on the now-melted outer metal coating rise to the top of the liquid bonding metal where they cannot affect joint integrity. The function of flux in welding is less critical than in soldering: it merely enhances the appearance of the finished connection by breaking down the discolored oxide that would otherwise remain on the surface of the joint. When the surface consists of tin or tin/lead, even Type R flux can perform this function.[14]

During true soldering, in contrast, the flux must:

1. Be carefully selected to be capable of breaking down the oxide bonds
2. Reach the surface and remove the oxides before the solder is applied
3. Prevent reoxidation between the time when oxides are removed and when solder flows over the surface

The requirements for flux application and activation are therefore much more stringent for soldering than for welding. Most significantly,

[14]An old adage among veterans of the original DoD soldering courses is that "nothing solders like solder." The sentiment is accurate; however, the statement should actually be "nothing *welds* like solder."

solderability is a concern only for actual soldering operations. Non-wetting will be observed only on surfaces that required soldering.[15]

7.10 Heat

Soldering cannot take place without heat. Heat is needed to activate the flux and to ensure the solder stays liquid (i.e., does not freeze) until flowing everywhere it is needed. During repairs, heat must be applied to melt the solder holding defective components. But for its usefulness, heat is the most destructive factor in the electronics assembly industry. The number of product failures caused by excessive heat during assembly is typically greater than the effects of all other failure modes combined.

7.10.1 Heat-Induced Failure Modes

Every electronics assembly employee is familiar with visible forms of heat damage such as lifted pads or tracks on PCBs, delamination of multilayer boards, or discoloration of the laminate itself. Far more common and serious damage occurs at temperatures significantly lower than the temperatures responsible for obvious damage like lifted pads. That damage takes place inside components where it cannot be seen. Moreover, the hidden damage can only occasionally be detected by quality control procedures such as test. Generally, heat-induced damage manifests itself as component failures after the product reaches the customer—not infrequently several years after the unit goes into service.

7.10.1.1 Destruction of Internal Bonds. Section 7.8 described the formation of intermetallic bonds between solder and the base metal. The destruction of components such as microprocessors results from the same physical phenomenon, in this case the growth of intermetallic layers between wire bonds and pads within components.

Microprocessors can properly be visualized as miniature printed wiring assemblies. Microscopic conductive tracks run between miniature pads to which wires are bonded. The wires—much finer than human hairs—connect the pads to the component lead frames. The techniques by which the

---

[15]Industry veterans will recall seeing "dewetting" on tin-plated surfaces in years gone by. This demonstrated that plating had been applied over oxidized or contaminated surfaces. During reflow, the solder drew back from the re-exposed contaminated or oxidized surface. Fortunately, this condition is very rare today.

wires are bonded to the pads vary but always must result in interatomic contact between the bonding metal and the pad. Although the contact does not involve soldering, the result is nonetheless an intermetallic bond at the interface of the bonding metal and the pad. Formation of an intermetallic bond is inevitable since, as Sec. 7.7 explained:

1. Electrical continuity requires direct contact between the filler metal and the conductive surface. This means that no barrier layer—whether oxide or contamination—can exist between the two metals.
2. When two metals are brought into direct contact, intermetallic bonding must occur between the two metals.
3. Neither metal need be liquid for intermetallic bonding to take place (see footnote 8, the bonding of bearing surfaces in space).
4. The rate at which the bond develops depends on the affinity between the two metals but always accelerates at a greater than linear pace as temperature increases.

As we have seen, the intermetallic layer of a solder joint grows over time. A thinner intermetallic layer was preferred to thicker layers because the intermetallic is brittle and less conductive. Within components, the intermetallic layer will also grow between the metal bond and the pad. Again, the thicker layer is not desirable because it is less conductive than either the bonding metal or the pad and will thus alter switching characteristics of the device. More significantly, the thickness of the pad is only a matter of angstroms (unlike the very thick metal surfaces encountered in the discussion of soldering). Thus, as the intermetallic layer inside a microprocessor grows, the pad essentially disappears into the much larger base of bonding metal. Over time, the loss of pad brings about a discontinuity in the device's electrical circuitry and failure results. Prior to absolute failure, intermittent failures of the device will manifest themselves as operating conditions cause expansion and contraction of the affected connection.

No electronic component lives forever. Eventually, deterioration of the component's internal connections will lead to failure. However, the life expectancy of the component in the absence of extreme forces should be on the order of several decades. But when the component is subjected to extreme force in the form of excessive heat, the life span is reduced anywhere from slightly to abruptly terminal. The extreme heat refers not to elevated temperatures during operation of the unit but to thermal energy conducted up the component lead and along the internal connecting wire into the bond. If the path along which the heat travels is an effective electrical conductor (as would be the case in every imaginable electronic device), heat will also be conducted efficiently so that the internal bond quickly approaches the temperature of the solder joint itself.[16] Heat being a function of temperature and time, the intermetallic growth will increase with the temperature applied and the time for which it is applied. Failure analysis employing electron microscopes shows that application of a soldering iron tip heated to roughly 370°C (700°F) for as little as 3 seconds has catastrophic consequences.

Often, the pad inside the microprocessor is composed of aluminum with gold used to bond the connector wire. The intermetallic of gold and aluminum has a distinctly purple hue, with the result that component engineers typically refer to the dissolution of the aluminum into the gold as "purple plague." Other metals are frequently found in today's microprocessors, but the reference to purple plague lingers on.

This heat-induced failure is extremely common, accounting for 80 percent or more of all assembly failures found at test and in field use. Awareness of this problem has been blunted by the tendency of the affected assembly companies to blame either the component manufacturers or ESD damage. Companies that have compared the statistically probable failure rate against their actual failure rates using the method presented in Sec. 3.14, however, realize that they are experiencing many more failures than can be attributed to faults in component manufacturing. Of those more enlightened companies, a handful have undertaken the electron microscopy analysis of failed joints to determine the true cause of failure. They have discovered the failure mode noted here. ESD represents a very real hazard in modern electronics assembly, but its importance has been exaggerated by erroneous attribution of heat-induced failures to ESD damage.

Failures from excess heat do not follow predictable time lines. All that can be said for certain is the components will fail prematurely. Some but not all heat-induced failures occur before the unit reaches test. While unwelcome, pre-test failures are certainly preferable to the much larger number of failures that occur after the units have been received by customers. Many devices will never be seen to fail because the product's usage life is short.

7.10.1.2 Other Heat-Induced Damage. Although in principle there is little difference between a PCB assembly and the interior of a microprocessor, the quantum difference in relative sizes makes some real-

[16]Of course, the only incontrovertible statement about the component's internal temperatures is that they must be somewhere between ambient temperature and the temperature at the solder joint.

world physical distinctions inevitable. One such distinction was seen in the propensity for pads inside a microprocessor to dissipate into the ball bond; in contrast, the intermetallic of a PCB assembly never destroys the pads or other soldered surfaces. Similarly, tiny particles of dust or other airborne contaminants that would have no effect on PCB assemblies have catastrophic consequences if they make contact with the circuitry inside a microdevice. In one instance, we found dendrites growing inside thousands of LED arrays sent to North America by sea from Asia.[17]

To prevent all contaminants from reaching the circuitry, component manufacturers encapsulate their microdevices in plastic or ceramic packages. The package forms hermetic seals around leads of the device. All is well—so long as the hermetic seal remains intact. However, thermal stresses rupture the seals to open gaps sufficiently large for easy entry by destructive particles.

The thermal breakage mechanism is the same for both plastic and ceramic packages, but the manifestation of the damage differs slightly between the two package types. In both cases, application of heat during soldering causes the metal component leads to expand and apply pressure on the surrounding package.

In the case of plastic packages, the heated leads soften the plastic, which is then easily pushed aside. When the heat is removed, the leads return to their original size but the plastic remains distorted; the hermetic seals no longer exist.

When leads of ceramic packages are heated, they also expand. However, the ceramic does not become pliable. Cracks appear in the ceramic around the leads. Repeated heating and cooling of the leads can cause the ceramic to fracture, again exposing the interior circuitry to the atmosphere.

One effect of losing the hermetic seal is accelerated oxidation of surfaces inside the packages. Where oxygen would otherwise be able to enter the package only by migrating through the plastic or ceramic, openings around the leads allow complete access to oxygen. Other gases such as chlorine in the operating atmosphere can also affect the circuit's performance and life cycle. While these are serious dependability issues, even graver consequences result from post-solder cleaning. The wash solution can penetrate the gaps around leads and deposit

[17]The entire shipment was returned to the manufacturer, supposedly to be destroyed. A few weeks later, another client called to ask advice about failures of LED arrays just received from the same manufacturer. The offending LED arrays in both cases were identical. Rather than destroying the components, the manufacturer had simply shipped them off to another customer, apparently hoping the second recipient would be less vigilant than the first.

ionic particles in the microprocessor. The probability that wash solution will penetrate the device package increases with larger gaps around the leads and lower surface tension of the wash solution.

It is worth noting that the surface tension of solvents in general but water in particular decreases as its ionic content increases. Therefore, the solutions that are potentially most hazardous to component dependability because of high ionic levels are also the solutions best able to enter small cavities.

7.10.1.3 Sources of Excess Heat. "Excessive" (i.e., capable of causing damage) heat is very rare in machine soldering. Even with the solder pot set to a recklessly high 260°C (500°F), it is virtually impossible to damage components during wave soldering. Other forms of automated reflow are also unlikely to approach the temperatures at which components age appreciably. The use of soldering irons, however, does expose components to dangerously high temperatures. Thus it is through manual soldering and desoldering operations—especially touchup—that heat-induced failures are born.

The destructive capabilities of soldering irons have been recognized for decades. Soldering specifications from the 1960s require application of mechanical heat sinks such as alligator clamps between the soldering iron's point of contact and the body of the component (the interior of the component being most vulnerable to heat damage). As components grew smaller (and more heat-sensitive), however, the use of mechanical heat clamps became impractical and operators were simply instructed to work quickly. Obviously, "working quickly" has no precisely quantitative properties and there is no way of knowing whether a component has been overheated.

In fact, the sequence of steps taught in defense soldering courses actually maximizes the probability of heat damage. Operations indoctrinated in defense systems are instructed to apply the iron's tip to the connection first, wait a moment before applying the wire solder, remove the unmelted wire solder after sufficient solder has accumulated on the connection, and only then remove the soldering iron. What temperature do conductive surfaces linked to this heat transfer reach? There is no way of knowing, except they will be somewhere between the melting temperature of solder [which, depending on the alloy, will typically be between 183°C (361°F) and 198°C (390°F)] and the temperature of the soldering iron tip.

7.10.1.4 Control of Heat during Hand Soldering. Heat damage during manual application of solder can be prevented by using the solid solder itself as a heat sink. The technique depends on the physical property of latent heat of fusion under which the amount of energy required to transform any given amount of a solid at its melting temperature into a liquid is a substantial multiple of the heat required to raise that same amount of solid matter 1°C without changing it into a liquid. As solder melts, it requires roughly 29 times as much energy to transform from a solid to a liquid as it takes to increase 1°C without changing state.

Applying the wire solder to the connection first (placing it between the component body and the point where the soldering iron will make contact), then applying the heat source, removing the heat source as soon as sufficient solder has flowed, and then removing the unmelted solder wire will prevent the temperature from exceeding the melting temperature of the solder by more than roughly 10°C, thus keeping the temperature of the parts safely below the 232°C (450°F) point at which component damage begins. Readers are encouraged to prove this in their own facilities by experimentation.

7.11 Scientific Management of the Soldering Process

From information presented in this chapter, we can see how scientific quality management would approach a perennial issue within the electronics assembly industry: How much solder is necessary before a connection is acceptable (or better)? This familiar question has long given rise to heated but subjective responses covering a variety of positions. However, we now possess enough information to resolve the matter objectively (i.e., scientifically).

7.11.1 How Much Solder Is "Enough"?

Quality personnel disagree on the minimum amount of solder required for a reliable solder joint. The debate predates introduction of printed wiring boards, but extremes of opinion became most evident with the arrival of single-sided circuit boards; in this context, the argument took the form of what percentage of a lead hole was required to be filled by solder. With the arrival of double-sided and multilayer boards, the dispute changed to the extent by which solder should rise through the hole from the bottom side to the component side. The same disagreements continue in this era of surface mount assemblies, now taking the form of how much skew is acceptable in a component lead relative to the pad and how high solder should flow up the component lead. The debates may appear to be different from one technology to the other, but there really is no difference in substance.

The issue of solder flow for both single-sided and double-sided (or multilayer) assemblies can also be analyzed according to the wetting forces presented in Sec. 7.6. Through-holes on single-sided circuit boards display incomplete solder fill when the effects of surface tension and gravity exceed the interatomic attractive force between pad and component lead. Increasing the lead-to-hole ratio and / or maintaining leads in the center of the hole during soldering solves the cosmetic "problem." With plated-through holes (presuming sufficient thermal energy is applied to prevent the solder from freezing before it can flow to the very top of the circuit board), the interatomic attraction between solder and metal surfaces must be sufficiently strong to overcome the three negative wetting forces of solder surface tension, gravity, and friction.

The question of how much solder constitutes a reliable connection was answered to our satisfaction more than a decade ago when a client hand-produced an almost invisible fillet between the component lead and the pad of a single-sided board. The circuit was isolated and an electric current applied. The current was steadily increased until it was more than 20 times greater than the specified operating power of the device. At that point, the circuit failed when the printed circuit track lifted away from the board and cracked. The minuscule solder connection remained intact.

Although the nature of the failure initially surprised us, the outcome of the experiment should have been obvious even before it was undertaken. Solder, being approximately two-thirds tin by weight, is an excellent conductor. The current-carrying capacity of even a small solder fillet is vastly greater than the capacity of the printed circuit track. The electrical bottleneck, therefore, is found in the track rather than the joint.

7.11.2 Top-Side Flow

The question of whether solder must flow to the top of a multilayer circuit board is only slightly more complicated. Only three possibilities exist:

1. That solder flow was prevented by discontinuities in the through-hole plating; in such cases, electrical flow would also be prevented by the discontinuous plating, and the circuit would not be dependable.

2. Contaminants or oxides—on the through-hole plating or the component lead—stopped the flow of solder, in which case the electrical continuity depends on the integrity of the plating.
3. Insufficient heat allowed the solder to freeze before reaching the component side of the board; again, electrical dependability is determined by the integrity of through-hole plating.[18]

Insufficient heat is the most common cause of partial fill of plated holes. This is particularly likely if substantial heat sinks like heavy components or ground planes are present. Insufficient heat is the most likely cause if the condition recurs in the same spot on other units of the assembly type. If the condition does repeat in a predictable manner, the presence of large thermal masses attached to the hole will verify that underheating was the cause.

Random occurrences of partially filled through-holes indicate plating discontinuities, contamination of surfaces, or failure to deoxidize. Random partial fills should be of greatest concern because they reflect variations in process management. However, partially filled through-holes should always concern the dependability-oriented quality professional—even if the solder has flowed more than 75 percent of the way up the hole and thus meets the standard MIL minimum. Partial fills caused by insufficient heat can easily be avoided by increasing preheat temperature and duration so that the solder remains molten until reaching the top of the through-hole barrel.

7.12 Soldering as an Example of Scientific Quality Management

The choice of soldering to illustrate the scientific forces controlling the process was deliberate. Soldering (or welding, as the case may be)—like all electronics assembly processes—is governed entirely by forces of nature. Most of those forces, it should now be apparent, are relatively basic. However, basic or not, the forces that determine the outcome are not widely known within the electronics assembly industry.

Every soldering outcome has an underlying physical origin. In fact, all processes are controlled by nature so that nothing happens in pro-

[18]This is the only condition that can properly be termed a "cold solder joint." Soldering defects commonly described as "cold solder" are actually instances of oxide or contamination barriers preventing wetting.

duction without being scientifically determined. Only by identifying, studying, and harnessing those natural forces will consistently dependable output be achieved. However, when we work *with* those forces rather than (whether consciously or unwittingly) fighting them, faulty products disappear.

Soldering is typical of the way natural forces control the results of every manufacturing process. Science dictates the consequences of every action (or inaction) in the electronics world. From its origins on a design pad to its last moment of service to the user, the destiny of every electronic product is governed exclusively by nature. The only question is whether we manage to distinguish the dominant forces and their effects on output; once the forces are known, the appropriate process management systems can be devised and put in place.

An example of how readily a pernicious natural force not related to soldering can be tamed once understood is found in the case of electrostatic discharge. Failures brought on by ESD were long believed to be natural fallout of defective components. Today, however, the destructive effects of ESD are widely appreciated and all prudent electronic assembly plants employ antistatic safeguards.

7.13 Final Words about Scientific Quality Management

As the company becomes increasingly aware that it must develop and manage its production systems in harmony with the forces of nature, profound implications for the role of quality management surface. There is only one opportunity to make a dependable product; any efforts to "correct" mistakes in the first-pass assembly will always result in a product of questionable integrity.

Clearly, therefore, the traditional functions of quality control must give way to scientific quality assurance. Resources expended on post-production qualification of output are largely avoidable costs; while post-production quality control activities may identify units that have already failed, they cannot identify units on the verge of failure nor can failed units ever be turned into consistently dependable product. On the other hand, resources allocated to quality assurance—i.e., verification of design and management of pre-production and production processes—prevent failures and raise the overall dependability of output. Given the choice between spending to catch defective product and investing less in preventing production of defective output, the wise company always elects the up-front investment. That is, quality assurance is always preferable to post-production quality control.

The argument that quality assurance must replace quality control is not new. But the nature and magnitude of the adjustments in quality management techniques to make the transition from quality control to quality assurance are still not widely appreciated. The next few chapters examine some of the more significant adjustments.

7.14 Summary

While it is impossible to understand quality management in the electronics assembly industry without taking better insights into scientific soldering than currently prevails in the industry, encyclopedic knowledge of soldering is not necessary. Accordingly, this chapter was not designed as a scholarly scientific treatise on soldering principles and sciences. Students of the physical and chemical sciences will therefore notice that some complex natural phenomena have been explained in highly simplified fashion and a few important factors ignored altogether. Also, automated soldering techniques—wave soldering and the various forms of reflow—have not been discussed at all.

Nonetheless, though the presentation may be simplified, it is by no means simplistic. Even the longest-standing and hotly debated issues of solder quality can be settled quickly and unambiguously when analyzed using the scientific principles presented here.

Important points raised in this chapter included:

- Electronics assembly consists entirely of processes governed by natural forces. Only by understanding the scientific principles of those natural forces is it possible to differentiate between acceptable and unacceptable output.
- True quality can only be achieved through scientific quality management, often described as quality assurance.
- The core process of electronics assembly is soldering. When the soldering process is properly managed, the number of failures will be very small and the causes of those failures can more easily be identified and resolved.
- Dependability of soldered connections cannot be determined visually.
- Many—generally most—product failures are caused by improper quality practices. Each time an assembly is subjected to thermal or mechanical stresses, its dependability suffers.

- Solderability management is critical to achieving successful output from any soldering operation.
- Most "soldering" in the electronics assembly industry is actually welding. The procedures for achieving perfect dependability from welding differ in important respects from soldering. Welding rarely produces defective output; virtually all problems that are not related to improper operation of equipment involve surfaces that must be soldered rather than welded.

8 Solderability Management

8.1 Chapter Purpose

Many of the faults found by post-assembly quality verification activities relate to soldering. As previous chapters have shown, soldering process problems lead to faults ranging from poor wetting (opens) to ionic contamination and even destruction of components by application of too much heat. In a substantial number of plants, faults resulting from causes other than soldering are almost unknown. Most soldering problems can be prevented by a pre-production activity known as "solderability management." This chapter explains:

- The meaning of solderability management
- Responsibilities of the quality department in solderability management
- How effective solderability management reduces the post-assembly workload of the quality department

8.2 Solderability Management

As was seen in Chap. 7, soldering consists of interacting natural forces. Real processes designed and managed in conformance with the requirements of those natural forces produce consistently defect-free output.

Processes that violate the natural forces also produce consistent output; however, that output will be consistently defective. Ultimately, the outcome of every soldering operation is most profoundly affected by the solderability management practices of the company.

8.2.1 Solderability Defined

"Solderability" is the most important word in electronics assembly. Quality and production personnel refer to "solderability problems" frequently in virtually every electronics assembly plant we have ever visited. Unfortunately, "solderability" is almost invariably employed incorrectly. Most members of the electronics assembly industry use the term "solderability problem" to describe any situation in which solder has failed to flow, although solderability—as properly defined—may not be (and generally is not) the issue at all.

Solderability relates *exclusively* to the conditions required for removal of oxides from a given surface. It can most usefully be defined as "the strength of flux required to deoxidize the surface in a given period of time at a given temperature." If surface A can be deoxidized by a given flux in a given period of time but surface B requires a stronger flux and/or longer time for deoxidation, surface A can be described as more solderable than surface B. Alternatively, it can be said that surface A has better solderability than surface B.

8.2.2 Flux Strength

The strength of flux required to remove all oxide from a given surface within a specified period of time is determined by two factors:

1. *The type of oxide.* The attractive force (i.e., the bond) between a metal's atoms and oxygen varies from metal to metal. The stronger the bond, the greater the acidity required to break it.

2. *The amount of oxide.* Thicker oxide layers are more difficult to remove than thinner layers of the same oxide.

In other words, solderability is determined by the type and thickness of oxide on the surface to be soldered.

A unit of copper oxide is more readily reduced than is a unit of tin oxide. But tin is frequently employed as a solderability enhancement layer over other metals (including copper), an act that may seem irrational given the fact that tin-oxygen bonds are stronger than copper-oxygen bonds. Use of tin over other metals is rational because tin oxidizes rather slowly in comparison to other metals normally found in electronics assembly. The net effect is that while copper surfaces are initially more solderable than tin surfaces, the tin surface will gradually become relatively more solderable than the copper surface, provided that both surfaces are stored in the same oxygen-bearing environment. Though the tin's solderability becomes better than copper's over time, the absolute level of solderability is declining for both surfaces; in other words, neither the aged copper nor tin surfaces are as solderable as freshly produced copper. Thus, solderability is a dynamic rather than static condition.

8.3 Solderability Management Defined

Solderability management is the science of ensuring prior to soldering that the flux employed will remove all oxides from all metal surfaces to be bonded.

Historically, solderability management has been approached from the flux side. That is, no attempt is made to control the solderability of the surfaces, and fluxes are chosen on the basis of their abilities to deoxidize the most heavily and/or tenaciously oxidized surfaces.

More recently, progressive electronics assembly companies have recognized that stronger fluxes increase the probability of product failures. Taking into account the greater risk of ionic contamination failure when stronger fluxes are used, a more enlightened approach has been developed.

8.3.1 Two Steps to Solderability Management

The enlightened approach involves two steps:

1. Select a mild flux whose ionic residues pose no risk to product dependability.[1]
2. All surfaces to be bonded are chosen on the basis of their abilities to be readily deoxidized by the chosen flux.

We will refer to the "enlightened" approach to solderability management as "scientific solderability management" and earlier approaches as "traditional solderability management."

8.3.2 Implications of Traditional "Solderability Management"

The traditional approach to "solderability management" is, of course, more appropriately seen as refusal to take responsibility for solderability management. Under the traditional approach, the question of whether the flux will deoxidize any given surface is not answered until inspection after the soldering operation. Inevitably, under this approach some surfaces enter the assembly mix that cannot be deoxidized by the flux even if the company uses a highly activated flux.

Companies that "manage" solderability in this fashion are always drawn to stronger and stronger fluxes, unwittingly accepting ever higher ionic contamination in exchange for fewer cosmetic defects. No scientific process control can be found in this form of process "management." Accordingly, the term solderability management can properly be applied only to the form of process control that holds the flux strength

---

[1]When in doubt about the strength of a flux, it is safest to use only rosin-based fluxes whose strength is determined by conductivity and corrosion standards managed by the U.S. Department of Defense. DoD standards exist for three rosin-based fluxes: Types R, RMA, and RA (in increasing order of conductivity and corrosion). Type R, the weakest, contains no activators; its limited ability to remove oxides derives from the mildly acidic properties possessed by rosin between approximately 66°C (150°F) and 232°C (450°F). Type RMA (i.e., "rosin mildly activated") contains small amounts of chemicals that become acidic when heated. The activators in the strongest flux for which rating standards exist—Type RA ("activated rosin") flux—become more acidic when heated than those used in RMA flux. Some rosin-based fluxes—often known as "superactivated rosin" fluxes—contain stronger activators than those found in Type RA. No restrictions exist on the strengths of superactivated rosin fluxes or fluxes that use material other than rosin to prevent reoxidation during soldering. In other words, the strengths of all synthetic and water-soluble fluxes are unregulated.

Historically, MIL standards have restricted the choice of flux to Type R or RMA. Under certain waiver conditions, Type RA flux may be approved for use in presoldering. Rarely is RA flux permitted for use on printed circuit assemblies. This is one MIL requirement that we endorse wholeheartedly.

European ratings of fluxes differ from DoD R, RMA, and RA flux standards even though the same terminology is used. Sometimes (as with Germany's DIN flux specifications), the differences are very significant. For example, a flux that would be rated superactivated by DoD standards may qualify for RA or even RMA designation according to DIN standards.

at a constant safe level and requires that all surfaces sent to assembly be capable of deoxidation by that flux. In other words, only "scientific solderability management" is truly management of solderability.

8.4 Steps in Solderability Management

Solderability management begins with selection of the flux. After the flux has been specified, it is necessary to ensure that all surfaces to be soldered can be deoxidized by the specified flux.

8.4.1 Choosing the Flux

Over many decades of use, it has been proved that there is little risk of failures in even the most sensitive assemblies if the flux conforms to RMA specifications. When fluxes stronger than RMA are employed, matters are considerably less clear.

Not all assemblies will suffer meaningful degradation of dependability if fluxes stronger than RMA are employed. Typical consumer-grade assemblies that will be used in low-humidity benign environments such as homes or offices can generally be exposed to RA flux without concern. Dependability of some simple assemblies will not be compromised by the use of fluxes even stronger than RA, though use of such highly activated fluxes is not a strategy to be undertaken without full awareness of the potential hazards. To determine the maximum strength of flux that can safely be used on their products, many companies build test assemblies using various types of flux and subject those assemblies to rigorous environment screening.

Stronger fluxes provide more latitude to the production department with respect to control of surfaces to be soldered. However, exercises to find an upper limit of acceptable flux strength are pointless and wasteful. Two reasons for adopting a solderability management program restricted to the use of RMA-level fluxes are particularly notable:

1. Verifying that a flux will not degrade dependability is costly, time-consuming, and prone to erroneous findings. Time and again, companies suffer excessive field failures because of operating conditions that were not envisaged during flux verification.
2. Type RMA flux works perfectly well when rigorous solderability management is practiced.

Therefore, it is both safer and realistic to reject fluxes stronger than Type RMA. This does not mean that the selected flux need be actual

RMA, a designation that can be applied only to rosin-based fluxes demonstrating specific conduction and corrosion properties. Increasingly, modern fluxes use materials other than rosin to prevent reoxidation of surfaces. The flux characteristic relevant to dependability is the strength of the activators, not the reoxidation prevention agent.

8.4.2 Managing Surfaces

Ensuring that all surfaces entering the assembly process can be deoxidized by Type RMA flux (or its acidic equivalent) does require more effort on the part of the production, quality, and stores personnel. However, by eliminating existing expenses such as flux verification and post-solder inspection, solderability management is cheaper overall than historic approaches to verifying solder joint integrity. The most significant saving comes in the form of substantially greater product dependability.

The logical first step toward ensuring that surfaces to be soldered can be deoxidized by the specified flux would appear to be issuance of a solderability specification to all component suppliers. Any purchaser of components should be able to provide all suppliers with a solderability specification that the supplier would honor. The solderability specification would say, in essence, "Surfaces to be soldered (leads, circuit boards, etc.) must be free of contamination and can be fully deoxidized by type (insert the type or strength of flux used in production) flux at 232°C (450°F) in less than 2 seconds."[2] That doesn't seem like much to expect.

Sadly, it *is* too much to expect. We have never found a supplier who could guarantee solderability of parts. Some vendors have claimed they could comply with a solderability specification but failed to live up their promises. The harsh reality is that suppliers—whether original component manufacturers or regional distributors—do not monitor the solderability of their own stock. Few electronic components are coded with their date of manufacture. ICs and bare printed circuit boards aside,

[2]The question that immediately arises with such a statement is how oxide removal is to be determined. Positive wetting force, as measured by a wetting balance, is the least subjective method. Note that this solderability specification differs profoundly from the well-known MIL Specification that evaluates solderability according to whether at least 95 percent of the surface to be soldered is covered by solder when Type R flux is applied and the surface inserted into a bath of liquid solder. This specification can be challenged on several technical and logical bases, including the arbitrary designation of 95 percent coverage as the appropriate minimum. The most significant limitation to the MIL solderability specification is that human judgment is required. Each inspector sees coverage differently so that what one person might interpret as 95 percent coverage could be viewed by others as 93 or 97 percent coverage. The loss of productivity caused by arguments over the extent of solder coverage is astoundingly high.

components provide no indicators of their age. How, then, could a vendor know the solderability of its shipments?

Compounding the solderability problem, component manufacturers occasionally accept returns from large customers whose needs change. Those returned components are normally sold to other customers, either end users or distributors, without notification that the components are of dubious age and may have been stored in hostile environments. Consequently, distributors are unable to determine if the parts they receive from component manufacturers are fresh or badly aged.

Distributors, too, may accept returns from customers without knowing the components' age or the storage conditions to which they have been subjected. Although they may be highly oxidized or contaminated, the returned components are resold to other unwary customers without any background history. For the most part, returns simply go back into the supplier's general stores and blend in with fresh stock.

Manufacturers and distributors of components would happily end parts returns if they could. Instead, component suppliers find themselves in a no-win position. If they refuse to accept returned parts, suppliers anger the customers attempting to make the returns. If it does take back components, the supplier must either resell the parts to other customers or discard the returned items and raise prices of all other stock to offset the losses.

To some extent, purchasers of components can protect their interests by dealing only with distributors who practice rigorous FIFO inventory control and turn over their inventories several times a year. Finding and verifying such suppliers is hardly a simple task, however. Purchasing agents—or, better yet, quality personnel—should conduct inspections of vendor warehouses to ensure that components are kept cool and dry even during the most sweltering summer conditions. Since sulfides can be even more tenacious than oxides, components should always be stored away from sources of sulfur such as most cardboards and papers. Ideally, components would also be stored in oxygen-free environments.

8.4.3 Enhancing Solderability

Electronics assembly companies can improve the prospects of receiving solderable parts by taking a few other basic precautions, including:

1. Specify that all leads or other surfaces to be soldered must be either (1) reflow plated with tin (or tin/lead) or (2) presoldered during manufacture. Presoldered surfaces are often called "hot dipped"; such surfaces have actually been soldered and therefore possess a heavy coating of sol-

der and an intermetallic bond beneath the solder surface. Reflowed plated surfaces are effectively soldered themselves; the interface between the plating and the base metal forms an intermetallic bond exactly like soldered surfaces, and reflowed surfaces often have excellent solderability as long as several years after their date of manufacture. Surfaces that have been plated but not reflowed are porous; oxygen can reach the metal or intermetallic surface below the plating and create serious solderability problems in a very short time—as little as a few weeks.

2. Avoid surfaces plated with gold or other costly metals. The thickness of gold plating is rarely adequate to prevent intermetallic compounds from growing to the surface in a very short time. Further, gaps in the plating coverage are very common. Combined with the propensity of gold to create brittle soldered joints, these factors dictate that gold plating should be employed only for leads and sockets that will be connected by friction fittings rather than soldering. Moreover, gold contaminates tin/lead solder; therefore, gold plating must be removed by presoldering prior to final reflow. Any solderability benefits—real or imagined—of the gold plating are more than outweighed by the additional handling steps prior to reflow.

3. Buy small lot sizes more often. The savings realized by bulk purchases that will meet the assembly company's needs for more than 6 months are easily offset by costs of managing solderability between the time parts enter and leave stores in the assembly plant.

4. Order circuit boards with attached snap-off coupons by which solderability can be tested at any time.

Fortunately, the solderability of today's components is vastly improved over conditions of just a few years ago. Reflowed plating and presoldered surfaces are quite common, with the result that solderability of components upon receipt at the plant is generally quite good—despite general neglect of solderability management by suppliers.

8.5 Incoming Receiving and Solderability Management

Electronics assembly companies regularly check incoming components to verify that they are the right model, have the correct electrical values, and otherwise conform to the order specifications. Auditing the solderability of incoming materials is equally important and can readily be integrated into the existing activities. Statistically based sample testing of compliance with the solderability specification of Sec. 8.4 is not onerous.

8.5.1 The Solderability Test

The method for testing solderability consists of just a few simple steps:

1. Select the surface to be tested for solderability (in Figs. 8.1 to 8.3, it is assumed that the surface to be tested is a component lead).
2. Immerse the surface to be tested in the same flux used in the plant's soldering operation.
3. Slowly insert the fluxed surface into liquid solder (a small solder pot is ideal).
4. Wait 2 seconds.
5. Observe the interaction between the surface being tested and the surface of the solder. The shape that the solder assumes is the meniscus.
6. If the solder climbs the surface—i.e., a positive meniscus results—as in Fig. 8.3, the surface is solderable using the plant's flux.

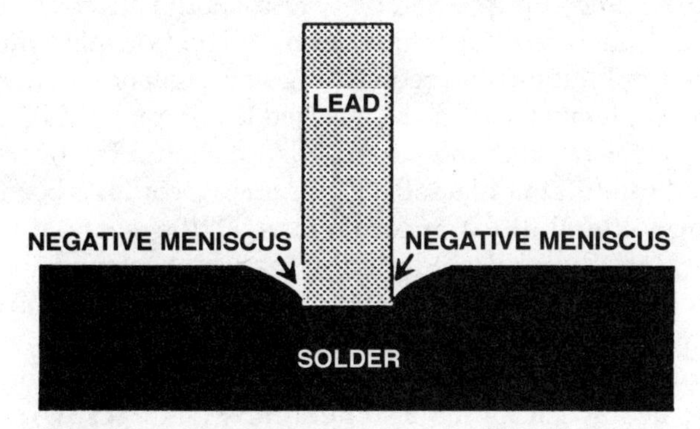

Figure 8.1. Negative meniscus (unsolderable).

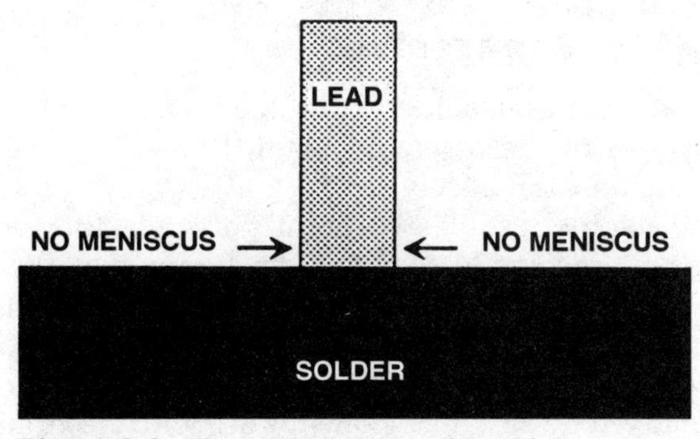

Figure 8.2. Flat meniscus (unsolderable).

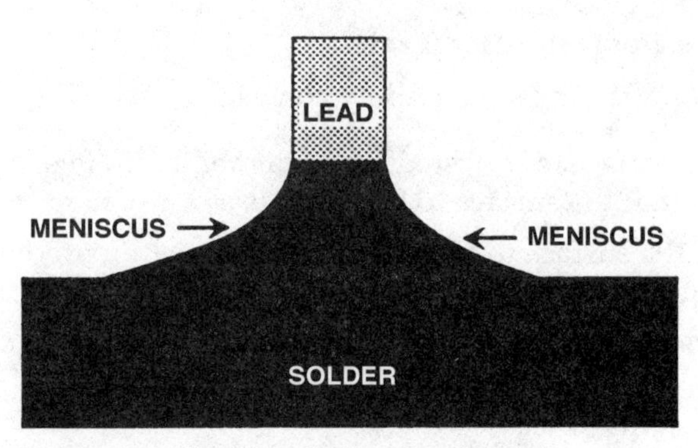

Figure 8.3. Positive meniscus (solderable).

7. If the surface of the solder is depressed (a negative meniscus) or flat (neutral meniscus) as in Figs. 8.1 and 8.2, respectively, the surface is not solderable using the plant's flux.

The results of incoming solderability tests should always be carefully documented and analyzed. Over time, certain components will show themselves to be troublesome and require ongoing vigilance. Other parts will be consistently solderable and can be checked less frequently. Parts with which the company has no solderability experience should be considered suspect and checked carefully until a satisfactory track record has been established.

Verifying solderability at incoming inspection is valuable for two reasons. First, lots that fail solderability can be returned to the supplier immediately and replacements obtained before production runs short of components. Second, if the components cannot be returned, it is essential that solderability problems be identified so that prompt action can be taken to achieve acceptable solderability (see Sec. 8.8).

8.6 The Role of Stores in Solderability Management

Surfaces are at their peak solderability the moment they are produced. For the purposes of the assembly plant, however, peak solderability exists when the parts are received.

Time is the enemy of solderability. As parts age, they acquire thicker oxide layers and become less solderable. Meanwhile, the surface coatings are combining with base metal to form intermetallics which can reach the outer surface of some thinly plated parts. If the intermetallic does reach the outer surface, it will react with oxygen to form oxides that cannot be removed by any fluxes safe for use with electronic assemblies.

The stores department normally has responsibility for ensuring that solderability does not fall below acceptable levels before assembly. However, the quality department should establish the documentation and compliance procedures that stores personnel will follow. Regular audits of in-stores solderability management by the quality department are also important.

Effective stores management requires that the time any component has been in stores can be determined quickly and easily. This control can be maintained by computers or simple color-coding schemes on component containers. The color-coding system is usually easiest and most effective. Under this system, colors are established for each portion of the year, whether monthly, bi-monthly, or quarterly. When a batch of components is received by stores, the appropriate color sticker (also bearing the date of receipt) is placed prominently on the container. The containers are stored in such a way that the stickers face forward and are readily visible. If a previous lot or lots of that component type are already on hand, the new container of parts is placed under the existing stores; strict FIFO must be maintained so that components of different ages are not mixed in a single container.

At regular intervals, stores personnel conduct a visual check on the warehoused parts. Containers whose sticker colors indicate that the parts are more than 6 months old should be removed from the shelves and their solderability tested. If the parts fail the solderability test, they can be restored using the method in Sec. 8.8.

The final check of solderability takes place just as the components are being released to production. Again, the stores personnel inspect the color of the sticker on containers from which the components are drawn. If the stickers show the components have been on hand for 6 months or more, the solderability of those parts must be verified.

8.7 Methods of Extending Component Solderability Life

Some companies prefer to save the effort of constantly testing the solderability of their stored parts. Either as soon as components arrive at stores or when the "use by" date of the components expires, these companies presolder all the relevant component surfaces (see Sec. 8.8.1 for this procedure). The presoldered surface will remain solderable for a period of several years and color coding of stock will not be necessary.

Increasingly, components are stored in inert gas atmospheres until required for assembly. In the absence of oxygen, the surfaces cannot oxidize; similarly, no sulfur compounds can be created if sulfur is not present. This approach is costly but highly recommended whenever several years' usage of some components must be purchased at one time, as happens when production of a needed component will be discontinued. Complex printed circuit boards used in small volumes over extended periods are the parts most often stored in inert gas. Inert gas will not prevent intermetallic bleed-through, although the opportunities for those intermetallics to oxidize after reaching the surface will be restricted. Lowering the storage temperature slows down development of intermetallic layers.

8.8 Restoring Solderability

Sooner or later, every assembly company finds that specific components are required immediately to fulfill orders. It may well be the case that the only available components are not solderable using the assembly company's flux. The traditional strategy in such cases has been to issue the components to the production department, hope that some will solder acceptably, and touch up the remainder. Such action violates the fundamental principles of solderability management, however, and virtually guarantees the demise of future solderability management activities. Moreover, sending unsolderable parts to production conveys a graphic message to all employees that the company is not truly committed to product dependability.

8.8.1 Presoldering

Fortunately, the easy alternative known as presoldering exists. The steps of presoldering—in sequence—are:

1. Choose an active flux that can deoxidize the surfaces (usually component leads) in question.
2. Apply the active flux to the unsolderable surfaces (generally by dipping the leads into a container of the flux, taking care that the flux does not reach the component body).
3. Insert the tip of the component lead into solder heated to approximately 233°C (450°F).
4. Wait until the flux activates and begins deoxidizing the surface; then gradually insert the leads into the solder.
5. Withdraw the leads from the solder.

6. Rinse the leads in an appropriate cleaning solution to remove the flux.
7. The surface is now solderable.

As the name presoldering implies, this process of restoring solderability to surfaces is actually a soldering operation. Unlike assembly soldering which electrically connects two or more conductive surfaces, presoldering applies solder to only one surface. Presoldered surfaces virtually never lose their ability to be soldered using only the weakest of fluxes.

8.8.2 Use of Activated Fluxes in Presoldering

Use of highly active flux is acceptable in this presoldering, though it cannot be condoned for production soldering. The reason is that the flux in presoldering cannot coat the assembly and degrade the resistance between current-carrying circuits. Further, since there are no tight spaces where the flux can be trapped during presoldering, any cleaning solution can easily dissolve and remove the flux material at the end of presoldering. There is no threat to product dependability.[3]

Presoldering can be employed at any time, from receiving until immediately prior to production. Since it does involve an extra processing step, presoldering is normally most cost-effective when employed with components that have been on hand for an unacceptably long period of time (6 months maximum, unless the parts have been stored in inert atmospheres, in which case shelf life is much longer) or were shown to be unsolderable upon receipt. If demand has been properly forecast prior to components being ordered, the number of components requiring presoldering will be very small.

8.9 The Quality Department's Role in Solderability Management

Solderability management is a quality issue no matter which functional departments are actually involved in receipt, storage, and disbursement of parts.

[3]One prerequisite must be stipulated: During presoldering, flux must not be allowed to flow under insulation. Flux trapped under insulation cannot be removed and over time may corrode the lead.

8.9.1 Receiving

Most plants assign responsibility for receiving to purchasing personnel on the theory that the function of receiving is to ensure that incoming parts correspond to the parts ordered. A few plants turn receiving chores over to the stores department that will have custody of the parts until disbursement to production. Virtually no companies recognize that solderability testing must be conducted at receiving, and even fewer know that the task is best suited to quality personnel. The line responsibility of the personnel performing inspection is not terribly important provided the quality department specifies the solderability test procedures—including recording of findings—and audits compliance.

8.9.2 Storage and Disbursement of Parts

In most assembly organizations, stores reports to planning and materials or is an independent department. Management of inventory and disbursement of components (including kitting where appropriate) are time-consuming activities that can easily overwhelm already overworked quality departments. While not accepting responsibility for overseeing the entire stores department, the quality department should nonetheless specify the solderability management methodology and audit for compliance.

8.10 Summary

- Solderability management is necessary for preventing defects in the assembly process.
- Solderability is defined as the strength of the flux required to remove oxides from the surface to be soldered; stronger flux requirements equal poorer solderability.
- Solderability specifications generally cannot be applied to vendors; however, specifying reflowed tin or tin/lead plating will ensure components remain solderable for much longer.
- Storage in inert gas (oxygen-free) environments maintains solderability for long periods.
- Unsolderable surfaces can be restored to solderability by presoldering.

- The quality department must take an active role in setting up and auditing solderability management programs within the plant.
- By applying some resources before assembly, the quality department can greatly reduce its total workload—and improve product dependability—after assembly.

9 Quality Standards

9.1 Chapter Objectives

Standards are the guidelines against which the quality of output is determined. This chapter will show that:

- A condition known as process consonance must exist for output to be free of defects and failures.
- When process consonance is achieved, the output will be free of defects and failures.
- Critical standards have not changed since the vacuum tube era.
- The same standards are employed in both commercial and defense electronics.
- Those standards cause dependable product to be rejected, reworked, and made undependable.
- Visual standards are ineffective at best.

9.2 Process Consonance

It has been noted (Sec. 6.4.2) that three process states exist concurrently in any plant:

1. The real process—what is actually done—is the physically tangible state. We observe the real process at work around us. It can be seen and measured.
2. The specified process—what the process engineers decreed should be done—exists only on paper and in the minds of the individuals who laid out the parameters.
3. The natural process—what scientific forces require to be done—makes its presence felt only when the real process violates the conditions of the natural process.

When reference is made to "the process" in normal plant life, only the real process is normally under consideration. There is also an assumption that the plant performs in accordance with the specified process—a highly questionable assumption at the best of times.

9.2.1 Process Consonance Defined

Scientific quality management ensures that the three process states coincide. That is, the "real," "specified," and "natural" processes must be identical before a plant can truly be described as operating according to effective scientific quality management. Defects and failures occur whenever any of the three process states deviates in any manner from the other two. When the three process states are identical, a plant can be said to have achieved *process consonance.*

Process consonance requires that the three process states be identical in all respects. When two process states are identical but diverge from the third, process consonance does not exist.[1] It is certain that output will contain the fewest defects and fail least frequently when process consonance exists. But it is by no means clear that movement toward process consonance improves output in a continuous function. That is—illogical though it may seem—output may contain fewer defects and failures when the real and natural processes deviate substantially com-

[1]More accurately, it is only necessary that the real process be identical to the natural process. The specified process has relevance only as a means of guiding the real process. Whether the real process could coincide with the natural process if the specified process calls for different procedures is another issue.

pared to when the two process states are closer but not identical. If some variables of the real process are inadvertently set incorrectly (i.e., the values for those variables contradict the values in the natural process), better results are frequently obtained by deliberately adjusting other parameters away from the values they would have in the natural process. The deliberate adjustment of parameters away from optimal settings balances the effects of the rogue elements. In wave soldering, for example, the natural process requires that when the assembly enters a still solder wave, it is aligned in such a way that rows of through-hole pins (as in a DIP) are at right angles to the wave. However, if the wave is turbulent—a violation of the natural process—fewer defects are often obtained if the assembly is turned 90° from the position in the natural process and rows of leads enter the wave broadside.

This erratic behavior when process consonance does not exist has interesting implications. For example, we can see why different experiments to determine the optimal settings of various parameters result in contradictory findings: Some unrecognized but critical variables have different values in the various experiments. Similarly, processes that work well with circuit assemblies whose layouts break some natural laws often produce high levels of defects when applied to circuit boards designed in conformance to the natural process. Companies that employ contract assembly houses to lay out their circuits can readily find themselves with designs that are compatible only with that assembler's flawed processes.

9.2.2 Achieving Process Consonance

It has already been seen in Chap. 7 that the specified process can readily deviate from the natural process. Electronics assembly veterans also know from experience that the real process often follows much different paths from the specified process. Therefore, it is characteristic of electronics assembly that great divergence exists among the three process states. Scientific quality management is the ability to establish process consonance.

Although the objective in all instances is to achieve process consonance, the techniques for ensuring homogeneity between the real and specified process states are very different from the steps that must be taken to gain congruity between the specified and natural process states. The former results from proficient uses of social science techniques including motivation and communication tools. But only through adept application of the physical sciences can the specified process be brought into congruity with the natural process. The combination of skills and knowledge necessary for process consonance to be achieved distinguishes quality from most other plant disciplines that require either scientific knowledge *or* social skills but rarely both.

9.2.3 Workmanship Standards

Regarded superficially, the quality department apparently controls none of the three process states directly. The natural process, for example, cannot be affected by any individual—within the quality department or without. The specified process appears on the typical company organization chart as a responsibility of process engineering and manufacturing (or, in a few organizations, design personnel) rather than quality. And the real process consists of every plant activity, not just those within the quality department.

In another and more realistic sense, however, the quality department has at least the ability and usually the effect of dominating the real process. An effective quality department involves itself directly in definition of the specified process, at the very least by pointing out inconsistencies between specified process and natural process. Particularly in this era of ISO 9000 popularity, the quality department's influence in every facet of the real and specified processes is profound. More and more, the quality department has formal responsibility for demonstrating that the company says what it does (the specified process) and does what it says (the real process).

ISO 9000 has reaffirmed in the collective plant consciousness the importance of written procedures. Within the quality profession itself, however, documentation has long been standard operating procedure. Probably the most influential documents designed and implemented by the quality department are the company's workmanship standards. Typical workmanship standards consist of pictures—sometimes accompanied by descriptive text, other times without any explanation—showing various circuit assembly attributes that an inspector might encounter. A few such attributes commonly include diverse types of solder joints, mounting characteristics of components, appearance of the circuit board laminate and traces, condition of miscellaneous hardware, and so on. For each attribute, a range of possible outcomes—designated "Acceptable Preferred," "Acceptable," "Minimum Acceptable," and "Unacceptable" (or some similar terminology)—is shown.

All plant personnel are aware that workmanship standards exist in the plant, and most personnel are confident that they know the specifics of the company's workmanship standards. (How else can the dust collecting on the standards books be explained?) However, the general plant population seldom has true knowledge of the content and intent of the workmanship standards. Indeed, the number of interpretations of the workmanship standards can be as high as the number of employees in the plant. Whether the workmanship standards are accurately understood and followed reflects the management talents of the quality department.[2]

9.2.4 The Commonality of Standards

The workmanship standards among companies in the electronics assembly industry show little variation. Not even national borders or highly divergent types of customers cause much divergence in those standards. While customer requirements may cause the amount of paperwork or other bureaucratic rituals to differ from company to company, fundamentally an American defense electronics manufacturer employs workmanship standards remarkably similar to those of an American broadcasting equipment manufacturer, a U.K. computer peripherals manufacturer, or a Spanish automotive electronics manufacturer. Yet:

1. Each of them believes their practices to be unique.
2. Except for the manufacturer of defense products, almost none of the companies knows that the standards used in their plants are products of NASA and Department of Defense bureaucrats.

[2]Lines of authority often become blurred at this point. The quality department sets the workmanship standards (perhaps in consultation with the production and other departments) and verifies whether output meets those standards. However, production personnel conduct the actual assembly operations. When we refer to the quality department's management talents, we include the ability to enlist support from those other departments. The distinction between effective quality management and a quality department going through the motions boils down to the amount of influence wielded by the quality management. Good quality managers can be stymied by unsympathetic and/or unsupportive general management, of course, but the decision facing the quality manager in such circumstances is whether to remain as a figurehead to collect a paycheck or whether to take a strong stand that could jeopardize the manager's own job.

3. Even fewer of those companies realize that the key quality standards of modern electronics assembly originated almost 40 years ago when state-of-the-art electronics assemblies consisted of vacuum tubes inserted into hard-wired sockets—and have remained the same ever since.
4. All of them put forth too much effort and expense in exchange for too little dependability.

9.3 The Origins of Formal Standards

Precision electronics assemblies are a relatively recent development. The most sophisticated electronics products until the middle of this century were found primarily in broadcasting applications. Indeed, almost anyone over the age of 50 can remember when local variety stores carried vacuum tubes and testing equipment so users of radio or television receivers could restore their sets to working order before the end of a favorite program.

Accordingly, the rigorous process management and quality control standards common to more mature industries were almost unknown in electronics assembly until well after the end of World War II (although manufacturers of the vacuum tubes themselves employed their own internally developed standards). This absence of general industry standards prevailed until preliminary work was undertaken by the Redstone missile project team around 1955. When the National Aeronautics and Space Administration (NASA) was formed in 1958, the development of comprehensive standards was undertaken in earnest by NASA personnel, many of them veterans of the Redstone project.

NASA's concern about electronics standards was logical. The agency would be relying on leading-edge electronic systems for its spacecraft and ground control operations. Moreover, these systems would be produced by private companies under contract to NASA. How could the user—NASA—ensure that the electronic assemblies would be dependable?

9.3.1 Early Military Influence on Specifications

Insisting on comprehensive warranties from contractors was one option and, in hindsight, would have been preferable to the choice that was made. Warranties, however, have rarely been part of the military way of doing business, in large part because sending a unit back for repairs is not an attractive option when the product in question—artillery, for example—would be used only once and the consequences of failure could be loss of life. Although NASA was nominally a civilian agency, many of its founders had military backgrounds and attitudes. Unsurprisingly, the agency decided to manage electronics contractors in the same military tradition of detailed process and workmanship standards that already applied to mechanical and raw materials suppliers.

Senior management of manufacturing enterprises have seldom taken any interest in mundane matters of manufacturing processes. So, while NASA's leaders—including the scientific elite within the organization—recognized the need for process standards, they were not about to write those standards themselves. Consequently, the authors of NASA's electronics assembly standards were high school graduates who possessed no advanced education as scientists or engineers. Requirements affecting the integrity of essential assemblies were established by inexperienced young men with no real scientific credentials; their personal knowledge of electronic components and systems, if any, consisted of what they had observed as repair technicians in NASA facilities. All questions were decided according to readily observed mechanical considerations; chemical and physical factors that we now know to be critically important were almost entirely overlooked.

Given the authors' lack of education, the specifications they produced actually met the needs of the day surprisingly well. For example, the era's state-of-the-art technology—vacuum tube sockets, solid wires, metal chassis, and similarly massive rugged parts—demanded high temperatures applied for long periods of time during soldering. Without application of great heat, the thermal energy would dissipate in the massive metal components and the solder would freeze before completing the connection. Compounding the difficulties of applying adequate heat, soldering irons were much less efficient than their modern counterparts. Bearing in mind the need for considerable thermal energy to form solder connections between very large parts—and well aware that heat damage was impossible at less than blast furnace temperatures—the NASA authors specified the processes in such a way that underheating could never occur.

9.3.2 Stagnation of Processes

Gradually, however, new components displaced the vacuum tube technology. Transistors appeared first, followed not long after by primitive solid state devices. Printed circuit boards (originally known as printed wiring boards) replaced the copper wires, socket terminals, and metal chassis. The componentry changed with astounding speed, but the process and workmanship specs stood still.

Before long, therefore, a reliability problem that had never concerned the authors of the original standards appeared. With rare exceptions such as multilayer boards containing substantial interior ground planes, underheating ceased to be a concern. Overheating (see Sec. 7.10) became the dominant cause of system failures. The MIL Specifications and other standards based on them guaranteed heat damage. Worse yet, almost every electronics assembly company, consultant, and trainer today—defense and commercial alike—relies on some form of those original NASA/MIL standards.

9.3.3 Dictating Both Process Parameters and Output Characteristics

The most basic principle of industrial engineering is that real process determines output. A consistent real process will result in consistent output (although, if the process is flawed, the output will be consistently defective). If the output is imperfect, the imperfections can be eliminated only by correcting the process.

Fundamentally, the visual standards are reasonable portrayals of the results that would be seen with process consonance. In other words, most of the attributes identified by the MIL standards as acceptable are indeed desirable. On the other hand, the desirability depends entirely on the conditions under which the appearance was derived. Reworked connections with the appearance of perfect dependability may or may not be reliable; there is no way of knowing.

Further, the appearance reflects the level of integrity in the process used to achieve those results. From the output, the integrity of the process can be deduced. Alternatively, the outcome is entirely predictable from the process parameters (presuming, of course, that sufficient scientific knowledge is possessed to enable those determinations). The process and outcome are mutually dependent; when one is set, the other is also unilaterally established.

It follows from the condition of mutual dependence that *either* the process *or* the outcome—but not both—can be specified. In mathematical terms, there is only one degree of freedom. The MIL standards violate the condition of mutual dependence by dictating *both* the process parameters *and* the characteristics of the items produced using those specified processes. Moreover, there is conflict between the process specifications and the output requirements. The required output is consistent only with a perfect real process, but the process specifications violate many requirements of the natural process. Therefore, the MIL standards ensure that output produced in accordance with the process specifications can only meet the visual criteria through considerable rework. It can fairly be said that the MIL standards involve two separate production operations—one for the initial output and a second to make the output conform to the visual requirements.

9.4 The Spread of NASA Standards

With wider reliance on electronics products in the early 1960s, the U.S. Department of Defense (DoD)—like NASA before it—perceived a need for standards to manage contractors' production and reliability. Various branches of the armed forces adopted core sections of the NASA documents under a range of MIL reference numbers. Each of the military services issued its own standards, although the differences were more superficial than real. The NASA roots became obscured by the proliferation of MIL documents.

9.4.1 Defense Standards in Commercial Plants

The arrival of solid state technology made formal quality assurance standards necessary in commercial electronics assembly as well as defense and aerospace contractors. Nondefense members of electronics industry trade associations looked to the associations for guidance on workmanship standards for use in their own plants. Trade associations responded by creating standards-setting committees composed of their members' employees. The committees were given the mandate to create process guidelines and quality standards that the associations published for implementation by their members. The IPC and IEEE standards originated in this manner.

Only rarely would the associations pay the travel expenses and wages of standards committee members; the cost was shifted to the companies where those committee members were employed. While some commercial electronics companies saw participation by their technical personnel in the setting of industry standards as a form of civic duty, most such companies could not justify the substantial cost. Defense contractors, on the other hand, had great incentive to place employees on industry committees of all types, including standards committees; unlike private sector manufacturers, defense contractors worked to "cost plus" contracts so their profits actually increased as costs rose. Inevitably, defense contractor personnel dominated the committees of industry associations and set standards corresponding to those with which they were already familiar—i.e., the MIL Specs—with the result that quality procedures of the commercial electronics industry became identical to the defense quality standards in every matter of consequence.

MIL Specs now dominate quality attitudes and practices in the vast majority of U.S. electronics manufacturing plants, although few of those plants actually produce for defense consumption. Many of those same standards have been exported to other countries, primarily America's European allies. Ironically, as the relative importance of defense contracting in the electronics industry has waned, the MIL Specs of several decades ago are more pervasive than ever.

9.4.2 The Costs of Using Outdated Standards

If American electronics assembly companies were to reassess the relevance of the workmanship standards used in their plants, they would find that the standards have not kept pace with changes in product technology. The standards employed today may have been satisfactory at a time when vacuum tubes represented state-of-the-art electronics but are totally incompatible with modern delicate, heat-sensitive solid-state components. Consequently companies continuing to use those standards—and training programs derived from them—suffer excessive costs and inferior product reliability. Just a few of the consequences include:

- Massive amounts of non-value-added labor
- Invisible defects leading to the mistaken belief that quality problems do not exist
- Large numbers of avoidable product failures
- Lack of scientific process control knowledge
- Unhappy customers
- Loss of markets
- Management and engineering time wasted trying to "fix" symptoms rather than problems
- Loss of competitiveness and market share

Most of the damage and non-value-added work cannot be seen. Components damaged as a consequence of rough handling mandated by the most common workmanship standards, for example, tend not to show up until the product reaches the customer. Consequently, there is little understanding in the electronics manufacturing industry as a whole that the workmanship programs may be the problem rather than the protection. With the problem lacking visibility, the critical need for corrective action has not been widely recognized. Thus, far from improving, the situation continues to deteriorate as increasingly delicate components replace older, more durable, parts.

Certainly not every aspect of the MIL Specs or industry standards misses the mark. Although revisions to long-standing activities such as soldering have a pernicious tendency to perpetuate past errors, specifications written for the first time after multilayer printed circuit boards became common are more likely to help than hinder. An outstanding example of a newer technical issue that has been well addressed by standards can be found in protection against damage from electrostatic discharges (ESD). Earlier generations of components were too robust to be affected by bursts of static electricity, so the issue was not widely addressed until the early 1980s. A powerful secondary reason for development of extensive ESD control standards was that manufacturers of antistatic systems and packaging saw a market opportunity that warranted large-scale investments to publicize the problem and, by extension, the need for their products. Good standards for judging certain aspects of printed circuit board fabrication also exist, in no small measure because the mechanical difficulties presented by warped or poorly plated boards are relatively obvious and easily measured.

Unfortunately, a mix of mostly good standards together with a few disastrously wrong specifications translates into almost as much trouble as totally defective measures. The antiquated practices such as visual inspection and touchup that virtually guarantee damage to modern components mingle freely in the DoD and industry standards with the acceptable procedures; only a highly experienced technologist with an open mind can separate the good from the bad.

9.5 New Components Need New Processes

The changes in component technology demanded equivalent changes in assembly processes. However, most industry standards and practices remained firmly rooted in vacuum tube assembly ideology. The training courses for electronics assemblers today are identical in every meaningful sense to the courses taught during the Kennedy administration. Far from being the solution to reliability concerns, both DoD and general industry standards compromise component integrity.

Not surprisingly, MIL and IPC/IEEE advocates dispute this contention. However, hard field evidence proves the point. Proof of this thesis came early in the 1980s when Ford Motor Company's Electronics Division abandoned the usual industry process and quality verification standards. In their place, Ford implemented systems that deviated 180° from the electronics assembly industry's usual practices. Some of those changes included new manual and machine soldering techniques to ensure that defects would be very uncommon, new quality manuals to replace the familiar IPC pictures (which are almost identical to the Martin-Marietta picture standards common in defense contracting plants), eliminating some test procedures, abolishing visual inspection, and ending burn-in. During the first year using the unorthodox procedures, the company's warranty claims fell by more than 80 percent.

9.6 Other Flaws

While the original NASA standards worked much better in the vacuum tube era than today's versions meet modern needs, serious scientific and human resource flaws undermined the standards right from the start. Unquestionably the most serious failings are the requirements to conduct visual inspection, touchup, and rework. These requirements reflect four strongly held but erroneous doctrines:

1. The bureaucratic belief that people (particularly civilians) cannot be trusted to perform dependable work without intense supervision
2. The opinion that dependability can be determined by appearances
3. Faith that inspection will identify undependable connections and only undependable connections
4. The assumption that touchup and rework can restore the integrity of a defective connection

Several consequences, all of them serious, result from these errors in judgment. Basically, the consequences can be grouped into two categories: (1) inefficiencies and (2) needlessly high failure rates.

9.6.1 Inefficiencies

Mainstream workmanship standards waste resources in many ways involving both labor and equipment. Costs increase while throughput slows, two of the most disastrous liabilities that can befall any modern manufacturing business but especially ruinous in the rapidly changing electronics industry.

9.6.1.1 Inspection. Missing, reversed, or wrong components are quality attributes visible to the human eye. Solder shorts and (to a lesser degree) opens can also be seen. No other common aspect of dependability can be determined by inspection. Moreover, as component sizes shrink and densities expand, whether a failure mode *can* be seen is less relevant than whether it is realistic to expect that it *will* be seen. The visual criteria that make up the preponderance of electronics assembly standards are not just irrelevant—they lead directly to some of the most bloated unnecessary costs in this industry today.

A human being paid to find defects will find defects. Like work, the number of "defects" identified expands to fill the number of labor hours allocated to the activity. Moreover, "defects" will be found even where none exist. This can easily be proved through the experiment described in Sec. 10.3.1.1.

The costs of inspection-based quality have been the subject of endless industry complaints. Yet, despite the almost universal antagonism to inspection, most electronics assembly plants still make extensive use of visual quality control. Defense contractors visually inspect because of their contractual obligations; commercial assembly plants inspect because that's the way quality has always been measured. Remarkably few members of the industry seem to realize that the labor costs and bottlenecks created by inspection operations are probably the least meaningful liabilities of those activities; in addition to increasing costs, inspection results in dependable products being sent off for rework that will make them much less dependable (see Sec. 9.6.2).

9.6.1.2 Hidden Rework. The real level of touchup and rework in the typical electronics assembly plant is much higher than the official statistics indicate. Wherever manual solder assembly operations are employed (soldering components for which automated reflow is not an option, for example), the operator normally remakes some connections several times until being satisfied with the appearance. While criticizing and reworking their own work, operators are also likely to inspect and rework some of the joints generated by the automated reflow system. This hidden rework is often missed by the statisticians tracking defect rates and rework activity. Because hidden rework goes unnoticed, corrective action to eliminate its cause(s) is less likely than actions to reduce the amount of formal inspection and rework. Of course, hidden rework is every bit as costly and undesirable as formal rework but, being invisible, is much harder to rectify.

Hidden rework occurs for two main reasons:

1. Operator training classes tell the operators that changing the appearance of solder connections makes them more reliable.

2. The presence of inspectors farther down the production operation makes the operators nervous. The operators are afraid of the embarrassment they will suffer if an inspector finds fault with the work. This negative reinforcement poses a dilemma for quality management who are told that the only way to improve the operator work is by feeding back information about how the inspectors assessed that work. If the assessment is accurate, the advice to keep the operator informed of the inspection findings has merit. Where the inspectors' interpretations of work quality is capricious, information feedback only makes the operator confused and more inclined to rework joints before they can reach inspection.

9.6.1.3 Wasted Classroom Time. The industry's most widely used training courses for solder assembly operators are worse than useless. The official courses in which operators are certified to MIL standards are actually hazardous to the product's dependability and cost. ("Hi reliability" assembly procedures are even more destructive than the garden variety procedures.) The problems include:

1. Training develops motor skills but not knowledge. While manual dexterity is certainly necessary in assembly, lack of knowledge means the employee is incapable of recognizing real process problems when they occur. Naturally, an employee who cannot identify problems cannot bring the existence of those problems to the company's attention.

2. The techniques in which the trainees are drilled—the application of excessive heat, for example—damage modern electronic devices.

3. Even if participants did take some meaningful process improvement knowledge away from the classes, no systems exist for translating class lessons into operations improvements.

4. Productivity of a training department is typically judged by the number of hours spent in classes, although true productivity equals the amount of improvement in personnel relative to the amount of time spent in classes. Greater improvement in less time means higher training productivity.

9.6.2 Degradation of Dependability

We know that visual approaches to quality increase costs by imposing unnecessary extra activities ranging from inspection to rework. Additionally, it can be proved that dependability varies inversely with the amount and type of handling to which the assembly and its components are subjected. Visual "quality" imposes not just more handling but also the types of physical stresses that cause the greatest deterioration in product dependability. Among the more important of those extra stresses that result from managing quality according to visual standards are:

1. *Heat damage.* Although the amount of thermal energy to which components are subjected can be rigorously controlled during initial assembly (even if manual soldering is involved) and roughly managed during repair using the technique presented in Sec. 7.11.1.4, it is impossible to control the heat applied during touchup. Touchup exists only where visual criteria dictate "quality."

2. *Electrostatic damage.* Although effective—in many cases, excessive—ESD controls are now found in every competent electronics facility, the chances of breakdowns in the prevention system rise each time a component or assembly is handled. Soldering and desoldering tools also tend to be well shielded, but the constant use of such tools by assembly plants makes some tool-related ESD damage a common occurrence.

3. *Mechanical damage.* Each time an assembly is handled, it encounters physical forces ranging from jolts to scrapes to pressure from hand tools. If assemblies are moved from station to station in boxes or carts (which means repeated loading and unloading of assemblies), the damage can be quite severe. The amount of handling to which products are subjected at each stage, from components to finished goods, must be minimized, yet visual standards ensure that cannot happen.

9.6.3 Trends in Visual Quality

For years, industry leaders have predicted that the emphasis on cosmetic standards would soon disappear. It has not happened. On the contrary, as the size of components and their connections have shrunk, more elaborate equipment for looking into joints has come into extensive use. When surface-mounted components replaced the larger through-hole devices of two decades ago, there was an opportunity to move away from the traditional workmanship standards; instead, the historic through-hole criteria were applied to the new devices.[3]

The DoD in recent years has shown more awareness than the com-

[3]For example, the rise of solder up the lead of a surface-mounted component is identical to the through-hole criterion of solder flow to the top (or "component") side of the assembly. The requirements for component alignment can be traced back to the stipulation that solder fill the entire hole of a single-sided circuit assembly.

mercial sector of the need for change, though admittedly the long-standing DoD standards require greater change. Unfortunately, the standards have not evolved as rapidly as appreciation within the industry itself that new ways of managing quality of defense electronics are necessary. The MIL STD 2000 family was praised as a step back from reliance on visual criteria in defense contracting in favor of process control. Certainly, MIL STD 2000 did bring about some much-needed improvements but still falls back on cosmetic criteria and specify production processes unsuited to modern electronics technology. Meanwhile, billions of dollars are spent each year on new units of systems that may have been in production for two decades or more.

One other peculiarity of the DoD standards system should be noted. New—and, presumably, improved—standards rarely apply to existing contracts. It is customary to require that the systems be built in conformance to the specifications in force when the first order was placed, even if the standards prevailing at that time have long since been revised and superseded. In addition to defying logic—why introduce improved standards if they are not better for all product, existing as well as new—this habit greatly increases the difficulty of managing production and quality. It is not unusual to find half a dozen or more differing requirements applicable to the various products built in the same plant; naturally, information overload has become the dominant characteristic in quality management. New standards arrive at regular intervals, but the old standards linger on as well. The standards manuals in some plants now take up scores of linear feet. Naturally, no one knows exactly what is in each standard, nor do they have time or inclination to search through the pages when in doubt. Assembly verification is dominated by faulty memories, limited visual acuity, and prejudices. Scientific quality management remains all but unknown.

9.7 Some Obvious Questions and Not So Obvious Answers

The people associated with assembly specifications may have made mistakes, but they are certainly not stupid. Recognizing the intelligence of the individuals involved in the development of standards over the decades, some obvious questions demand answers. Two questions in particular stand out:

1. Why would defense contractors not have caused the Department of Defense to change its standards?
2. Why would commercial companies use quality standards created in the defense electronics industry?

9.7.1 Why Defense Contractors Supported MIL Specs

Until recently, defense contractors had several solid business reasons for supporting the MIL standards:

- At least some advocates of the MIL approach believed that MIL processes brought about more dependable products. The advocates were wrong, but they cannot be criticized on moral grounds.
- Cost-plus contracts meant that the inefficient MIL standards ensured higher profits for defense contractors. DoD specs required vast amounts of expensive record keeping, inspection, rework, test, and retest, all of which drove up costs and increased earnings.
- MIL standards removed much of the contractors' decision-making burden and risk. If the contractor followed the process parameters dictated in the purchase contract and met the necessary cosmetic standards, the government was obligated to purchase the product *even if the product was potentially unreliable.* The contractor was not required to do the job right the first time; touchup and rework were sanctioned ways to achieve the stipulated cosmetic criteria. To show the extent of the havoc wrought by this ridiculous procurement philosophy, it is only necessary to look at the reliability records of nuclear missiles. As recently as the late 1980s, nuclear missiles were powered up regularly for electronic diagnostic checks. While the Pentagon never officially revealed the test results, individuals working for the suppliers of missile electronic systems have admitted that on a good day fewer than one-third of the missiles were operational. The bad days were unspeakable.
- The bureaucratic overhead required to meet the standards prevented new competitors from entering the field. In addition to the cost of setting up the verification operations, prospective entrants into the defense contracting business also had to arrange for bureaucratically specified training and certification of their personnel. Generally, the certifications were demanded before a company could bid on contracts.

The days of cost-plus have largely disappeared, and in recent years we have worked with increasing numbers of defense contractors who do need to find ways of cutting costs. In a quirk of justice, however, the defense contractors' past negligence has caught up to them. Now that they have reason to lobby the DoD for less costly quality (i.e., workmanship) standards, defense contractors lack the technical knowledge to know what to recommend and how to substantiate the technical validity of their recommendations. Decades of inbreeding eradicated the industry's ability to look at matters scientifically.

9.7.2 Why MIL-Based Standards Still Prevail in Commercial Electronics Plants

Every commercial company rants about the gross inefficiencies in defense electronics plants. Why, therefore, does the nonmilitary electronics industry allow itself to be governed by the defense sector? Among the reasons:

- Lack of technical expertise—not to be confused with memorization of standards—in most electronics assembly companies means there is no one with the ability to identify scientific errors in the standards.
- Limited awareness of where and how the standards were developed.
- Lack of confidence to rock the collective boat; if everyone else in the industry shares well-known beliefs, it is easier and safer to follow.
- Shortages of imagination and initiative that inspire the search for additional knowledge.

9.8 The Challenge of Change

The decision to modify the company's workmanship standards can be taken without fully appreciating the magnitude of the difficulties implementing those changes will involve. The great obstacles to implementing a new regimen of quality standards generally do not involve the creation of replacement standards and procedures. Indeed, if the scientific knowledge about requirements of the natural process is at hand, the hardest part about writing new standards is finding the time to compose the necessary documents.

On the other hand, reprogramming employee minds to follow the new standards and procedures is an altogether more formidable challenge. Compared to changing the work habits of operators who have often been conditioned to MIL-type ways by training programs and yearly reindoctrination, developing new standards and procedures is trivial.

The technological challenges, in other words, are simple relative to the effort required to alter human attitudes. Old habits that regard inspection and touchup as viable approaches to cost-effective production of reliable product *must* be broken. Comprehension that doing the job right the first time is the only feasible survival option *must* be cultivated. And neither shift in attitude will be accomplished easily.

9.9 Summary

- Output that is free of defects and failures occurs only when the real process, the specified process, and the natural process are identical. This condition is known as process consonance.
- Achieving process consonance requires skillful application of both social and physical sciences.
- The quality (or workmanship) standards found in most plants originated with NASA four decades ago. Under administration by the Department of Defense (DoD), they became known as the MIL Specs.
- The same standards are employed by commercial and defense electronics assembly companies.
- Although output and process are mutually dependent, MIL Specs dictate both the process parameters and the appearance of the output.
- The consequences of using MIL Specs are high costs and low dependability.

10
Non-Value-Added "Quality"

10.1 Chapter Objectives

This chapter will:

- Identify quality practices that compromise product dependability and increase costs.
- Examine the reasons these undesirable quality practices exist.
- Suggest alternative strategies that do improve product dependability and reduce costs.

10.2 Value-Added, Non-Value-Added, and Negative-Value-Added Work

A few important points from previous chapters deserve to be repeated here. The purpose of the quality department is identical to the purpose of any other department—to help the company increase long-term profits. Two tactics are available to the quality department: (1) improving product dependability and (2) reducing costs of quality. When quality is managed effectively, the combination of lower costs and greater dependability will be achieved. These results allow the company to sell higher-quality goods at lower prices, which in turn attracts more customers and enhances long-term prosperity. On the other hand, when quality is not managed effectively, the company is forced to increase prices for lower-quality goods, and customers are lost.

From Chaps. 7 and 9, it should be apparent that many electronics assembly quality practices cannot be justified economically or scientifically. At best, those activities are unnecessary; they represent avoidable costs and prevent the company from progressing closer to the goal of maximizing long-run profits. At worst, they degrade product dependability. In far too many cases, the quality department has dragged down rather than boosted the company's fortunes.

Readers who compare their company's actual failure rates to predicted failure rates using the techniques presented in Sec. 3.12 will prove that the actual failure rates are considerably greater than can be explained by inherent faults in the components. Clearly, some aspect(s) of in-plant activities cause the failure rate to rise. By means of basic (though admittedly tedious) statistical experimentation, it can also be demonstrated that dependability varies inversely with the amount of handling to which the unit and its components are subjected.

10.2.1 Handling and Dependability

Every time a component or assembly is touched, moved, packaged, unpackaged, prodded, probed, covered with chemicals, or cleaned, dependability suffers. There is no such thing as "good" handling. Even unavoidable handling may damage the assembly; while a handling step can be necessary under our existing technological know-how, it need not be innocuous—a point explored with respect to test in Chap. 11. Some little-known examples of handling-induced failures were discussed in previous chapters,[1] and all electronics assembly personnel are already aware of other dependability problems caused by handling.[2] One of the most satisfying ways to improve the quality of electronic products is by eliminating unnecessary handling of components, subassemblies, and finished goods. Both costs *and* failures go down when unwarranted handling is abolished.

---

[1]Touchup, for example, was identified in Sec. 7.7.2 as a leading cause of dependability degradation.

[2]e.g., static electricity discharge.

10.2.2 The Hidden Factory

Several decades ago, Feigenbaum coined the phrase "hidden plant" to describe the amount of avoidable work found in the normal manufacturing organization. According to Feigenbaum's definition, the hidden plant is "the proportion of plant capacity that exists to rework unsatisfactory parts, to replace products recalled from the field, or to retest and reinspect rejected units."[3] In other words, the hidden plant (often known today as the "hidden factory") consists of the resources required to "fix" things that were not done correctly the first time.

Presented during an era when the Martin Company was earning applause for its Zero Defect program, Feigenbaum's hidden plant departed radically from mainstream quality thinking. In the Martin environment, inspection and repair were considered highly productive work; it was acceptable to the company that its Zero Defects program relied on intensive inspection and repair to prevent faulty product from reaching the customer. Feigenbaum, on the other hand, recognized that product verification and repair were penalties for failing to do the job properly during production. No new output results from the detection and correction efforts; the number of shippable units may increase but the additional units should have been manufactured correctly in the first place. In economic terms, detection and correction are non-value-added work. In contrast, resources that cause output to rise are classified as value-added work.

10.2.3 Negative-Value-Added Work

For many years, electronics assembly functions could be neatly segregated into just those two categories of value-added and non-value-added work. Finding and repairing faulty product did not create additional output, but neither did it cause failures. With modern heat- and static-sensitive electronic components, however, conditions have changed. Detection and correction of faults continue to be non-value-added work. In addition, though, post-production activities such as inspection and rework compromise the quality of otherwise dependable product.

---

[3][Feigenbaum 1983], pp. 46–47.

In other words, a third category of work has appeared that is more to be feared than non-value-added work. By reducing dependability, these activities *decrease* value. They can properly be described as "negative-value-added" work.

The three modern classifications of work can therefore be defined as:

1. *Value-added;* expending these resources increases total output.
2. *Non-value-added;* additional inputs of these resources do not alter total output.
3. *Negative-value-added;* each additional unit input reduces the quantity of dependable output.

10.2.4 Locations of Negative-Value-Added Work

Inspection, touchup, and rework are the best-known constituents of the hidden factory. They are also prominent examples of negative-value-added work. The reasons these activities degrade dependability are discussed in Sec. 10.3.

Inspection, touchup, and rework are regarded by most electronics assembly professionals as quality functions. Certainly they are driven by the quality department. Yet some or all of these activities are carried out by the production department. This observation has great significance to the electronic assembly plant: policies originating in and/or administered by the quality department result in non-value-added and negative-value-added work for departments other than quality itself. Indeed, when a list of prominent negative-value-added operations is compiled, the work will be seen to involve almost every department; many of those other departments expend more effort on quality-mandated negative-value-added work than does the quality department. Touchup and rework, for example, are driven by quality standards yet typically carried out by the production department and reflected in the company ledgers as charges against the production department.

10.2.5 A New Challenge for Quality Management

The existence of insupportable activities carried out in the name of quality is not a recent discovery, of course. Quality managers have long been familiar with the hidden factory concept. With few exceptions, the profession understands that its mandate extends to assisting in defect prevention that allows reduction of hidden factory activity. The controversial question has never been *what* to change but *how* to change.

Attacking the hidden factory by employing quality assurance methods is distinctly preferable to policing output after production. As increasing numbers of firms have proved by implementing quality assurance in place of quality control, the results of proactive quality are impressive indeed. One need only recall that late in the last decade, defect rates of many thousand parts per million were considered the sign of a well-managed electronic assembly plant to realize that enormous advances in process management have been made in a very short time. Today, plants with defect rates of fewer than 100 parts per million are common—and believed to need further improvement.

Nevertheless, depressingly large amounts of resources continue to be spent throughout the plant on reactive "quality," particularly the negative-value-added activities dictated by unreasonable quality specifications. The greatest opportunities for profitable quality improvement initiatives today involve identification and elimination of quality-driven negative-value-added work. The quality director who redesigns quality strategies from a scientific perspective can achieve astoundingly high dependability and companywide efficiencies.

10.3 Examples of Negative-Value-Added Work

A partial inventory of negative-value-added work in the normal electronics assembly plant would reveal some or all of the following activities:

- Visual inspection
- Touchup
- Repair
- Rework
- Flux removal
- Cleanliness testing of soldered assemblies

Test and burn-in, two other common quality activities, should also be reconsidered. Unlike the activities itemized above, however, the case against test is not absolute; at times test is desirable, while under other circumstances it is only an expense from which no benefits derive. In our opinion, choosing the appropriate test strategy is sufficiently important to warrant a dedicated chapter of its own. Accordingly, Chap. 11 is dedicated to test practices and strategies.

10.3.1 Inspection

Inspection dominated quality control practices for decades.[4] The Martin Company's Zero Defects program of the early 1960s demonstrates the extent to which some sectors of the electronics industry have relied on inspection as a useful quality practice. However, there has never been universal acceptance that inspection was a satisfactory method of ensuring quality. In recent years, there has been increasing acceptance that scientific process management (i.e., "quality assurance") is the only way to obtain product dependability. The positions taken in this book are predicated on experience that only quality assurance produces quality products—and quality control cannot be defended.

10.3.1.1 Visual Inspection. Countless studies have proved that visual inspection of solder joints and other cosmetic criteria follows no consistent pattern. This can easily be proved in any plant by the following simple experiment:

1. Randomly select a batch of circuit assemblies generally representative of the company's products; for statistical purposes, a batch of 20 assemblies is satisfactory.
2. Code each of the units with its own unique mark, such as a number.
3. Have the assemblies evaluated by regular inspection personnel who are not informed that this is a special exercise.
4. Instruct inspectors to mark the locations and nature of defects on the assemblies. Arrow-shaped labels can be used to point out the defect location, and the color of the arrow can indicate the nature of the defect.
5. For each assembly, record the location and type of defects found during inspection. The easiest way to record the findings is by taking photographs of the assemblies after the labels have been applied.

[4]Quality control differs from quality assurance in these pages. For the distinction, see Sec. 3.13.

6. Remove the arrows and return the assemblies to the same inspection personnel, who remain unaware of the exercise.
7. Repeat steps 4 and 5.
8. Compare the results of the two inspections.

With rare exceptions, there will be no statistically significant correlation between the two inspections.

A primary reason why visual inspection yields wildly variable results will quickly become apparent to anyone who undertakes visual inspection for the first time. Simply spending a couple of hours looking at circuit boards causes most managers to appreciate that everything quickly starts to look the same. Veteran inspectors suffer the same sort of visual impairment but because of long exposure to the practice fail to appreciate the extent to which their decisions depend on guesswork.

The dissimilar natures of the reference materials and the actual connections being assessed constitute a second leading contributor to confusion in determining joint integrity. Typically the reference materials consist of photographs or pictures representing acceptable and unacceptable conditions. However, two-dimensional reproductions of reflective surfaces bear little or no resemblance to the shiny three-dimensional objects that confront the inspector's eye.

10.3.1.2 Automated Inspection. Over the years, quality managers have regularly considered employing automated equipment to verify solder joint integrity. Such equipment ranges from machines that evaluate the surface appearance of solder joints to those that attempt (by x-rays, infrared sensors, or other technologies) to look inside the joints. Although neither type of inspection machine is particularly useful, the equipment that looks inside joints can cripple a plant's throughput and impose enormous evaluation costs. These unpleasant consequences of internal inspection of joints occur because most solder connections naturally contain cavities (sometimes called "voids") that suggest to the uneducated evaluator lack of structural integrity in the joints. Far from reflecting problems with the connections, those cavities indicate perfectly healthy joints.

This seeming contradiction—that internal cavities are consistent with perfect joint integrity—is a function of solder's thermodynamic behavior. When heated, solder expands; when cooled, it contracts. For this reason, the volume of a liquid solder mass is greater than the volume of that same mass after solidification. If the solder tried to retain the same dimensions after solidification that it possessed in liquid form, strong internal stresses would be working inside the joint attempting to contract the solder to a smaller, stable shape. As it cools and shrinks, the molten solder draws back. The result of this withdrawal may occur on the surface in the form of "pinholes" or beneath the surface where cavities form. (The pinhole phenomenon can be seen in exaggerated form by looking at the surface of a solidified solder pot.) Pinholes are natural and even desirable; historically, however, they have been cause for rejection and enormous effort has been wasted trying to find a "cure" for these surface blemishes. Companies that install on-line equipment to scan joint interiors invariably find themselves busy trying to "fix" the subsurface equivalent of pinholes. In electronics assembly quality, there are times when ignorance is bliss.

10.3.1.3 Insertion Issues. Inspection is not restricted solely to post-solder verification. Visual checks for missing or reversed polarity components are often used to spot insertion errors before soldering. If this technique often reveals misplaced components, more effort should be invested in prevention rather than inspection. (It is not uncommon to find that plants experiencing meaningful numbers of component placement errors will post inspectors immediately after soldering to catch the mistakes. Obviously, the post-solder positioning of inspectors comes too late in the operation to help the units affected.)

10.3.2 Touchup

Touchup invariably accompanies visual inspection. But it is also clear from Sec. 7.10 that touchup degrades product integrity by causing heat damage. Therefore, touchup epitomizes negative-value-added work. Every quality professional professes to hate touchup, but few actively or effectively discourage it.

Much touchup occurs before quality personnel even see the units affected. However, absence from the scene of the crime does not mean the quality personnel are entirely innocent. The readiness of assembly operators to touch up even unrejected work is based in large measure on a history of product rejection by inspectors that fosters expectation in the production department of further rejection. Knowing that the workmanship of their output will be checked by quality inspectors—and having previously experienced the humiliation of rejected work at inspection—production operations equipped with soldering irons touch up joints over and over before passing the units along to inspectors.

Touchup is more pervasive than generally recognized because so much of it takes place informally and almost invisibly. Careful but discrete study of hand-solder operators usually reveals that many (sometimes all) of their connections are touched up and reworked several times until the operator is satisfied with their appearance. This is perhaps the purest form of hidden factory because the value-added work is mixed with non-value-added and negative-value-added work, with all the effort appearing to the uneducated eye to be expended on behalf of value-added assembly.

The complete range of forces responsible for the needless touchup and rework as the product passes through the production department is not solely attributable to arbitrary or capricious behavior by quality inspectors. Although the regular rejection of dependable product by inspectors does confuse production operators, lack of clarity (and excessive extraneous information in the form of multiple grades of "acceptable" product) in quality standards is often even more important. Since quality standards are not fully understood even by inspectors, it should not be surprising that production personnel less schooled in the content of those standards are even less confident when assessing the quality of their department's output. When in doubt, the human inclination is always to attempt to make the product "better"—even if better is only a cosmetic concept.

Therefore, merely abolishing formal touchup departments will be insufficient on its own to eliminate the plant's non-value-added and negative-value-added work. Workmanship standards must be stripped to their essentials. Also, thorough education in those slimmer standards (and the futility of touchup) must be provided to all production personnel in addition to quality department employees. More information about the characteristics of effective quality standards can be found in Sec. 10.5.

10.3.3 Repair

With the exception of components that were dead on arrival at the plant, any repair activity reflects damage caused by in-plant handling and processes. Consequently, when repair is necessary, investigation and correction of the processes is generally appropriate as well. While methods do exist to minimize the chances of further damaging an assembly during repair, repair must always be recognized as evidence of failed process management and considered the approach of last resort. Although a repaired unit will work better with repairs than without (the unrepaired unit, after all, does not work at all), refurbished assemblies always suffer a greater field failure rate. Repaired units will never be as reliable as those that have never been exposed to soldering irons or desoldering equipment. Consequently, a busy repair department is a sure sign of higher warranty claims and customer complaints down the line.

10.3.4 Rework

Rework should not be used as a synonym for repair. Repair properly means replacement of defective components or mending of damaged circuit boards. Rework consists of changing incorrectly specified components, adding jumper wires, and otherwise carrying out activities that would not be necessary except for design, stores, and operator errors. Like repair, rework results in less reliable product and should be a serious quality concern. Although most rework can be prevented by scrapping the flawed circuit boards once the design error is recognized, it is not unusual for companies to try salvaging the boards through rework instead. Partly because the printed circuit boards themselves are complex and expensive and partly because of a culture that accepts rework as productive, some of the most flagrant examples of reworking circuit boards are found in the defense industry. The quality department should strongly oppose purchasing departments that order large quantities of circuit boards before the design has proved dependable.

10.3.5 Flux Removal

Until relatively recently, removal of flux residues after soldering was standard industry practice. But in the past 10 years—and particularly during the 1990s—substantial numbers of companies have adopted low-solids fluxes which are left on the assembly (giving rise to the terminology "no-clean fluxes"). Despite the proven reliability of no-clean fluxes even in exceptionally harsh environments such as automotive electronics, many quality departments continue to oppose the use of these new fluxes. This position should be reconsidered, taking into account the greater sophistication of modern flux chemistry.

10.3.5.1 Cleaning after Wave Soldering. Contact between the assembly and the flux is different under wave soldering than in other cases.[5] The difference is that wave soldering applies flux to the entire bottom of the assembly while other forms of soldering apply flux only to the individual discrete joint areas.

Depending on the type of flux employed, cleaning after soldering may actually be detrimental to the assembly's dependability even if the

[5]Drag soldering involves full coverage of the board surface with flux just as with wave soldering. This note aside, no explicit reference to drag soldering is contained in these pages. Among other liabilities, drag soldering subjects assemblies to excessive heat and has no place in modern electronics assembly.

flux does not belong to the low solids family. Rosin, for example, cures to form an excellent nonconductive barrier to moisture (i.e., a conformal coating). Recalling (from Sec. 7.9) that the effects of ionic contamination are greatly magnified by the presence of humidity, it follows that removing rosin only enables moisture to reach any ionic residues that lie below the rosin.[6]

Ionic matter poses a threat to reliability because it is conductive and reduces the insulation resistance in the area where it is present. Current leakage as a result of ionic residues can adversely affect the performance of the assembly. In extreme cases, the ionic residues reduce the resistance to current flow so greatly that a short-circuit condition results. The conductivity of the ionic residues and thus the extent to which they compromise dependability is determined by the flux activators.

There is widespread belief that organically activated (so-called halide-free) fluxes are preferable to fluxes containing inorganic activators.[7] This supposition involves serious traps for the unwary, however, as demonstrated by a flux based on citric acid that attracted considerable attention when introduced by Delco-Hughes in the early part of this decade; the flux easily deoxidized surfaces that could not be reduced by RA fluxes, reflecting the relatively low pH of citric acid relative to activators found in more traditional fluxes.

The primary motivation for flux removal is the belief by quality departments that visible flux indicates potential dependability degradation from ionic bridging and/or corrosion of the joint. In reality, the visible constituents of flux—such as rosin—are harmless and, as we have just seen, often protect the assembly from moisture that could cause failures. The activating chemicals—i.e., the ionic content—never constitute more than 1 percent of liquid flux and often make up only 0.25 to 0.5 percent of the total volume. Thus ionic residues—the only aspect of flux that should cause any concern—are invisible. Requiring removal of any visible flux residues is equivalent to requiring fumigation for termites in a building simply because wood is present.

---

[6]A common objection to leaving rosin on assemblies is the tendency of dust to collect on its sticky surface. In high heat and humidity, molds have been known to grow on rosin. Both of these results are possible. In practice, however, the rosin quickly hardens and supports neither dust nor organisms. The current trend toward lower solids fluxes further weakens the case for flux removal. Thus, even with the ban on CFC cleaners, rosin-based fluxes can and should continue to play a prominent role in electronics assembly formulated as "no-clean" fluxes.

[7]See, for example, Germany's DIN FS-W-32 standard, which suggests organically activated fluxes are universally equivalent to American RMA fluxes.

Admittedly, two quality problems can arise from failure to remove flux:

1. Flux residues may clog the heads of test contact pins, resulting in false failure readings. This becomes less onerous when the amount of testing is reduced in response to decreased failure rates accomplished by following scientific process management as recommended in Chap. 6. Additionally, test operations normally learn to recognize potentially false failure readings and adopt an efficient strategy for coping with such situations: Set the failed assembly aside while a few other assemblies are tested (which normally will clean the clogged test nail); then retest the failed unit. While the additional test clearly represents non-value-added work, the net benefit after accounting for savings earned by eliminating post-solder cleaning is highly positive.

2. Conformal coating may not adhere to flux residues. A possible solution: Try different combinations of conformal coatings and flux compositions to find a compatible combination. Where low solids (i.e., "no clean") fluxes are employed, the severity of this particular problem is generally but not always limited.

So eliminating post-solder cleaning is not entirely free of complications. Offsetting the new challenges, however, elimination of post-solder cleaning provides two very real benefits:

1. The most obvious benefit is that elimination of cleaning also eliminates the costs of cleaning agents, machine maintenance, and operating expenses such as electricity, deionization, and labor.

2. Less obvious but significant nonetheless, eliminating post-solder washing will sharply reduce or eliminate residues (generally white or brown in color) often noticeable after cleaning. Three types of chemical reactions typically cause residue problems, and only one relates to post-solder cleaning. The first cause of residues is a chemical reaction between the flux activators and the solder mask; this should not occur if the recommendations on flux selection (Sec. 7.9) are followed. The second cause is excessive heat causing oxidation of the flux; the heat is encountered as the assembly passes over the solder wave and is rarely a problem if the solder temperature is below 250°C (480°F). Plants that are required to obey MIL standards demanding minimum solder temperatures of 260°C (500°F) will find scant comfort in this information, just one more example of the costs imposed by the antiquated mentality still permeating DoD specifications. Finally, residues are frequently the

product of reactions between flux remnants and cleaning solutions. By eliminating post-solder cleaning, this problem—which tends to be the most troublesome of the three residue-producing reactions—also disappears.

10.3.5.2 Flux Removal after Other Soldering Processes. Facilities that perceive a need for washing assemblies after wave soldering generally believe that flux should also be removed after other forms of soldering. Actually, even less justification exists for requiring removal of flux after hand soldering than after wave soldering. Unlike wave solder, in which flux and its ionic content is spread across the entire circuit board, properly conducted hand-soldering operations apply flux only to individual joints. Unless the flux is applied too generously or carelessly so that it flows between neighboring conductive paths, the flux used for hand soldering cannot cause ionic bridges across insulating surfaces.

Attempts to remove fluxes by brushing or wiping a solvent across the surface of the circuit assembly actually undermine dependability. The operator merely spreads the activators over broad areas that otherwise would have remained perfectly free of ions. Any improvement in appearance can be attributed to removal of the anti-reoxidizing portion of the flux (such as rosin) that has previously been seen to have desirable properties as a de facto conformal coating.

Manual cleaning of assemblies may even increase the total ionic material on the assembly. Unless fresh brushes or cloths are used for each assembly, ionic residues from previously cleaned assemblies can be transported to other assemblies.

The observations that flux residues increase false failure readings by clogging contact pins in test fixtures and may be incompatible with certain conformal coatings apply equally to residues left by hand soldering.

10.3.6 Cleanliness Testing of Completed Assemblies

"Cleanliness" is rather misleading because it implies a search for such things as dirt, grease, or other filth. The actual purpose of cleanliness testing is to find the amount of ionic material (the activators in fluxes) left behind after the assembly has been washed.[8]

---

[8]There is, of course, no reason why ionic levels of uncleaned circuit assemblies cannot be tested as well. At the same time, it is difficult to conceive of a scenario under which there would be incentive to check unwashed assemblies.

Checking the ionic residues of assemblies has a certain inherent appeal. Even when restricting flux usage to RMA strength or milder, it would be comforting to know if potentially troublesome ionic residue conditions exist on an assembly. Unfortunately, the only available method for measuring ionic contamination—cleanliness testers[9]—cannot provide that information. Indeed, cleanliness testers constitute wasted resources because ionic matter left behind after washing will primarily be trapped in places that cannot be reached by the cleanliness tester's solvent.

The problem involves the relative surface tensions of flux, the washing solvent, and the solvent employed in the cleanliness tester. Flux has very low surface tension which becomes even lower when the flux is heated (as is the case in all soldering operations). This low surface tension allows some flux to flow into tight spaces, an action that is reinforced by capillary action.

In order to remove trapped flux residues, whether for cleaning itself or cleanliness testing, the solvent not only must have lower surface tension but must also exert a stronger attraction on the flux residues than the capillary forces holding those residues in place. All solvents have surface tension equal to or greater than that of flux and therefore are unable to penetrate narrow spaces where flux residues remain. Water has very high surface tension, and deionized water's surface tension is higher still. The probability of water washes removing flux residues from under components lying close to the circuit board (a common feature in pure or mixed technology[10] surface-mounted assemblies) is very low. Assemblies with surface-mounted components will almost certainly have flux residues remaining after cleaning.

Devices for testing cleanliness of assemblies use solvents—typically a mixture of deionized water and isopropanol—having surface tension greater than or equal to the flux. The assembly being checked is immersed in a tank filled with the mixture of deionized water and isopropanol. Any ionic matter taken into solution from the board will reduce the solution's electrical resistance. The cleanliness tester reads those changes and calculates *average* ionic cleanliness based on the surface area of the circuit board. Since the solution of deionized water and isopropanol has higher surface tension than flux, however, it is unrealistic to presume that the solution will reach and dissolve all ionic residues. For that reason, it is best to assume that:

---

[9]Two of the more common brand names for cleanliness testers in North America include the Ionograph from Alpha Metals and the Omegameter from Kester Solder, a division of Litton Industries.

[10]Sometimes referred to as "hybrid" assemblies, a terminology that invites confusion for anyone who has experience with thick- or thin-film hybrid circuits.

1. Whenever flux is applied to the assembly, some residues will remain behind no matter how thorough the washing process.
2. The cleanliness reading will underestimate the severity of the ionic residues.

Finally, it must be emphasized once again that the results of the cleanliness test are averaged for the submerged surface area of the assembly. However, the residues left on the assembly are not distributed equally; rather, they will be clustered in many small areas of dense contamination. Accordingly, assemblies that show acceptably low average readings may well have areas of heavily concentrated ionic matter that will cause failures under certain operating conditions.

Cleanliness testing does serve one useful function in plants that wash assemblies after soldering. By testing various batches of identical assembly types produced at intervals of several days, it can be seen whether the washing operation is well controlled. Since the cleanliness test basically reaches only those surfaces accessed by the post-solder wash, changes in the readings of the cleanliness test reflect changes in the ionic content of the wash solution. This information provides guidance on when the washing system requires maintenance.

10.4 Causes of Negative-Value-Added Work

Few of the activities identified here as either non-value-added or negative-value-added work will not have been questioned at one time or another by the quality professional. Most open-minded members of the electronics assembly industry have long debated the value of these activities. Why, therefore, are so many worthless and destructive practices still in place? Primary causes include:

- Intellectual inertia
- Lack of knowledge
- Excessive reliance on consultants
- Fear
- Inept designs
- Carelessness
- Expediency

10.4.1 Intellectual Inertia

The technology of products changes rapidly. Companies accept the inevitability of rapid evolution in product technologies, anticipating that components will become smaller and/or denser in very brief times. Inexplicably, however, the same companies have difficulty realizing that changes in component and assembly technology may require changes to processes.

Despite the obvious changes in component technologies in the nearly four decades since solid state technology began replacing vacuum tubes, many critical assembly operations and quality standards have been in stasis. We have just seen one example of this reluctance to change: the continued use of fluxes that require washing rather than taking advantage of no-clean fluxes to improve reliability and eliminate handling steps. Another example, found in virtually all assembly plants, is continued use of test operations long after process improvements have eliminated most of the failure modes that the test location in question was designed to identify. Instead of subjecting every circuit assembly to such a test, it would be much less expensive to combine that test with later functional tests and simply discard the assemblies that fail. (Admittedly, whether the assembly should be scrapped depends on the value of the assembly.) There is no point in improving process yields if quality evaluation expenses are not adjusted to take advantage of the improved reliability.

10.4.2 Lack of Knowledge

Today's components are more fragile than components in the past. Recognizing one danger to components, every sensible company has invested heavily in the prevention of electrostatic discharge damage. But components continue to be subjected to touchup, to scrubbing after manual soldering, and to other physical abuses that at times defy belief.

One of the most gruesome practices is solder-cut-solder (or, more commonly now, solder-cut). In this operation, a conveyor carries the assembly through reflow soldering (typically wave soldering), then over spinning rotary blades that trim protruding leads to the desired length.[11] After lead cutting, the conveyor may then transport the assembly through a second solder reflow step, or the exposed ends of the trimmed leads may be left unsoldered.

In earlier days, the deleterious effects of sawing off leads could easily

[11][Manko 1986], pp. 81–83.

be seen when glass diodes exploded; normally, after witnessing this type of component destruction a few times, production engineers realized the wisdom of trimming leads prior to soldering and put the cutting equipment in storage. Today, the damage occurs inside components where it is less visible (though no less real), and a new generation of process engineers who have never seen components explode when attacked by the cutting blade are happily resurrecting this disastrous practice.[12] Ironically, plants where solder-cut-solder is used spend vastly more money and time to test the abused assemblies than they save by not forming and crimping the leads prior to soldering. Since the leads must be trimmed to the proper length at some point in the assembly operation, solder-cut-solder can appear to qualify as value-added but is certainly negative-value-added by virtue of its destructive effects on components.

10.4.3 Excessive Reliance on Consultants

Paying for outside expertise is often the most cost-effective and fastest means of improving operations. In fact, companies that reject outside counsel rarely achieve levels of dependability and efficiency taken for granted by less xenophobic organizations. A relatively small investment in advice can return enormous dividends. Too often, however, consultants are hired for the wrong reasons, most often to justify actions that the department involved wants to take or to ensure that blame can be shifted if actions do not work out. Even when the decision to bring in consultants is taken for the right reasons, another problem presents itself: It is not always easy to distinguish good advice from lunacy. Over the years, the authors have spent a distressingly large amount of time attempting to revive clients of incompetent, careless, or unscrupulous consultants.

Some of the blame for the misfortunes of consulting clients must be laid at the feet of the clients themselves. Many are all too ready to accept without question advice that a moment's reflection would show to be ill-

---

[12]Some studies on the effects of cropping after soldering have concluded that the practice is harmless. These studies do not reflect the real world, however, because the blades were freshly sharpened and changed frequently. In real-life production environments, no one has time to change and sharpen blades every 4 or 5 minutes, which is the maximum amount of steady use that can reasonably be expected before the blade loses its edge and begins hacking rather than smoothly cutting the leads. Examination of the "trimmed" leads after this abuse reveals jagged edges. If the cropped leads are resoldered immediately after the cutting, the fresh coating of solder will cover the ends and conceal the evidence. However, resoldering cannot heal the fractured bonds inside the components.

founded. Again, the practice of solder-cut-solder (or solder-cut) serves as a good example. One of the most prominent North American authorities on soldering, Howard H. Manko, advocates solder-cut-solder (which he calls "inline cutting") for mass production in his popular book *Soldering Handbook for Printed Circuits and Surface Mounting*. "Cutting after soldering should not be permitted for high reliability work because of:

- Danger of cracked solder joints
- Exposed ferrous lead ends[13]
- Damage to components"

After this promising start, however, he adds: "For low cost assemblies, post-solder cutting is often used without resoldering. With the proper amount of care, the dangers listed above can be minimized or eliminated....In terms of the dangers outlined above, the inline cutting process with rotary blades is unique. It has the smallest mechanical damage impact and the revolving surface tends to smear some of the solder coating over the cut surface, giving it a limited protection."[14]

Manko's advice is far from clear. It is a classic equivalent statement. Even after careful scrutiny, Manko's message is easily interpreted as endorsement of solder-cut-solder provided rotary blades are employed. That interpretation has enticed many plants into adopting solder-cut-solder or solder-cut systems in the mistaken belief that there will be no meaningful compromise to the integrity of their products. In fact, of course, rotary blades normally lose their edge and require replacement within minutes, after which the operation endangers the integrity of the product.

10.4.4 Fear

Product failure, particularly in the customer's hands, is every plant's worst nightmare. Hoping to minimize failures, therefore, many companies subject assemblies to prodding, probing, fondling, banging, heat-

[13]Depending on the environment in which the assembly will be used, leaving the lead's core metal (often, but not always, ferrous-based) exposed rather than soldered can result in serious corrosion, primarily but not exclusively because of oxidation. One of the desirable properties of the tin-lead alloy used as solder in most of the electronics assembly industry is its resistance to oxidation. Soldered surfaces will ultimately become oxidized through to the core metal, but oxidation to this extent typically requires several decades rather than a few years in the case of an exposed lead core. For assemblies that will be used in low-humidity environments such as an office or home, exposed lead ends can be reliable for decades of use. But the salt air found in seaside areas will accelerate the rate of corrosion to a serious extent if the core metal is bared.

[14][Manko 1986], p. 82.

ing, cooling, and other bizarre abuse almost from the moment components arrive at the receiving door. The net effect is generally degradation of product dependability rather than enhanced reliability—and much needless cost. For a more complete assessment of the merits and demerits of test practices, see Chap. 11.

10.4.5 Inept Designs

Design engineers who provide artwork only when it is too late for any revisions to be made create enormous non-value-added and negative-value-added work in the plant. The blame cannot always be assigned solely to the design personnel, either; the mandated time for the product to move from drawing board to production may be unreasonably short. Companies that do not allow designers sufficient lead time for prototype builds and testing are equally guilty of causing non-value-added and negative-value-added work.

The rush to market with untried designs has forced large numbers of plants into chronic mods syndrome. Mods, as assembly modifications are widely known, once applied only to upgrading of units that had been serving in the field. Today, however, mods are commonly seen as standard operating procedure by plant personnel at the initial assembly stage. Jumper wires are run across boards. Circuit board tracks (or traces, depending on one's preference) are cut. Some plants even go so far as to rework tracks inside multilayer boards.[15] As previously noted, quality requires doing the job right the first time—and design is the very first time. Chapter 13 examines design issues in depth.

10.4.6 Carelessness

We never cease to be amazed by the number of plants that shove partially assembled units into racks, place them on foam rubber pads, and generally subject them to unspeakable abuse. Among other serious problems, this rough handling leads to misaligned and missing components which, in turn, mean post-reflow rework—and degraded dependability.

There is nothing mysterious about the effect of such careless handling on product quality: Dependability deteriorates sharply. The mystery is why quality-conscious individuals allow it to happen.

---

[15]Department of Defense training courses in repair even include techniques for repairing interior tracks. The practice is part of what they call "high rel" (i.e., high reliability) work. We've always believed DoD usage of "high rel" to be a non sequitur like "jumbo shrimp," "airplane food," or "government intelligence."

10.4.7 Expediency

The answer to the question raised in Sec. 10.4.6 is that bad practices survive because it is more convenient to live with those practices than work around them. Sloppy handling of assemblies is only one all-too-common sign of superficial quality management. Over the years, we have also: watched plants try to assemble systems using parts known to be unsolderable because no one was willing to accept responsibility for writing off obsolete inventory; try to get accurate placement of components requiring tolerances tighter than their equipment could achieve; employ temporary help because the wage rate is less than for permanent employees; bring in trailers to store defective output rather than shutting down the production line until the causes of those defects could be established and fixed; and on and on.

A perennial example of expediency in action can be found in the practice of assembling units before all the parts are available (i.e., "partial builds"). Partial builds require desoldering and resoldering with the inevitable consequence of compromised integrity and much higher costs. Alternatively, some form of solder resist may be applied to the area where the missing component should go; later, it is necessary to remove the solder resist and add the component by hand. No matter what technique is followed, partial builds always end up costing more and generally perform less reliably.

The driving force behind partial builds is the production department's desire to report output even when that output may not be in shippable form. But partially assembling units before all the parts are available accomplishes nothing aside from giving the production department a false sense of achievement and inflated performance reports. The products do not reach the customer any faster than if assembly had been suspended until all shortages have been filled. The labor content increases markedly without any compensating reductions in other resources used. And, of greatest concern to the quality department, partial builds increase probabilities of failure.

The partial build problem is widespread. Component shortages afflict almost every plant at one time or another. In recent years, as the costs of carrying inventory have figured more prominently in the industry's decision making, parts shortages have become steadily more common. ("Just in time" inventory management too often translates into "never in time.") While every company feels the same kind of financial pain when facing parts shortages, only a small percentage of plants respond to the challenge in the appropriate manner: by waiting until all components are in house before assembly begins.

Admittedly, refusing to begin assembly in the face of component shortages may mean that some or all plant personnel are left idle or must be sent home early, neither situation making life more pleasant for production management. The more prudent—albeit initially less comfortable—approach is to accept the short-term pain of idle workers. In exchange, very real long-term gains typically come about when support departments such as purchasing and stores realize that production management takes shortages seriously. A production manager who insists that all components be in-house before assembly begins will generally find that more senior managers suddenly become very aware of the need for tighter controls on parts management. The focus of executive annoyance switches from the innocent production department to production planning, purchasing, and stores, where it belongs. Conversely, production managers who take what appears to be the path of least resistance and carry out partial builds will find that they are constantly facing component shortages and unhappy executives.

Because of the detrimental effects on dependability, partial builds should be resisted by the quality management. However, quality departments traditionally have not recognized the need for their active involvement in eliminating the parts shortages bottlenecks. But whenever actions are taken that compromise dependability, it is the quality management's ethical and professional responsibility to intervene even though it means crossing traditional department lines.

10.5 Some Additional Recommended Changes

Clearly there are steps any quality department can take to reduce the cost of quality and increase dependability. Some of those steps, as this chapter has noted, entail becoming involved in matters that have traditionally been seen as out-of-bounds to quality personnel. We trust that the necessary changes in the quality department's priorities are obvious to the reader. The first remedial adjustments should be taken within the quality department itself, however. Revising the workmanship standards is the most urgent task.

Workmanship standards have historically been weak in two ways. First, the requirements too often contradict fundamental scientific principles. Second, the standards themselves are presented in ways that cause confusion and bad judgment. The workmanship standards will be positive rather than negative influences within the company if they are revised so that:

1. *Distinctions between "acceptable preferred," "acceptable," and "minimum acceptable" are banished.* Acceptability should always be based solely on whether the assembly conforms to the standards of dependability stated in Chap. 3. Therefore, a connection can possess only one of two possible states: Either it is dependable and therefore acceptable or it is not dependable and unacceptable. The range of "acceptable" conditions only invites destructive touchup and rework in misguided attempts to provide the customer with "better" (i.e., the "preferred") product.

2. *"Unacceptable" outcomes are based exclusively on dependability.* Appearances that do not relate to known failure modes cannot be considered unacceptable regardless of their cosmetic niceties. Cold, hard objectivity must prevail.

3. *Inspection manuals show only rejectable conditions.* This basically means only two soldering conditions: shorts and opens and perhaps a handful of mechanical issues along with wrong or reversed components. Extraneous "information" only clutters the user's mind and leads to improper interpretations of output quality.

4. *Workmanship manuals contain explanations of the reasons why conditions deemed unacceptable are faulty.* The reasons, it should go without saying, must be based on scientific consequences rather than "because we say so." This information is necessary for productive communication between quality and production personnel as well as meaningful quality evaluation.

5. *Standards distinguish between connections produced by machine soldering processes and those derived by touchup and rework.* Visual perfection obtained through touchup or rework is not the same as product dependability.

Above all, standards must be unambiguous and no deviation from those standards allowed. Any standard that is so unimportant it can be waived at times should be taken off the books completely. As Crosby said when criticizing "fitness for use," if a requirement is not enforced, either change the enforcement or eliminate the requirement.

10.6 Summary

While it is necessary for the quality department to verify the actions of other departments, there is room for improvement in the way we carry out our functions.

- Many quality practices in the electronics assembly industry neither improve quality nor promote efficiency.
- Costly activities are required by the quality department without scientific reason.

11 Test

11.1 Chapter Purpose

Test and its related activities, such as burn-in, can make valuable contributions to an electronics assembly company's quality if managed carefully and scientifically. Used to excess, however, test can easily become a negative-value-added function. This chapter examines the following issues:

- Origins of current attitudes toward test procedures
- Proper uses of test
- Improper uses of test
- The potential for test to damage product dependability
- How to manage test in the framework of quality optimization

11.2 An Introduction to Test Theory and Practices

Many quality issues in electronics assembly can be answered definitively and, as we have seen, with relatively little difficulty. Test, unfortunately, is not one of those issues. A handful of absolute statements—as will be shown—can be made about test. More often than not, however, the correct test strategy always involves some judicious reasoning by the quality management.

Although test is time-consuming and usually labor-intensive, most plants hold it in high esteem, and test figures prominently in their quality verification. The rationale for employing test is that faults can be fixed most cheaply when detected earliest. Further, the percentage of assemblies failing at in-circuit test is in general frighteningly high. An executive with one test equipment manufacturer claimed in 1995 that the normal pass rate at in-circuit test (generally the first test step involving board assemblies) is only between 70 and 80 percent.[1] That figure correlates well with our own observations in assembly plants around the world, although we have seen both considerably better and vastly worse results—on occasions in different plants owned by the same company.

The actions taken when a unit fails vary and are separate from the test activity. Reflecting the usual belief among quality personnel that a repaired unit is a dependable unit, a failed unit is typically repaired, returned to test, and—if it survives test—shipped to a customer. However, a small number of companies scrap failed units in the belief that a defective product can never be made fully reliable and the costs of a field failure far outweigh the money saved by refurbishing failed units. Of course, those companies that believe in scrapping rather than repairing must have exceptionally low test failure rates before they can turn their philosophy into a fiscally viable reality.

The subject of test as an investment has not been well documented in the literature about electronics assembly. Much has been written about the techniques of test, but very little attention has been given to the vital question: Does test represent good use of the company's quality budget? With test now representing as much as one-third of costs in electronics assembly,[2] a close look at the economics of test is essential to any company hoping to remain cost competitive.

The word "test" has a variety of meanings depending on the point in the operation where it occurs. Tests early in the product development cycle—including environmental qualification tests to determine how much abuse the product can withstand—validate that new designs work. Bare circuit boards are tested for electrical continuity. Tests are conducted to discover suitable materials, components, and production methods. Individual components may be tested for operating characteristics as they arrive from suppliers. Of course, test is found at the subassembly and final assembly points acting as validation of the product's integrity. And, perhaps most controversially, environmental stress screening (ESS) tests during production attempt to induce failures of weak parts.

---

[1]Gerd Lipski, director of global customer marketing for GenRad Inc., quoted in *Electronic Business Buyer*, July 1995, p. 93.

[2]Scheiber, Stephen F. "System Houses Pursue Many Paths to Ease Test," *Electronic Business Buyer*, July 1995, p. 87.

This chapter focuses primarily on the tests of subassemblies and final assemblies—which collectively can be designated "output verification tests"—and, to a lesser degree, ESS. Discussions of the other forms of test are located in the chapters dealing with the subject to which they are most closely related; design verification tests, for example, are considered in Chap. 13.

11.2.1 Nonelectronic Issues Not Covered

Quality in parts that are not purely electronic is not within the scope of this book for reasons found in Sec. 3.3.4. The "quality" of mechanical components may well be positively influenced by inspection and other practices criticized here as ineffective or harmful in an electronics context. Just as the meaning of "quality" varies from industry to industry, the most satisfactory methods for ensuring that quality also vary. The recommendations found here are absolutely applicable to electronics but should not be interpreted as suggestions for how quality should be managed with respect to any parts that are not electronic.

11.2.2 Test and Inspection

Test has long been perceived as a form of inspection. The view is so prevalent that discussion of test in [Juran 1988] is limited to a few vaguely worded pages in a chapter titled "Inspection and Test." Feigenbaum's contribution, also described as "Inspection and Test," consists of fewer than two dozen words. Other prominent figures in the quality world are strangely silent concerning this extremely costly quality verification activity.

Test and inspection may be distantly related, but considering them in tandem creates some thorny management problems. Although test and inspection share a common objective—keeping defective product out of customer' hands—the basic attributes are distinctly different. Some key distinctions include:

1. *Inspection is visual; test is performance.* Inspection assesses the product according to visual criteria which may but need not be cosmetic. Checking solder joint appearance is a classic case of inspection as a cosmetic activity while identifying missing or reversed polarity parts is wholly objective. The "vision" is not restricted to human eyesight; automated inspection which uses machines to find problems is based on electronic imaging.

Test spells out its findings numerically. The numerical information may be provided in strictly binary "pass/fail" format, in graphical format such as a plot of oscillations over time, in tables, and so forth. Test is totally objective: either the unit performs within specification (passes) or it does not (fails).

2. *Inspection does not add value, but test may.* As Chap. 9 discussed, inspection is never a means of improving quality. The vital characteristics (with the notable exceptions of missing or reversed parts, solder shorts, and opens) cannot be seen. Those that can be seen are likely to be overlooked as the inspector's eyes tire—which happens within minutes of starting inspection. The critical information is located inside the components where it cannot be detected by human or machine inspection. The most common result of inspection is rejection of perfectly dependable product that is then made undependable through touchup and rework.

Test does look inside the components, at least to the extent of measuring their performance at the moment of test. Some failures that could not be seen will show up in test. Stacked tolerances that cause performance to vary beyond the acceptance limits also show up. And, of course, some products require tuning of components, a task that is inseparable from the test itself.

3. *Test can detect the same items found by inspection.* The point of both operations is to identify those units that have failed. (Some test activities are intended to detect impending failures as well. The effectiveness of those activities will be considered later in this chapter.) The number of dependability problems that can be caught through inspection but not by test is either insignificantly small or nonexistent. In most plants, inspection is seen as an inexpensive method of catching problems before they reach test and slow down throughput. As we have seen, however, inspection is far from inexpensive because it pushes product back into touchup and rework. Touchup and rework are costly in their own right but also increase the subsequent failure rate at test.

4. *Test takes much more time.* Some test stages require very little time; a test for shorts and opens, for example, can typically be conducted as quickly as any inspection. As the assembly builds up through the coupling of subassemblies, the complexity of the test procedures increases. By functional test and final tests, the time involved can be very meaningful.

The actual amount of time for any given test is predictable within a very narrow range. Inspection, on the other hand, expands to fill the amount of time available. In the misguided belief that they are helping their company by being more diligent, some inspectors spend enormous amounts of time studying each unit. If the inspectors are paced by an assembly line, the time spent on each unit will be more controlled; but the extent of that control can be undermined if the company takes seriously inspectors' complaints that there is not enough time available to adequately review the units. The number of inspectors located along a subassembly production line can reach astoundingly high numbers.

The distinction between inspection and test should be apparent. Inspection is subjective and largely unquantifiable. It involves unpredictable and variable amounts of time and leads to serious damage being inflicted. In contrast, test is objective and quantifiable. The amount of time and expense for each test pass is highly predictable, and the activity is therefore readily managed. Most significantly, test catches faults that can never be detected by inspection.

For all those reasons, regarding test as just another form of inspection is a serious mistake. Test is a discrete, distinct operation that stands apart from other plant activities.

11.3 Test

The purpose of test is to identify defective products before they can reach the customer. The defective units can then be repaired or scrapped, depending on the company's quality philosophy.

11.3.1 Levels of Test

Normally several levels of test are employed, with the total number depending on the characteristics of the product and the company. Four types of test are most common, beginning with the individual circuit board assemblies and extending through to the completed product immediately prior to shipping.

11.3.1.1 In-Circuit Test. The first test during assembly is generally an in-circuit test which evaluates specific circuit characteristics: shorts, opens, polarity of capacitors and diodes, passive component integrity, and values. Complex active components such as microprocessors and programmable devices can also be functionally tested at this point.

The in-circuit test apparatus generally consists of spring-loaded pins making up a bed of nails.[3] The arrangement of the bed of nails typically requires some customization for families of circuit assemblies. Until recently, test equipment vendors and manufacturers of complex active components could supply bed-of-nails test pattern sets for most components. Today, however, the number and complexity of device types is so great and life cycles so brief that ready-made test patterns do not exist for many components.

A particularly attractive feature of this test is that the most common failures and performance aberrations can be determined in one brief test. Also, failure reports typically identify the exact location of the problem. The disadvantages are that:

- Test personnel (from test engineers to operators of the test equipment) expand the company's labor costs.
- A "brief" test for one unit when aggregated over all output usually means a significant slowdown in throughput.
- The purchase price per tester often amounts to several hundred thousand dollars.
- Each specific circuit board assembly (or family of assemblies) requires a specialized test program.
- High-density assemblies may lack sufficient space for test pins to reach the circuit board's test pads without damaging neighboring circuitry and components.
- Bed-of-nails testers require that test points be incorporated into the printed circuit board, but greater numbers of test points can increase electromagnetic interference to levels that are unacceptable in some applications.
- Test pins must be cleaned and aligned regularly.
- False failures are not uncommon; causes can range from the obvious (e.g., dirt on test pins preventing contact with the test pads, misalignment of test pins) to the esoteric (if test is carried out before com-

[3]Other test technologies that do not involve contact with test nodes—such as boundary-scan assessments—are still in their infancy but rapidly growing in importance. By early in the next decade, 50 percent or more of all semiconductors are expected to contain boundary-scan circuitry and so will not require contact-type testing.

ponents have cooled after soldering, the values of some components may differ substantially from the values they possess at normal operating temperatures).[4]

- Some failure modes—most seriously, unsoldered connections ("solder skips") on surface-mounted components—may not be detected.

Most important from a quality management perspective, faults normally found at this stage are all preventable by scientific process control. Thus, the number of failures identified at in-circuit test varies inversely with the effectiveness of the plant's process management.

The failure modes found at in-circuit test were historically dominated by shorts and opens. During the 1980s, manufacturing process improvements greatly reduced the incidence of soldering-related problems so that component failures increased in significance (though the absolute number of all faults declined). With the increasing use of fine-pitch (25-mil and narrower) surface mount components, the incidence of shorts and opens has once again become dominant. Today, in plants where fine-pitch surface-mount components are common, the typical failure mode Pareto analysis would show solder shorts and opens accounting for about 80 percent of the faults, component failures making up between 10 and 12 percent, and non-solder shorts and opens constituting the rest.

11.3.1.2 Functional Test. Either in addition to or in place of in-circuit test, individual circuit board assemblies may be subjected to a more sophisticated and thorough evaluation known as functional test. Typically employed prior to final assembly, functional test simulates the ultimate working conditions that the circuit board assembly will face. At this level, it is possible and usual to carry out a full operational check of the assembly by measuring and recording its performance against specifications.

The advantage of functional testing over in-circuit test is the ability to discover faults generated by stacked tolerances (i.e., accumulated variances in performance of several interrelated parts). Disadvantages include:

---

[4]This form of false failure demonstrates the necessity of carefully planning not just the type of test but also the location and environmental conditions of the test. Where component values can be temporarily altered by transient process-related forces (such as heat from soldering), logic dictates that test not be conducted until the unit has returned to normal operating environmental conditions. While this requirement would seem readily apparent, it is not unusual to find in-circuit test placed immediately after wave soldering so that the output is still at elevated temperature during the test cycle. In some cases, air jets or other forms of rapid cooling are employed even though abrupt changes in component temperature should always be avoided.

- Competent test procedures require considerable skill and experience on the part of test personnel.
- The diagnostics phase of the test is time-consuming.
- The failure messages require interpretation (whereas in-circuit test points unambiguously to specific components and board sites).
- Costs for each test unit are between two and three times those of in-circuit testers.
- Since each test cycle is several times longer than an in-circuit test, more functional testers will be required to handle the same volume of output (assuming that the in-circuit testers operate at more than half capacity).
- The test programs for each circuit board assembly can be extremely complex and require several months to write—a problem that is increasingly acute as the production life for products shrinks rapidly.

11.3.1.3 Complete Unit Test. After two or more printed circuit assemblies are electrically connected, test is used to determine how they perform as an integrated unit. Depending on the complexity of the ultimate product, the combined printed circuit assemblies may be the final products themselves or subassemblies—often referred to as line replaceable units (LRUs) by defense contractors—that will be integrated into a more substantial end product.

Adjustment ("trimming") of some circuit values may be necessary at this stage, and many of the attributes checked during functional test are commonly repeated here as well.

For "single box" products, this may very well be the final test stage before the unit is packed and shipped to the customer. In such cases, the manufacturer often terms the complete unit test a "final test."

11.3.1.4 Systems Test. For large and complex systems consisting of several interconnected units or boxes, the complete system will normally be exercised before shipping. Less complex systems (the "single box" systems noted in the previous section) may be run through a systems test as well as the complete unit test. Some plants refer to this stage as "final test," but the term is not used consistently throughout the industry.

The in-depth functional tests of individual circuits are not repeated here, but the system is normally required to operate under power for several hours or even several days. More complex tests involving external testing equipment cycle the system through a range of operating functions and conditions. In the latter case, extensive data may be collected on the operating characteristics of critical functions; as an example, the navigational performance of inertial navigation systems undergoes more careful scrutiny than less critical circuits.

Built-in system test (BIST, often abbreviated to built-in test, BIT) runs the system through test routines incorporated in the system's circuitry. Some BIT routines—particularly where defense equipment is involved—are quite elaborate and sophisticated; a radar system's BIT routine, for example, may develop synthetic targets and automatically exercise most of the receiving, computing, and tracking circuitry, including displays.

The nature of the tests varies among industry sectors, companies, and even site to site in the same company. Not all the test stages are carried out in every plant, and some plants employ additional test tactics described later in this chapter. In some instances—notably including defense electronics and civil avionics—the test procedures may be dictated by the customer and evidence of satisfactory performance (typically a printout by the test equipment) required.

11.4 Burn-in

Between 1956 and 1958, the Department of Defense Advisory Group on Reliability of Electronics Equipment (AGREE) developed a set of recommended test procedures by which new product designs could be qualified before release to production. The AGREE report of 1958[5] placed heavy emphasis on burn-in of components, board assemblies, and subassemblies of large systems. The burn-in era had officially begun.

11.4.1 The AGREE Pedigree

AGREE is virtually unknown outside military contractors—DoD contractors in the United States and MoD contractors in the United Kingdom—and rapidly lapsing into obscurity even within the defense electronics sector. It is important to keep AGREE's memory alive, however, because our understanding of the thought processes that originally led to burn-in remains intact only so long as our awareness of

[5]AGREE, Advisory Group on Reliability of Electronic Equipment (1958), *Reliability of Military Electronic Equipment,* Office of Assistant Secretary of Defense, Washington, D.C

AGREE. In particular, the AGREE specifications were never intended to be applied to units that would be put into field service. The AGREE concept of burn-in was to qualify parts and designs in an era when component dependability was highly suspect.

11.4.2 Definition of Burn-in

The meaning of burn-in is not homogeneous throughout the electronics assembly industry. In this book, however, burn-in is intended to mean operation of a subassembly or finished assembly for a period of time without application of thermal, mechanical, or electrical stresses beyond those that the product can expect to experience regularly in normal use by the customer. Often, operating tests conducted at temperatures above those that the product should encounter in normal use are referred to as "burn-in." In our minds, however, tests conducted at abnormally high or low temperatures fall into the category of environmental stress screening described in Sec. 11.5. Some of the sources cited in this discussion of burn-in clearly have in mind operation of the unit at temperatures beyond the rated environmental level; it is important for the reader to recognize the temperatures implied in those statements.

11.4.3 The Proliferation of Burn-in

Members of AGREE did not foresee burn-in being applied to shipping units. Over the years, however, burn-in mutated to the point where many companies now require burn-in of every unit before delivery to the customer. Even where the manufacturer would like to bypass burn-in, some customers require that all units be burned in prior to delivery. Computer assembly companies, for example, are notorious for demanding burn-in of subassemblies by their suppliers.

Burn-in has taken on a life of its own. Some high-volume plants devote as much floor space to their burn-in chambers as they use for all manufacturing operations combined. And the cost of those burn-in chambers—from capital cost to operating cost to labor—in a few cases approaches the total manufacturing cost (excluding materials). The reliance on burn-in can be taken to ridiculous extremes. For example, during preparation of this book, the mother board of our home office central computer failed, causing chaos in work flow. This was clearly a situation where every minute of downtime was hideously expensive to us. Despite the clear need for speedy replacement, restoration of the system was delayed by a day because the manufacturer would not release a replacement until it had been burned in for 24 hours.[6] We noted that the first (and now deceased) mother board had been burned in to no avail and that we were more willing to risk failure of the replacement than to endure an extra day of system downtime—but the quality manager could not be persuaded.

11.4.4 Advantages of Burn-in

Some components received by the assembly company may be inherently weak and likely to fail soon after being put into service by the customer. Running the unit for some hours or days provides an opportunity for the weak components to fail in the plant and be replaced by healthy parts before the system ships to the customer. That, at least, is the hypothesis, although the evidence does not support the assumption well.

11.4.5 Disadvantages of Burn-in

- The extra handling required to place circuit assemblies in the burn-in chambers and remove them provides additional opportunities for damage to healthy units.
- The burn-in chambers are expensive, and the cost of electricity to keep them running can be meaningful.
- Operators are required to load and unload the units as well as to monitor the test results.
- Burn-in is a bottleneck operation; delivery of product is delayed.
- Setting the duration of the burn-in cycle is necessarily an arbitrary decision.
- There is no convincing evidence—although anecdotal and statistically meaningless reports abound—that burn-in reduces the rate of field failures.

---

[6]The lack of an in-stock replacement may seem strange except to anyone familiar with the decline in customer service that manufacturing companies have accepted in the cause of Just In Time inventory.

- Most failures revealed by burn-in reflect poor workmanship in the assembly plant and should therefore be preventable by process modifications.

11.4.6 Does Burn-in Help?

Although thousands of experiments have been run and massive numbers of papers published about the effect of burn-in on field performance, the work with respect to electronics assemblies has been disappointing. Too often, the experimenters and authors fell into the trap of attempting to demonstrate rigorous arithmetic distributions of failures over time, failing to recognize that the numbers and timings of failures are largely functions of the specific assembly plant processes. In many cases, the amount of negative-value-added work performed varies widely and unpredictably from day to day and lot to lot, with the result that the fallout during burn-in will also be unpredictably variable.

Based on our experiences, we believe that burn-in is more likely to reduce than enhance dependability, although the consequences should not be of great import if the test is restricted to pure (i.e., without elevated temperatures) burn-in. The effect of "burn-in" with elevated temperatures—which should more properly be considered a form of environmental stress screening—is almost surely detrimental to dependability. Moreover, the risks associated with burn-in must be substantially greater as submicron devices become increasingly common.[7]

The best analysis of burn-in effectiveness has been performed by manufacturers of integrated circuits and other forms of microprocessors. A highly regarded paper[8] from the Engineering Academy of Denmark summarized the evolution of attitudes within the semiconductor manufacturing industry about the desirability of component burn-in. Among the points raised are:

- Semiconductor fabrication processes have improved so markedly that burn-in of "mature semiconductor components" is a waste of resources. "All-in-all, the consensus as regards burn-in of mature technology components from reputable vendors is: *don't do it,* it is just not effective."[9]

---

[7]Device technology of less than 0.3μ is relatively common in leading-edge components of 1996.

[8]Jensen, Finn. "Component Burn-in: The Changing Attitude," in Christou, A., and Unger, B. A. (eds.), 1990. *Semiconductor Device Reliability*. Netherlands: Kluwer Academic Publishers, pp. 97–106.

[9]p. 100; emphasis reprinted from the original publication.

- Burn-in is a negative-value-added activity. "More likely, the situation will be such that the *burn-in process itself is generating severe weaknesses that fail components early.*"[10]
- For "state-of-the-art devices," burn-in will be effective only in flagging devices that are so inherently flawed that failures will occur "using the burn-in times and temperatures commonly favoured in industry still, that is typically 24-168 hour [sic] at temperatures in the range 125-155 degrees centigrade."[11] The component manufacturer may wish to conduct such component validation burn-in on preliminary batches of new designs simply to ensure the absence of an inherent design flaw. However, the likelihood of such design flaws in major semiconductor manufacturing companies today is small (which is not the same as saying that the component design is consistent with meeting the performance specifications—an altogether separate issue from hard failures during burn-in).

Two years prior to publication of the Danish work, the U.S. Institute of Environmental Sciences published results of a study by its Environmental Stress Screening of Electronic Hardware parts committee. The report concluded: "The participating OEMs retemperature cycled more than 70,000 devices...rejects were due mainly to handling damage and electrical overstress. This study does not support retemperature cycling of military I.C.s, although retemperature cycling of large die I.C.s may be beneficial."[12]

Sammy G. Shina nicely summarizes not just the unlikeliness that burn-in will reduce subsequent field failures but also the clash in attitudes between typical production and quality departments over the efficacy of burn-in. "There are many disagreements between the manufacturing and quality departments of companies on the effectiveness of burn-in," Shina writes.[13] "The quality department will point to continuing failures during burn-in as proof that it is needed, while the manufacturing department will experiment with shipping products with and without burn-in and collect field failure data that will show that burn-

[10]p. 100. Again, the emphasis is from the original publication. For discussion of the dangers of burning in "mature" components, see Unger, B.A., "Early Life Failures," *Quality and Reliability Engineering International,* vol. 4, no. 1 (1988), pp. 27–34.

[11]p. 102. The upper end of the temperature range described by Jensen as "commonly favoured" is much greater than customary. The temperature of 125°C (for 168 hours) is the usual MIL STD—and even that is much too high for most devices, in our opinion.

[12]Institute of Environmental Sciences, "Integrated Circuit Screening Report." November 1988.

[13][Shina 1991], pp. 260–261.

in does not statistically influence the field failure rate. The argument is made more confusing since the clever use of statistics can bolster either of these conflicting arguments."

Interestingly, the most telling comments about the dubious value of burn-in can be found in the *Reliability Design Handbook* of the Rome Air Development Center (1976), a document derived from MIL-HDBK-217B. "Many manufacturers provide a 'burn-in' period for their product, prior to delivery, which helps eliminate a high portion of the initial failures and assists in establishing a high level of operational reliability," the RADC handbook says.[14] "Examples of early failures are:

- Poor welds or seals
- Poor solder joints
- Poor connections
- Dirt or contamination on surfaces or in materials
- Chemical impurities in metal or insulation
- Voids, cracks, thin spots in insulation or protective coatings
- Incorrect positioning of parts"

The reader will doubtless have noticed what is missing from the list: anything that could be construed as a component problem. The entire list consists of workmanship flaws that should have been prevented by better process management and/or detected much earlier in the quality cycle than during burn-in. The RADC handbook is now more than 20 years old, however, and components have changed substantially. If RADC were compiling its list of early failure modes during burn-in today, it would almost certainly include component failures near the top of the list. But the addition of component failures to the problem list does not alter the list's basic character; the components fail almost exclusively because of overheating during assembly (as seen in Chap. 7), and thus the list continues to be composed of preventable workmanship errors.

11.5 Environmental Stress Screening (ESS)

For many quality practitioners, burn-in at temperatures and power variations similar to those the unit would experience in the field is not con-

[14]RADC, *Reliability Design Handbook* (1976), p. 6.

sidered adequate test of dependability. They advocate much more aggressive testing known as environmental stress screening (ESS). One concise statement describes both the general motive for using ESS, the basic technique, and the result: "Environmental Stress Screening (ESS) can eliminate some problems by subjecting electronic assemblies to extreme conditions and weeding out resulting failures."[15] The motive is questionable, but ESS certainly does subject assemblies to extreme—even abusive—conditions. And some units do fail.

11.5.1 ESS Activities

Exactly what constitutes the "extreme conditions" of ESS depends on who defines the company's test procedures. No single concept of ESS exists; its meaning varies substantially among companies. Common elements of ESS include thermal cycling, power variations, and vibration testing. Some plants also incorporate elevated humidity, mechanical shock, and other quite eccentric forms of punishment in their ESS. Units that survive this brutality are deemed reliable while failed units are typically sent back for repair—after which the survival tests begin again.

11.5.2 ESS vs. Burn-in

Some companies regard burn-in as a part of ESS while others do not. Other companies subject products to extremes of temperature and power cycling that should more properly be considered ESS yet call the operation burn-in.

In our minds, burn-in should be considered separate from ESS if the system is burned in at normal power under normal operating environment conditions. When "burn-in" pushes the unit beyond its regular operating parameters, the border into ESS is crossed. Therefore, when reference is made to elevated or cycled temperatures and/or power variations during "burn-in," the discussion is not actually about burn-in but rather about ESS.

11.5.3 AGREE and ESS

Like burn-in, ESS first gained widespread attention in the AGREE report of 1958. As with the other elements of AGREE, the original intent

---

[15]Scheiber, Stephen F. "System Houses Pursue Many Paths to Ease Test," *Electronic Business Buyer*, July 1995, p. 88.

was to assess durability of new designs and, occasionally, revalidate manufacturing processes through destructive analysis of production units.

In the context of validating designs—i.e., used as intended by AGREE—a strong case can be made for ESS, today much more so than when the AGREE report was published in 1958. Pushing prototype units beyond the worst conditions they would experience when used by customers according to specifications can reveal patterns of weakness that otherwise would not be detected until patterns became apparent in field failures. As design cycle time grows increasingly shorter, ESS of prototype units guards against catastrophes in waiting. (Ironically, although shorter design cycles increase the need for prototypes, the trend is away from pre-production builds. This serious issue is examined in Chap. 13.)

ESS was intended for design and process qualification only. The authors of AGREE did not anticipate that units surviving such ill-treatment would be sold to customers. Before long, however, forms of ESS became standard practice for evaluating units intended for customers. Through some bizarre distortion of good intentions, abusing product before it reached the customer became accepted as a way of ensuring better field performance.

11.5.4 ESS and Test Failures

Burn-in conceivably has merit under certain combinations of component technologies, product complexity, and consequences of field failure—especially while kinks are being worked out of unfamiliar processes and newer designs. On the other hand, any rational reason why ESS should be anything other than a costly exercise in degradation of dependability is certainly not clear.

If the success of ESS is measured according to whether failures are induced, then ESS is eminently successful. Failures occur regularly and often. At the same time, it seems obvious that pushing a system beyond its limits will cause failures—even if the failures occur in components that, in the absence of ESS, would have served dependably long after the customer lost interest in, or need for, the product.

Two observations about ESS should be noted:

1. ESS performed on units to be sold is analogous to auto manufacturers driving every new car for several thousand miles before delivering it to the customer; at that point, the customer would be buying a used car, not a new car. Some plants subject units destined for customers to such serious abuse that an equally good case could be made for drop-

ping units off the plant roof and shipping any units that survive. Rigorous ESS can be justified for destructive analysis of new designs—it is impossible for us to accept that a unit which survives being run over by a truck is more dependable for the experience.

2. Quality cannot be inspected or tested in. Quality must be *built* in. Therefore, ESS—which is an attempt to instill quality after the fact—violates one of the bedrock virtues of quality assurance.

11.5.5 One Company's Use of ESS

In 1990, a test engineer with power supply manufacturer (and subsequent Malcolm Baldrige National Quality Award winner) Zytec reported on that company's use of ESS during design and production.[16] The design procedures involved vibration, thermal cycling above and below maximum operating specifications, and electrical overstressing with internal protection circuits closed down; these tests of new designs are consistent with principles outlined by AGREE.

Zytec also subjected 100 percent of its output of switching power supplies to burn-in and all output of several models of power supply to vibration testing prior to shipment to the customer.

The vibration screen preceded burn-in. While powered up and connected to a measurement device (described in the report as a "black box"), each unit was vibrated over three axes at six *g* rms for 4 minutes. Any unit whose output drifted outside the specified limits was rejected. Not surprisingly, considering the mechanical nature of the test, "Vibration failures consist mostly of workmanship/mechanical defects."[17] Roughly 95 percent of units survived the vibration test.

The subsequent burn-in test conformed to our definition of "pure" burn-in. That is, the temperature and power inputs did not exceed the unit's operating specifications. With temperature of the burn-in chamber maintained "within 2°C of the max operating temperature of the supplies,"[18] the units were tested for "6 to 48 hours dependent on the time to failure data." The test itself consisted of cycling the unit hourly "with a 90 percent duty cycle." According to the report, "typical process

[16]Dalland, Donald. "Use of Environmental Stress Screening in Development and Manufacturing of Switching Power Supplies," *Institute of Environmental Sciences Proceedin* 1990, pp. 65–69.

[17]p. 69.

[18]p. 68.

yields in burn are 99 percent," which possibly reflects good product integrity or reinforces Jensen's contention that the probability of revealing component failures today is negligible even with temperatures as high as 155°C.

What did the ESS accomplish for Zytec? The company's report concluded that "Used in manufacturing, ESS can be...an effective method to identify weak areas in components, workmanship and design. Eliminating the defects that the ESS identifies can provide higher internal yields, lower cycle time, higher customer quality and reliability, and most importantly, higher customer satisfaction." Unfortunately, however, the report did not explain how defects were "eliminated" except to indicate that failed units were reworked before shipping to customers. No indication can be found in the paper that the company obtained useful information about causes of failure that it then was able to use in refining the processes so the faults would not recur.

All told, the Zytec report provides little or no useful data. Despite the company's belief that vibration and burn-in screens improved the reliability of product delivered to customers, there is no data on field performance to support any such conclusion. The report is not at all helpful in establishing that ESS increases or decreases dependability but is no less satisfying than any of the other studies published over the last decade. The case for ESS remains to be proved.

11.6 Choosing a Test Strategy

Test activities result in numbers (data). Some of those numbers include:

- The amount of electrical, thermal, and mechanical stress required to generate failure in a product
- The percentage of units free of failures—including shorts, opens, dead components, and reversed components—at various stages of production
- The operating characteristics of individual components and assemblies under varying environmental conditions

Numbers are useful, even essential, for scientific management of quality. At the same time, numbers are dangerous when they convey an unwarranted sense of science and control. In developing a test strategy, it must always be kept in mind that any unit of a product can be caused to fail if subjected to sufficient stresses. Assessing the effectiveness of test procedures by the number of failures produced—like measuring the productivity of inspectors and touchup operators by the number of defects, real or imagined, rejected—generally ensures development and implementation of steadily more abusive testing. When developing a test strategy, it helps to stop occasionally and remind ourselves that the purpose of test is to prevent failures from slipping through to the customer. Forcing dependable product to break down should never be interpreted as a measure of success.

11.6.1 Positive and Negative Tests

As was seen earlier in this chapter, two very different types of tests exist:

1. Tests that determine whether the unit works properly in its anticipated operating environment
2. Tests that push the unit significantly beyond normal operating conditions in the hope that undependable components can be identified and prevented from reaching customers

Tests in the first category—including in-circuit, functional, and complete unit tests together with "pure" burn-in—are fairly safe and may therefore be considered "positive tests." Assessment of their role in the company's quality strategy should be conducted on the basis of test costs vs. benefits received. Guidelines for carrying out such cost-benefit analysis are found later in this chapter.

Tests in the second category—the ESS family, including burn-in at elevated temperatures—can and frequently do degrade product dependability. It is reasonable to designate such activities "negative tests." There is no doubt that many failures found through ESS test methods occur only because of the stresses subjected by testing. Of course, advocates of ESS point out that encouraging failures is the precise objective of the exercise; far from regarding inducement of failures as undesirable, they see failures as a welcome outcome. In our opinion, however, the ESS disciples overlook a simple yet profoundly important corollary: ESS capable of inducing premature failures at test is equally capable of reducing the life expectancy of other components and thus increasing the number of field failures during the reasonable usage life of the product.

11.6.2 ESS in Design Qualification

ESS (and environmental screening, which determines the suitability of the product for operation in specified conditions) can be shown to have positive effects in qualification of new designs as well as components with which the company has little experience. Designs involving technologies with which the design personnel have a history of successful applications should require much less qualification testing. The amount of ESS testing employed in design qualification should be determined using the same cost-benefit principles suitable for appraising when and how much to test production output without applying artificial stresses.

11.6.3 Costs and Benefits of Test

The underlying assumption of the decision to employ test is that some units will fail despite the best efforts of the production department. The company therefore spends money on test in the expectation that money will be saved by preventing faulty units from reaching the customer. Further, the least expensive point in the product's life—other than in the design phase—at which to detect failures is the assembled circuit board level.

11.6.3.1 Cost of Failure in Relation to Progress through Assembly. The cost of diagnosing and repairing faults rises sharply as the product progresses through the assembly cycle. One source[19] contends that it is 100 times more expensive to detect and repair a fault at final test compared with the cost if the fault is detected at in-circuit test. The same source also states that the cost of a field failure is 10 times greater than the same failure found at final test. The cost of field failures does not include customer dissatisfaction, which can be very hard or even impossible to quantify. Like most figures cited to compare costs of failure at different points in manufacturer and use of the product, these numbers are extremely arbitrary and poorly documented. It is certainly true that cost is greater the later in the production cycle failure is found, but it is unlikely that the increase in costs occurs in neat multiples. Further, the amount by which costs of finding and repairing failures increases will vary with the

[19][Bateson 1985], p. 37.

value and complexity of the product; repairing a unit that has been conformal coated will surely be many times more costly than repair of a unit that is not conformal coated.

11.6.3.2 Assessing the Relative Benefits of Test. The decisions of when and how much to test should be based on careful analysis of costs and benefits associated with the testing. The company equipped with complete information about costs and benefits of test can readily determine whether a given test procedure can be justified by the economic criteria. However, exact knowledge of those costs and benefits is highly unlikely. The reasons why accurate quantification of costs and benefits is unlikely include:

1. The possibility that test itself is responsible for failure of inherently dependable parts. This possibility is very high where ESS testing is employed.

2. Repair of failures found at test may result in units that can pass test but will subsequently experience a high field failure rate. The question must be asked: Can repair itself reduce product dependability? A small number of companies have invested in bar-coding systems to answer this question by tracking dependability of each unit from first test onward. Those companies have found that units requiring repair before reaching the customer also fail much more frequently in the field in comparison to units that have never required repair. Their findings indicate that net benefits of locating failures through test are overstated by failure to take account of increased costs for warranty service. Detailed tracking of subassemblies, LRUs, and complete systems of inertial navigation systems for Tornado aircraft revealed that items causing problems during assembly and test continued to give trouble in the field despite apparently satisfactory rectification prior to shipping.

3. Lack of a satisfactory method for measuring the value of customer dissatisfaction in the event of a field failure. Quality departments frequently estimate those costs, but the accuracy of the estimates cannot be proved.

4. The failures being caught at test might be preventable by inexpensive modifications to manufacturing processes. Since tracking down failures after production is typically much more expensive than prevention through scientific process management, the net benefits of test are generally overstated.

5. False failure readings and inabilities of repair technicians to identify the failed component from the test equipment readings lead to

replacement of dependable parts. Technicians commonly replace many components in a circuit before finding the device responsible for the failure.

For these reasons, the benefits of test have frequently been exaggerated. More reliance has been placed on test than accurate cost-benefit analysis would warrant.

11.6.3.3 Test in a Low-Volume Plant. The number of failures arising at test—especially the in-circuit test, where most failures reflect in-plant process flaws—can be reduced to negligible amounts by scientific process management. When other progressive quality management practices such as eliminating inspection, touchup, and rework are employed, the rate of component failures and variances will also be greatly reduced. In some cases of mass-produced low-value products, no level of test will make sense.

High-volume production is always more easily controlled than small-batch output of nonhomogeneous products. As the range of products increases and the output of each shrinks, the probability of failures rises substantially. The greater incidence of failures in batch production compared to mass production[20] has two causes:

1. The need to change production setups frequently. Each time machine settings must be altered, the opportunity for error arises.
2. The greater reliance on semi-automatic or manual systems. Humans are more flexible than machines but also less consistent.

Additionally, the value per unit of product often varies inversely with the size of the production run. Thus, the need for test and probable net benefits will normally be greatest in plants where lot sizes are smallest.

Flawed cost accounting has caused a peculiar restructuring of American manufacturing. Mass production involving little labor input but substantial capital equipment per unit output goes to offshore plants while labor-intensive batch production has been retained. Only a minority of American plants now engage in what might reasonably be

[20]In mass production, output should be more consistent. This often means consistently defective output rather than consistently perfect, however. Mass production is easier to control only where the manufacturing personnel possess the requisite process management knowledge. Without that knowledge—and rigorous application of processes developed in accordance with that knowledge—the failure rates in high-volume production environments can be enormous.

called high-volume production. Consequently, the average American plant does require one or more test stages.

11.6.4 How Much Test Is Enough?

The amount and type of test cannot be established arbitrarily; what and how to test will be dictated by the plant's success in quality assurance. As more effective processes are implemented upstream, the need for downstream verification activity such as test is reduced.

Rigorous, informed process management invariably reduces true defects and failures to insignificant levels. But it is difficult to acquire sufficient hands-on experience to develop the informed process management techniques when production volume of a product is small.[21]

Just as the decline in flawed output makes it feasible and desirable to bypass inspection so that output moves straight from assembly to test, it is advantageous to eliminate test steps that do not discover meaningful numbers of failures. Depending on the value of the product, the difficulty of replacing defective modules, and the probability of in-plant failure, it is often entirely practical to test only once—after system assembly. Some outstanding plants have such low failure rates that they do not even attempt to fix failed products; any unit that fails at final test is scrapped.

11.6.5 Test Is Always a Cost

The best test strategies are reached when planning begins with the recognition that test, like inspection, is always a cost. Test may find faults but can never increase output beyond the level that would be achieved if production and pre-production activities were performed correctly the first time. A failed unit identified at test can be brought back from the dead by repair but cannot always be restored to perfect health. Once a unit has failed, its probability of failing again and again—not necessarily in the same manner each time—goes up.

The objective in devising a test strategy should therefore be to find the

[21]Plants producing customized modifications of the same basic product are fairly common. Though the numbers of any one flavor of the product may be rather small, the plant's production volume of the product family is high. For practical purposes, there is little or no difference between volume production of a single homogeneous product and volume production of a product family. A true low-volume assembly plant works with widely diversified product technologies.

minimum amount of test that provides acceptable probability of catching failures. Once test activity has been reduced, the forward-looking plant then seeks ways of further curtailing test. The obvious parallels between constantly seeking ways of reducing test and working to eliminate flaws in output are not coincidental: The need for all types of post-assembly product screening varies directly with the quality of the output. When output has inherent integrity, the need to seek out deficiencies ends.

11.7 Defining the Test Strategy

Four decisions must be reached in defining a test strategy:

1. Why test?
2. What attributes should be tested?
3. How should test be performed?
4. When should test be conducted?

Once those questions are answered, identification of the appropriate test strategy rarely presents serious problems. However, it is often more difficult to understand the questions than to answer them.

11.7.1 Why Test?

Superficially an obvious question, determining the reason(s) for testing—or not testing—a given product or products requires some thought. The reasons for test should be more sophisticated than the rote answer "to ensure customers do not receive faulty product."

Reasons that could support a decision to test include:

- Serious liability exposure in the event of product failure. Defective pacemakers are clearly cause for more concern than faulty radios. Defective thermostats or smoke detectors carry more liability than defective clocks. Without demeaning the importance of having satisfied customers no matter what the product, possible serious injury or loss of life entail far more serious issues for both customer and manufacturer.
- Massive inconvenience to customers in the event of a failure. This is different from safety issues. A defective control circuit for a washing machine can be a serious irritant to the busy householder, but failure

of a broadcast transmitter or an insurance company's master computer can close down the entire business.

- Customer requirements. Some customers—defense and aerospace customers in particular—often want to play a major part in determining the test strategy and dictate test procedures to suppliers.
- Undependable components. In a perfect world, flawed components would be detected before they ever reach the shop floor. In real life, batches of defective components occasionally slip through the best production defenses. A manufacturer that cannot trust the quality of its inputs must resort to screening the outputs.
- Unproven production processes. As processes mature, the failures generated by those processes decline to insignificant levels. Testing that may have been necessary early in the plant's learning curve should become dispensable as the manufacturing technology becomes more familiar.
- Designs exceeding the capabilities of the production department. Although patently unacceptable in a properly managed company, designs that exceed the tolerances of placement equipment, that try to place too many components into too little space, and that violate good design rules in other ways still show up. One way to combat this is by charging expenses such as test back to the design department.

Conversely, reasons for minimizing test could include:

- Recognition that most failures at test could be prevented by improved process design and management that would be considerably less costly than test.
- No meaningful consequences in the event of failure. This applies most often but not exclusively to inexpensive, disposable products.
- No exotic or customized components. The dependability of mainstream components today is extremely high.
- Mature designs and processes.
- Realization that test costs more than the customer is willing to pay. Defense contractors who were quite happy to perform every imaginable test under cost-plus contracts are less favorably disposed today to operations that increase costs without providing benefits.
- Failure to find failures during test.

- Knowledge that test means additional handling steps that can cause failures.
- Expense. The combination of capital equipment, equipment programming time, operation, maintenance, and increased cycle time makes test an exceptionally costly activity. In today's electronics assembly environment, test represents the largest single discretionary expense—and every dollar saved goes directly to the clichéd "bottom line."

The analysis that would seem so obvious that it need not even be mentioned is, in fact, rarely carried out. That analysis consists of tracking field failure rates with test and without test. If no statistically significant difference can be tracked, no justification for test exists. Our experiences with such analyses have invariably shown that test does not improve dependability of output for a company with firm control of its assembly and design processes.

11.7.2 What Attributes Should Be Tested?

Hard failures such as shorts, opens, and nonfunctioning components are readily identified as causes for rejection. But when the unit operates, it is not always obvious what characteristics are critical to acceptance or rejection. Minimum signal-to-noise ratio may be in the specification, but why should a unit that is well within the tolerance limits under normal operating conditions be rejected if it displays slightly too much noise under an extreme operating condition? Do operating characteristics not listed in the manufacturer's specifications matter? At what point do test criteria become too stringent, like the visual inspector who rejects an "acceptable" connection because it is not "preferred"?

The crucial attributes vary among industry sectors and companies within each sector. No ready-made formulas exist for universal application in this area. Ultimately, quality management will be forced to rely on educated judgment more often than they might like.

11.7.3 How Should Test Be Performed?

The nature of the test will typically derive from the characteristics to be tested. Shorts and opens, for example, can only be tested in a small number of ways, with in-circuit test being easiest and fastest. But the issue can quickly become complicated. For example, what is the appropriate test strategy when all shorts must be detected but the incidence of shorts is minuscule? Should in-circuit test be perpetuated or displaced by an all-inclusive final performance test? Cost-benefit analysis proves very valuable in this case because the costs of waiting until system test to find in-circuit failure modes can easily be calculated and compared to the costs of operating in-circuit testing. The complicating factor is that some failure modes cannot be detected at system test.

The most contentious issues in test arise when ESS is contemplated. These predicaments are easily avoided by rejecting ESS as a worthy test methodology.

11.7.4 When Should Test Be Conducted?

Quality directors often say that their objective is to make their job unnecessary. Test is one of those quality functions that fit neatly into this objective. The effective quality department will always be looking for test activities that can be eliminated. Realistically—at least, realistically in the absence of customers demanding otherwise—the test strategy should be directed toward elimination of all test except a final system check. Initially, failure levels at earlier test stages may be so high that deferring all test until the system is built would be more expensive than maintaining the earlier test stages. Such a situation only demonstrates the necessity for aggressive refurbishing of pre-test processes to eliminate the causes of failure. Where failure levels at intermediate test points are too high to allow consolidation of all test in a single stage at final assembly, the company clearly requires attention to upstream operations.

A number of companies have abandoned all test without encountering field performance problems. Those companies are highly qualified in design and scientific process management. A safer strategy—at least as an interim measure—would be to test random samples. Sample testing is highly recommended as a means of monitoring stability in designs, processes, and components. If unanticipated problems should arise, they will be detected by sample testing and universal testing reinstated until process and product integrity have been demonstrated once again.

11.8 Summary

Points raised in this chapter included:

- Test is the single most costly discretionary activity in the plant; moneys saved by limiting test go directly into profits.
- The original intent of burn-in and ESS was to evaluate designs.
- Many test activities—especially those classified as ESS—can cause damage and should never be applied to product intended for sale.
- Test is useful for monitoring stability in processes and products; for these purposes, random sample testing coupled with scientific process management works quite acceptably.
- Where test reveals large numbers of failures, investment in process redesign and supervision is called for.
- The effectiveness of test is not measured by the number of failures detected at test but by its impact on field failure rates. If field failure rates when test is employed are not meaningfully lower than field failures without test, test is entirely a cost without providing any benefits.

12
Human Tools

12.1 Chapter Purpose

In this chapter, we will look at what has gone wrong in the past, obstacles that any company hoping to tap into the plant worker's heart and mind will encounter, and how to overcome these challenges. Whereas quality departments promoting various forms of total quality see their role as encouraging teamwork, it will be shown here that providing employees at all levels with knowledge of technology, problem-solving, and general program appraisal skills is essential before widespread employee involvement can work effectively. Analysis of real and imaginary personnel development methods in Japan is also presented.

12.2 The Worker's Role in Quality

Manufacturing quality theorists over the years have agreed about relatively little. But on one point there has rarely been dissent: Workers are the company's most valuable assets. Moreover, in recent years quality departments have increasingly perceived catalyzing employee involvement as a primary quality mandate. While quality assurance requires knowledgeable involvement by all plant personnel, making the most of employee participation requires enormous effort on the part of all quality personnel. Obtaining worker participation in teams and other such employee involvement activities is the easy part; the truly formidable task entails transforming participation into significantly enhanced company development.

In discussing the worker's worth to the company, a distinction must be made between *potential* and *real* worth. *Potentially,* the employee's worth to the company is immense. But all too often the workforce *in reality* becomes the company's worst quality liability. As quality theorists correctly emphasize, the problem seldom lies with the workers; they generally perform their work more conscientiously and willingly than their employers deserve. However, while the workers tend not to fail the company, the company does fail the workers.

In particular, the company passes along misconceptions about quality and workmanship that severely handicap the workers' abilities to turn out the best possible product. Those misconceptions include the erroneous beliefs that:

- Defects occur because the worker lacks the proper ability. In fact, as previous chapters have shown, defects and failures are caused by faulty processes over which the worker has no control.
- Defective solder joints can be improved by touchup. Of course, we now know that touchup actually increases the probability of product failures.
- Reliability can be determined by cosmetic appearances. This belief is also wrong and leads to rejection of dependable product—and paranoid workers.
- When in doubt, it is best to rework. Not surprisingly, rework flourishes and the failure rate soars.

The hourly workforce did not invent those beliefs. The company's quality standards and training programs indoctrinated the workers' minds and shaped their attitudes. Only substantially revised quality standards and thorough reeducation will enable the workers to live up to their potential.

12.3 Education vs. Training

Management consultants—quality specialists and generic business consultants alike—emphasize that workers require training to perform their jobs properly. The value of training is at the very core of quality's conventional wisdom.[1] In keeping with this conviction, most companies invest heavily in training. Many of today's leading electronics companies allocate 40 to 80 hours of classroom time per worker every year. Unfortunately, while the intentions are good, *training* on its own rarely delivers meaningful benefits. Truly effective personnel development emphasizes *education* over training.

The distinction between education and training is not widely appreciated. The confusion between the two words is reflected in the frequency with which training and education are used interchangeably. Further evidence that education's importance is poorly understood in the quality world can be found by examining the content of randomly selected quality books. While it is a rare book indeed that does not contain extensive discussion of training, reference to education is virtually unknown.

12.3.1 Training

Training is the development of motor skills or instinctive responses to specific situations. Through training, for example, hand soldering operators develop the manual skills of bringing the tools and materials together in consistent ways. Training imprints *what* to do on the trainee's mind. Through repetition, the trainee acquires the ability to carry out the exercise as second nature. However, nothing in training helps the individual understand the reason the acquired skills produce specific results.

The question of *why* performing the operation in exactly that manner is necessary never comes up in training. Indeed, the trainee is actively discouraged from asking the reasons for any element in the job. Training therefore reduces the individual to a living machine, capable only of performing the same operation on assembly after assembly without variation. For the trained individual, as for the trainer, the principles—

[1]One of the most thorough—and misguided—statements about the value of training may be found in [Deming 1986], pp. 248–251. "Anyone, when he has brought his work to a state of statistical control, whether he was trained well or badly, is in a rut," Deming observed (p. 249). "He has completed his learning of that particular job. It is not economical to try to provide further training of the same kind. He may nevertheless, with good training, learn very well some other kind of job." Deming failed to recognize the difference between training and education. Far from learning in the classic sense of comprehension, the individual has been programmed (by training) to perform the physical operations consistently in the identical manner and sequence from one assembly to the next. *Training*, not learning, is the reason Deming can say the individual has "brought his work to a state of statistical control."

most notably the underlying scientific laws—governing the operations have no relevance.

12.3.2 Education

Education deals with the *whys* ignored in training. In the hand soldering example, education would instill understanding of considerations that determine what *should* be done. Education explains the natural laws dictating the selection of materials, tools, the sequence of events—and the consequences of breaking any of those natural laws.

Previously (Sec. 6.4) the existence of three processes was noted. The natural process is the sequence of activities that physical forces dictate for perfect output; this is what *must* be done. The specified process is written by plant personnel, usually process engineers, and is the official statement of what should be done. Finally, the real process is what actually happens during production. When the real process diverges from the natural process, defects result.

If the techniques in which the individual is trained coincide with the requirements of the natural process, the output will be dependable. But the chapters on soldering and solderability management showed that process requirements are not static. The solderability of surfaces, for example, deteriorates over time. Operators who are not familiar with the natural forces they face are unprepared when defects appear (i.e., the techniques in which they have been trained are inconsistent with the requirements of nature). Rather than recognizing the existence of a solderability problem that requires pre-assembly corrective action, the trained operators assume that they somehow lack sufficient "skill." Through a combination of excessive heat and perseverance, the operations succeed in producing cosmetically perfect connections. As we have seen, joints produced in this manner may be good for the operator's ego but seriously degrade product dependability.

Soldering is only one example of how lack of education hinders the employee's ability to do what is best for product dependability. No employee can do the job better than that individual's education allows. This fundamental fact of human behavior must be kept firmly in mind by the manager who advocates pushing responsibility down to the employees closest to the activity.[2]

---

[2]If management does not know the natural process any better than the lower-level employees, the company loses little by cutting back on administration (unless, of course, paring away management layers leaves a leadership void—a condition that is more common today than generally recognized).

12.4 Theories X, Y, and Z

When North American companies first began to pay serious attention to the Japanese manufacturing "miracle" in the late 1970s, they noticed three things in particular about Japanese industry:

1. Japanese companies organized their factory communities differently from the plant structures in North America.
2. Ordinary Japanese production workers played leading roles in operational decisions while their Western counterparts had no voice.
3. Productivity in Japanese plants was soaring while American productivity was leveling off.

The societal difference between Japanese and American factories was so pronounced that many Western observers believed cultural factors must account for the huge advantages the Japanese seemed to enjoy in productivity and quality. With that belief in mind, American companies began replacing traditional authoritarian management with "participative" management. Executives, business schools, and consultants developed intense interest in Theory Y versus Theory X management techniques. Some companies bypassed Theory Y entirely in favor of the new Theory Z management style.

The terms Theory X and Theory Y management first appeared in a 1960 book[3] by Douglas McGregor, a professor at MIT's Alfred P. Sloan School of Industrial Administration. Although the terminology was new, the underlying concepts had been more or less in the mainstream of American management thinking for several decades. For example, Theory X is McGregor's terminology for Frederick W. Taylor's "scientific management" system[4] that, while devised in the early 1900s, remains to this day the prevailing approach to organizing and managing a factory workforce. Taylor (who himself worked as a lathe hand before becoming a mechanical engineer and accordingly has more first-hand knowledge of production worker attitudes than the academic theorists who came later) contended that workers:

- Inherently dislike work.
- Lack the creativity and discipline to make meaningful contributions to the company.

---

[3][McGregor 1960].

[4]For the best presentation of Taylor's views in his own words, see [Taylor 1947].

- Are motivated only by money.
- Are interchangeable and disposable.
- Have no greater intrinsic value than any other easily replaceable commodity.
- Will always perform best when assigned simple, repetitive tasks defined by management or engineering specialists.

Since the workers will shirk their duties at every opportunity, Taylor argued, the role of management must be to define and enforce the company's expectations of its workers. These management activities included:

1. Breaking production down into many mindless, repetitive tasks.
2. Paying more to those workers who perform best while penalizing the less productive workers.

Taylor's philosophies epitomize the popular perception of "autocratic management" that Deming would attack so vehemently. It is often said that under Taylor's system, managers are "thinkers" and "police" while workers are closely monitored mindless "doers."

"Theory Y," meanwhile, was McGregor's designation for the management style that logically grows out of work done in the late 1920s by behavioral scientist Elton Mayo. Beginning in 1927 at the Hawthorne Works of the Western Electric Company (that is, at the same time and in the same plant where Walter Shewhart was developing modern statistical process control, Joseph Juran was helping write statistical training courses based on Shewhart's work and Edwards Deming was working summers while finishing his doctorate at Yale), Mayo conducted experiments to see how changes in plant lighting conditions would affect productivity. To his surprise, he discovered that productivity increased each time the lighting was changed, regardless of whether it was intensified or made dimmer.

The concept that any change in the work environment would increase worker productivity became famous as the "Hawthorne Effect." However, Mayo was not satisfied with that conclusion and investigated further. Through interviews with workers, Mayo discovered that the experiments had not been completely unbiased. Supervisors, being aware that the experiments were taking place, treated their workers better. The workers themselves became aware that they were experimental subjects. Knowing about the experiments, the workers felt their menial tasks had acquired greater stature, and their supervisors' behavior reinforced this belief. Mayo therefore concluded that the real lesson to be learned from the Hawthorne experiments was that productivity is a function of the worker's sense of involvement and self-esteem.[5]

McGregor postulated that the Taylor approach had no place in the postwar world. Mayo's studies had satisfied McGregor that making workers feel more responsible for their work would increase productivity. Moreover, the 1960 worker was better educated than the workers Taylor studied at the turn of the century—or even the Hawthorne employees studied by Mayo just before the Great Depression. Consequently McGregor concluded that the contemporary worker not only had the *ability* to identify obstacles to better performance on the shop floor but also possessed a strong *desire* to take part in company improvement. From these theorems, McGregor argued that workers in a Theory Y company:

- Constitute the company's most valuable assets; they must be maintained and improved like any other valuable asset.
- Will perform best and most happily when presented with constant challenges and the opportunities to influence their work quality and conditions.
- Are both able and eager to demonstrate their creativity and discipline in a work environment.
- Perform poorly when deprived of the opportunities to contribute to their work quality and conditions.

Accordingly, the Theory Y manager sets only the basic requirements and allows the workers to determine how best to perform their work. While final decisions remain the Theory Y manager's prerogative, workers are encouraged to provide input. The Theory Y manager's ultimate task is removal of barriers placed in the paths of the innately responsible, industrious employees.

Once McGregor's Theory Y became popular, a Theory Z was inevitable. The only surprise is that it took two decades before finally appearing.[6] When it did appear, Theory Z proved disappointing. Whatever faults may have existed in Taylor's scientific management and McGregor's Theory Y, they were at least well defined. Theory Z management, on the other hand, is an amorphous, ambiguous, and ultimately unsatisfying concept that appears to leave even its own creator, UCLA Graduate School of Management professor William Ouchi,

---

[5][Silbiger 1993], p. 248.

[6][Ouchi 1981].

uncertain of both the theory and, more importantly, the application. Theory Z management evidently consists of a Theory Y–oriented manager working for a progressively minded company that offers lifetime employment. However, in Ouchi's own words, "Each Type Z company has its own distinctiveness—the United States military[7] has a flavor quite different from IBM or Eastman Kodak."[8] Perhaps the best way to judge the usefulness of Theory Z management is by noting how far his business paragons—IBM and Eastman Kodak—fell in the decade after Ouchi presented his thesis.

The most successful management style, sometimes called the "contingency approach," combined the best elements of Taylor, McGregor, and Ouchi. It is probably fair to say that Hewlett-Packard—which considered but rejected a "Japanese" style of management—best exemplifies this pragmatic but humanistic manner.

12.5 The Japanese Way

The sudden interest in consensus management among American companies resulted—as always seems to be the case when Western companies decide to copy Japanese tactics—from mistaking Japanese appearances for Japanese reality. Japanese workers did participate in decision making at lower levels of their companies. However, a look at autobiographies by Japanese executives indicates that consensus management in Japan means executives decide and underlings concur.

12.5.1 Decision Making at Matsushita

In *The Panasonic Way,*[9] former Matsushita Electric Co. Ltd. president Toshihiko Yamashita scathingly addressed the question of consensus management. "A mountain climbing party suddenly beset by bad weather has to reconsider its route and make other plans," he notes. "To continue or go back? To take the route to the right or to the left? A group decision, where everybody expresses an opinion and the choice is a consensus based on compromise, is frequently wrong. An experienced

---

[7]This is not a misprint. Ouchi evidently believed that the armed forces operate according to consensus management.

[8][Ouchi 1981], p. 71.

[9][Yamashita with Baldwin 1987], p. 36.

climber should be in charge and make the decisions. This was the conclusion of a series of fascinating experiments on decision making at high altitudes conducted by Nagoya University's Research Institute of Environmental Medicine....The group leader functioned best and scored highest in the exercise because he had enormous experience and was accustomed to low-pressure conditions.

"...That holds true for business, too. When a fundamental policy choice must be made, a majority vote or consensus invites failure. If groupism worked, why have a president?...If a majority thinks a plan has a chance, you can be sure another company is already doing it. Rejecting staff proposals isolates the president, but the loneliness of command comes with the territory.

"Of course, I listen to relevant opinions, but I make the final choice. That means also taking any flak that follows. Unpopularity is also sometimes part of the territory."

12.5.2 An Example from Sony

Sony founder and recently retired chairman Akio Morita expresses similar thoughts in his autobiography *Made in Japan*.[10] Describing the invention, refinement, and marketing of the Walkman portable tape player, Morita unreservedly takes full credit for the original idea and subsequent refinements. "I thought we had produced a terrific item and I was full of enthusiasm for it," he writes,[11] "but our marketing people were unenthusiastic. They said it wouldn't sell, and it embarrassed me to be so excited by a product most others thought would be a dud. But I was so confident the product was viable that I said I would take personal responsibility for the project." Morita was traveling when the marketing department created the Walkman name. When Morita returned, he was horrified.[12] The marketing department, however, informed him that it was too late to change the name. Advertising had been prepared and, even as they spoke, thousands of tiny cassette tape players bearing the Walkman name were coming off Sony's production lines. But Sony U.K. and Sony U.S. had different ideas, fearing "they couldn't sell a product with an ungrammatical name like Walkman but we were stuck with it. We later tried other names overseas—Stow Away, in England, and Sound About in the United States—but they never caught on. Walkman

[10][Morita, Akio, with Reingold, Edwin M., and Shimomura, Mitsuko 1988].

[11]p. 89.

[12]p. 90.

did. And eventually, I called up Sony America and Sony U.K. and said, 'This is an order. The name is Walkman!' Now I'm told it's a great name."

12.5.3 Changing Decision-Making at Honda

Similar stories abound, including a *Fortune* account of a management shakeup at Honda.[13] During 1991, Honda's CEO, Nobuhiko Kawamoto, toured the company's plants in Japan. He discovered that the world-famous participative, conciliatory management style just wasn't working well anymore. The workers had slowed down and were starting to spend more time on their personal lives. Kawamoto decided that the time had come to switch to a more American management style, to get things done faster.

The problem for Kawamoto was that the system he wanted to put in place directly contradicted the dictum of company founder Soichiro Honda. "First there are people," Honda had often said, "then there is work, and a minimum necessary organization follows so the people and work are effectively managed. An organization exists to serve its members, not the other way around."

According to Kawamoto, he went to see Honda himself and said, "I'm sorry to say it, but not everything you said is correct now."

As Kawamoto foresaw, Honda was not happy. Nonetheless, demonstrating the intelligence that had enabled him to create first the world's most successful motorcycle company and then a giant auto company, both times against the wishes of MITI, Honda did agree. And so Kawamoto launched a corporate restructuring that by Western standards was ambitious and by Japanese standards remarkable. He started with the executive floor at corporate headquarters. The tenth floor housed 32 executives with titles of vice president or higher, 20 of whom had no specific responsibilities. Or, to put it another way, everyone felt responsible for every decision. So, whenever a problem arose, all 32 senior executives got together and debated the issue until they could arrive at a consensus. More often than not, this took a very long time. So Kawamoto assigned a specific responsibility—such as purchasing or customer relations—to each executive. When a problem arose, it was to be decided solely by the executive responsible for the territory into which the problem fell. The result: faster decision-making and no more

---

[13] "A U.S.-Style Shakeup at Honda," *Fortune,* Dec. 30, 1991.

waiting for consensus. Moreover, the decisions were implemented faster because the same individual was responsible for making things happen.

Kawamoto also concluded that many of the company's middle managers weren't performing up to his expectations, either. While no one was actually fired, many had their independence reduced and some had their management positions eliminated. In other words, Kawamoto emulated the traditional Big Three approach of centralized control. Again, he had violated a Honda dictum: "If you succeed 1 per cent of the time, 99 per cent failure is acceptable." (While there is no record of Honda ever being put to the test on that statement, we suspect that he would not have been quite as content with a 1 percent success rate as his words suggested.)

12.5.4 *Tatemae* and *Honne*

If the Japanese themselves use consensus management only when it suits the chief executive, why do Americans believe otherwise? The answer is found in two interrelated Japanese words: *tatemae* and *honne*. Although references to *tatemae* and *honne* can be found in many books about modern Japan, the best explanation is found in van Wolferen.[14] "In their daily lives, the Japanese are very helpful to one another in minimizing embarrassment, and will make very clear that what they have just said may not refer to factual reality," van Wolferen explains. "All one has to do is catch the signal. Foreigners are expected to do this also, for sometimes the excuses or 'explanations' given in international dealings are simply too crude to be taken seriously. To make things even easier, the Japanese have a relevant terminology: *tatemae*, or the way things are presented, ostensible motives, formal truth, the façade, pretense, the way things are supposed to be (often wrongly translated by Japanese as 'principle'); and *honne*, or genuine motives, observed reality, the truth you know or sense.

"This *honne-tatemae* dichotomy is constantly referred to, and it is usually considered an ethically neutral if not positive aspect of Japanese society. But it also provides a frame of reference in which many forms of deceit are socially sanctioned. The Japanese can be honest about their fakery to a degree that Westerners could not possibly be. They

[14][van Wolferen 1989], pp. 234–236.

are allowed to pretend honesty without fear of being chided for dishonesty."

The emphasis on consensus is a prominent example of *tatemae*. Within their companies and in dealings with foreigners, Japanese businessmen can observe the rituals of consensus. The Japanese participants in the consensus dance recognize the exercise for what it really is (the *honne*): a social ceremony by which subordinates save face. Western, however, accept the *tatemae* at face value.

When dealing with foreign competitors (in business or politics), however, the value of *tatemae* to Japanese negotiators extends far beyond a simple social nicety. When bargaining with foreign visitors who hope to sell them an idea, license a product, or enter into a joint venture, Japanese companies capitalize on the knowledge that foreign visitors are subject to time constraints. If they can avoid making commitments before the visitors must return home, the Japanese can examine the situation more closely and decide at their leisure how best to take advantage of what they have learned.

Western businesses have been exceptionally gullible in their dealings with Japanese companies. The consensus ruse has served the Japanese exceptionally well. More than a few inexperienced Western negotiators have been overwhelmed by Japanese hospitality that occupies every free minute. At the last minute, as they are being put on the plane home, the visitors become acutely aware that their hosts have not committed to the proposal that brought the Westerner to Japan in the first place. When a commitment is requested from the Japanese escort, the Westerner is politely informed that it will take some time for the Japanese team to reach a group decision.

Time and again, American managers touring Japanese plants have mistaken *tatemae* for *honne* in ways that benefited the Japanese tremendously. Western visitors to Japanese plants have long been encouraged to believe that the secret of Japan's industrial success is automation; by coincidence, the Japanese host company happens to own a subsidiary that manufactures automation equipment available for sale to their guests. Company songs, lifetime employment, smaller differences in remuneration between senior executives and rank-and-file workers, total quality programs, and much more—all *tatemae* earnestly presented as *honne*, the real reason why Japanese control their domestic markets and do so well in exports. The reality doesn't measure up to the mythology. Indeed, the most profound example of *tatemae* is the reputed robust health of Japanese industry. By the standards of Western industry, the financial performance of Japanese companies has been horrendous.

12.6 Quality Circles

As companies sought ways to demonstrate their new-found Theory Y or Z worker-oriented attitudes in keeping with the new desire to emulate the Japanese, they usually turned first to quality circles. Western visitors to Japan had reported that quality circles had proved highly successful in Japanese plants. In fact, quality circles were working well in Japan, but the American observers had failed to identify the reasons for that success.

In a pleasant contrast to many quality concepts of the past two decades, the term "quality circle" was defined with considerable precision. Almost without exception, quality circles meet the following criteria:

- The circles consist of small groups of shop floor workers—certainly fewer than a dozen and normally around eight individuals—assembled to find and solve problems in the workplace.
- The group members may all come from the same department or contain representatives of several departments.
- Typically, the personnel are hourly workers, but participation by supervisors is not uncommon.
- Membership is voluntary.
- Once a group has been formed, its membership does not change unless required because of external forces (such as personality clashes or members leaving the company).
- The circles select the problems they wish to pursue and, while the normal guidelines encourage circles to concentrate on problems within their own departments, there are usually no formal barriers preventing them from looking at issues beyond their departments.
- Circle meetings are held regularly (weekly is the norm) on company time, although many circles also meet before or after regular working hours (and without receiving additional pay).

12.6.1 The "Quality" in Quality Circles

The most misleading aspect of quality circles concerns the name; quality circles do not necessarily attack quality problems. Although the teams are free to select what types of problems they will pursue, true quality issues historically have played a minor role in the activities of typical quality circles.

On paper, quality circles looked very promising. Their pedigree was good: Quality circles were widely regarded among Western businesses as the secret weapon in Japan's industrial successes. Managers who supported quality circles could feel virtuous and progressive. Meaningful improvements in quality and efficiency were widely anticipated. Workers were expected to be happier and more productive. It all turned out to be *tatemae.*

12.6.2 Quality Circle Failures

Quality circles never worked well, although not all companies that went down the quality circle path admitted it was a dead end journey.[15] The groups turned out not to be capable of solving many issues of consequence. Significant support from managers and supervisors was needed. The costs were much greater than anyone had expected. Far from strengthening their companies' competitiveness, quality circles undermined financial abilities.

12.6.3 Unrealistic Assumptions of Quality Circles

That quality circles failed should have surprised no one. They were based, after all, on two unrealistic assumptions:

1. *Most workers naturally want to eliminate problems; they only need the opportunity to contribute.* Worker desire and ability to solve the company's problems if management would just stand aside is the core belief of Theory Y management. When McGregor published his Theory Y research in 1960, this belief may have had some merit. The economy was expanding steadily. Companies had not yet begun exporting jobs in massive numbers. The worker expected to stay with the same company right through to retirement age. If the company was more efficient, workers had some reason to anticipate higher wages.

During the 1980s—and even more in this decade—the conditions that had made workers willing to sacrifice for their companies evaporated. Reengineering, downsizing, and a lot of other euphemisms for exterminating jobs destroyed whatever trust may have existed. Workers per-

[15]Some companies still use quality circles and claim to have derived real benefits. Third-party audits of many of those plants compared the alleged results to measurable benefits. The company satisfaction in almost every case results from inept monitoring of circle results by management.

ceived improvements in efficiency and quality as ways for their employers to eliminate more jobs. Few workers expected to spend their entire careers with the same company; the majority were simply hoping to stay employed somewhere. Inefficiencies like "busy work" became one of the few ways workers saw to postpone the inevitable loss of their jobs.

2. *Workers, being closest to the actual production operations, have the best knowledge of company problems.* Workers are aware of what happens on the jobs to which they are assigned. However, failures by the school system and the company to provide technical knowledge and reasoning skills fatally impede the workers' abilities to differentiate between positive and negative actions—or to correct the failings that they do identify.

Long before entering the job market, workers are victimized by a school system that ignores or dilutes the language, math, and science knowledge they will need to recognize, analyze, and solve real problems on the job. The company's contribution to their knowledge typically consists of training programs that require memorization and unquestioning observance of instructions written by others. The workers are trained rather than educated, so the instructions have no meaning for them.

Additionally, as previous chapters have showed, the techniques in which electronics assembly personnel are trained are generally undesirable. Instructions—even in companies not connected to the defense industry—are taken directly from antiquated MIL Specs. So the workers are conditioned to build in higher probabilities of premature product failures by touching up solder connections that fail to meet cosmetic standards.

Further complicating matters, the interests of separate departments often conflict. Workers in one department have little appreciation of how their work affects other departments. Indeed, the customary reaction when problems show up in one department's work is to blame other departments—normally design and purchasing—even though the problems may be totally self-induced. None of this is deliberate undermining of the company as a whole; it is simply human nature.

In recent years, it has become fashionable to blame managers for preventing the rank and file from performing their jobs properly. In many cases, however, the greatest management error is *overestimating* the abilities of the workforce. In the case of quality circles, management wrongly assumed that the workers possessed more technical knowledge than was actually the case. This erroneous assumption stopped companies from providing the critical technical and problem-solving education that would have equipped teams to identify and resolve important issues.

12.6.4 The Life Cycle of Quality Circles

Accordingly, the life cycle of quality circles tended to follow a predictable pattern. In the first stage, the circle activities concentrated on work-related practices that were important to some or all of the members but not necessarily useful in strengthening the company. But sooner or later (and usually sooner), the teams either fixed the issues of concern to their members or gave up in frustration. Since the companies seldom set up systems for feeding information to the teams about organization problems that the circle members had not themselves identified, the teams' idea banks became exhausted.

At this point, a rational system would have caused the circle to disband. However, the circle was expected to meet regularly and could not disband. Accordingly the circle entered the second stage in which the issues were unrelated to ways of working better. Circles in this stage of their evolution frequently debated trivialities such as the color of paint in the washrooms or the taste of coffee from the vending machines.

Ultimately, the members depleted their knowledge of even these trifling matters. This brought them to the third stage in which the circles assembled only to fulfill the requirement for regular meetings. Workers began to find reasons for avoiding the circle meetings. After some period of time in this state of paralysis, the circle faded away from lack of interest on the members' parts.

12.6.5 The Destructive Consequences of Quality Circles

Quality circles were conceived to motivate the workers as well as improve the company. Ironically, their experiences as circle members turned even enthusiastic workers into cynics. The company may have learned about some workplace issues that would not otherwise have come to its attention, but the price of gaining that knowledge was very high. Moreover, in many companies, the ideas put forth by the circles were never acted upon because the mechanism for evaluating and implementing circle recommendations simply did not exist![16]

---

[16]Although it seems hard to believe, companies regularly set up programs for soliciting employee input without having any structure for reviewing suggestions and acting on those with merit. The employees work hard to come up with ways to make the company stronger, put in the effort to write up the suggestions—and never hear another word about what, if anything, happened. This is the sort of blinkered management that kills employee initiative.

12.7 Teams

The failure of quality circles forces the question: Are all attempts to gain input from the hourly workers destined to fail—and fail miserably? The answer can be stated without reservation or hesitation: absolutely not. Assembly workers are the front-line troops, closest to the action and the first to see when processes fail. Although the structures and operational modes of quality circles were tragically flawed, the original concept of tapping the workers' familiarity with the plant operations could not be faulted. When the weaknesses of quality circles became apparent, experiments with more effective forms of worker teams began.

The second generation of group worker involvement—now called simply "teams"—benefited from the lessons of quality circle failures. In particular, it was now clear that successful use of worker teams requires vastly more effort than a management decree saying: "Effective immediately, hourly workers will form teams, hold meetings to identify and solve production problems, and pass those recommendations along to management which will decide what, if any, action to take."

12.7.1 Three Minimum Conditions for Successful Teams

At a minimum, the following three conditions must be met for worker teams to succeed:

1. *The company must provide workers with in-depth education about the technologies of their jobs.* Without knowing the chemistry, metallurgy, physics, and mechanics that are involved in most electronics assembly problems, the workers cannot identify critical quality issues. Remember that education is not the same as training. Circus animals are trained to perform tricks; humans improve their abilities with education.

In designing an education program for electronics assembly personnel, it must never be forgotten that most of the worker-students will have been subjected to years (possibly decades) of indoctrination in the mind-numbing DoD processes and workmanship standards. DoD graduates are extremely proficient at making cosmetically perfect solder joints even when the processes are running out of control. They believe that electronics assembly is an art rather than a science and they are the ultimate artists. Naturally, converting these proud artists to the scientific method requires patience.

2. *Workers must be motivated to inform supervisors or managers* when, as a consequence of seeing day-to-day operations in the new perspective

of greater knowledge, they see problems in the processes, the organization, or the facilities. This *does not* mean forcing workers into meetings. It *does* mean ensuring that there is someone (ideally, every supervisor and manager) in the plant who can be trusted to lend a sympathetic ear when a worker has an idea for improving productivity or quality.

One fact about attitudes of first-line supervision cannot be stressed too strongly: The supervisory level of commitment to worker input determines whether their employees will (or can) participate enthusiastically and effectively. Production workers regularly identify problems and even suggest feasible solutions only to see their input brushed aside by lead hands and supervisors who do not want suggestions from subordinates. Even in companies where formal systems for listening to workers are nominally operating, supervisors find ways to block input. Academic theorists about plant community interactions are not aware of this widespread structural impediment to personnel involvement, but production workers have intimate familiarity with the problem of closed-minded bosses.

3. *Systems must be in place to ensure the company will act on the workers' ideas.* In the best case, the workers should be so competent that they not only make the suggestions but also have full authority ("empowerment" in the overworked quality jargon) to put those changes into effect without first obtaining approval from higher levels of management. Teams that reach this stellar level of ability are commonly referred to as "self-managed" or "high performance."

12.7.2 Harder Than It Looks

Although these three requirements may sound simple, few methods of improving plant performance present more challenges. Each requirement can be achieved only through:

- Hard work by management as well as workers
- Company willingness to pay for thorough education rather than superficial training
- Exceptional discipline through many months of painstaking preparation and implementation

We have worked in and visited several hundred plants where teams are employed. Of those hundreds, only a handful have team programs that pay their way. The others are best described as "feel good exercises"—management *feels* enlightened and workers *feel* important but no progress is made toward achieving the all-important company goal.

Companies considering implementation of teams should also be aware that employee involvement on matters of productivity, quality, and efficiency is illegal in non-unionized American companies. A 1935 law intended to thwart company control of unions[17] prohibits employees and managers in non-union companies from meeting to establish "conditions of work." Until 1991, companies ignored the little-known law. In the 5 years beginning with 1991, however, the National Labor Relations Board successfully sued at least a dozen companies—including prominent names like Polaroid and DuPont—for statute violations. The position of the AFL-CIO's labor law task force director in early 1996 was that workers should elect members of company teams rather than letting the company select the personnel it wants.[18] Of course, companies want the most knowledgeable workers, not the most popular, for teams.

12.7.3 High Maintenance and Expensive Teams

Results of a 1994 survey of *Fortune* 1000 companies by the University of Southern California's Center for Effective Organizations reflect the magnitude of the effort and insights required to create effective teams. The study found that 68 percent of the *Fortune* 1000 companies use "self-managed" or "high-performance" teams—but only 10 percent of workers participate in the teams. Edward Lawler, the USC management professor who supervised the study, explained the findings with an appropriate analogy: "People are very naïve about how easy it is to create a team. Teams are the Ferraris of work design. They're high performance but high maintenance and expensive."[19]

Apparently naïveté abounds. Another *Fortune* article[20] enthused that "scores of service companies like Federal Express and IDS have boosted productivity up to 40% by adopting self-managed work teams; Nynex is using teams to make a difficult transition from a bureaucratic Baby Bell to a high-speed cruiser of the I-way; Boeing used teams to cut the number of engineering hang-ups on its new 777 passenger jet by more than half." The same article quoted Boeing President Philip Condit, saying,

---

[17]That is, the kinds of unions found in all Japanese companies of consequence.

[18]Jones, Del. "Bill Would Make Manager-Employee Teams Legal," *USA Today*, Feb. 7, 1996, p. 1B.

[19]"The Trouble with Teams," *Fortune*, Sept. 5, 1994, p. 86.

[20]"The Trouble with Teams," p. 86.

"Your competitiveness is your ability to use the skills and knowledge of your people most effectively, and teams are the best way to do that."

Evidence that Condit represents the current vogue in management thinking is easily found by referring to the content of annual reports in recent years. Annual reports can provide fascinating insights into a company's future as well as its past by laying out management's strategic thinking for improving future results. And lately the strategic thinking portion of the annual reports tends to fall back on jargon-heavy terms and phrases like "workplace of the future,"[21] "empowerment,"[22] "teamwork,"[23] "an empowered work force,"[24] "self-managed work force,"[25] "boundaryless people, excited by speed and inspired by stretch dreams,"[26] "Quality is our number one operational imperative. Empowered people are making it happen."[27]

Not surprisingly, there's trouble in this new team paradise, just as quality circles failed to meet expectations. As *Fortune* cautions: "Forget all the swooning over teams for a moment. Listen carefully and you'll sense a growing unease, a worry that these things (i.e., teams) are more hassle than their fans let on—that they might even turn around and bite you." A quotation from Eileen Appelbaum, author of *The New American Workplace,* sums up the *Fortune* theme: "It isn't that teams don't work. It's that there are lots of obstacles." Or, in the colorful words of an anonymous team leader at American President Companies, "A team is like having a baby tiger given to you at Christmas. It does a wonderful job keeping the mice away for about 12 months, and then it starts eating your kids."[28]

12.7.4 Traps for Teams

As Appelbaum properly notes, the company embarking on the journey of teams—or any other form of participative management for that mat-

---

[21]AT&T 1993 Annual Report.

[22]Chrysler Corporation 1993 Annual Report, which includes 16 pages describing how the company is using teamwork and other recently fashionable quality approaches to improve their business.

[23]Chrysler Corporation 1993 Annual Report.

[24]Corning Incorporated 1992 Annual Report.

[25]Exide Electronics 1993 Annual Report.

[26]General Electric Company 1993 Annual Report.

[27]Motorola 1992 Annual Report.

[28]Labich, Kenneth. "Elite Teams," *Fortune,* Feb. 19, 1996, p. 90.

ter—faces many obstacles. The more common traps lying in the path of the unwary company include:

- *Lack of direction.* Since any group that does not have a predetermined goal will drift aimlessly, considerable effort must be expended to teach team members the skills of preparing and following agendas and measuring their progress.

  Management must help the team stay focused: An unfocused team will become satisfied by merely holding meetings, unaware that the value lies not in the meeting but in the results.

  Management must not use teams as an excuse to offload responsibility for (and consequences of) decision-making. Employee participation does not mean management abdication: Too many management attempts to give the shop personnel a voice in decision-making end up with the teams failing to act and the company afraid to undermine its own efforts. This situation can be referred to as "consensus paralysis." Left to develop unopposed, consensus paralysis can lead to tragic consequences: Operational problems that could have been nipped in the bud with little cost or difficulty turn into giant headaches; team members feel betrayed when management needs to step in and take back control; quality drops and costs climb. Companies that have followed the guidelines for setting the goal laid out in Chap. 3 will not be troubled by these traumatic outcomes.

- *The rudderless manager.* All too often, employee involvement is the refuge of the weak manager. If plans don't work out, the manager knows there is refuge behind the shield of employee involvement. Whereas too many American managers believe teams are an excuse for them to become detached from daily involvement, Japanese managers typically devote between 30 and 40 percent of their time to coaching, finding additional resources and otherwise smoothing the way for teams under their wings.

- *The passive team member.* Some few workers will sign up for teams without sense of commitment or recognition of the effort involved for success. These passive members fail to carry their own weight and shift extra burden onto the team members who signed on for the proper reasons. There is a growing belief that the entire team will rarely be stronger than its weakest member; in other words, one or two superstar performers can do less to pull up a team than a single passive member will drag it down by damaging morale.

  What makes workers passive? Often, they regard problem solving as management's responsibility. In their minds, workers who solve problems should be paid for their contributions. The passive workers—who can often be more appropriately described as silently hostile—"volunteer" for teams because of real or perceived and/or management pressures.

- *Personality conflicts.* In the real world, friction between individuals is almost inevitable. In teams, the potential for conflict is even greater than normal. Effective team managers always anticipate that individuals have different priorities than the company. What is good for one department may mean more work for another, giving rise to disagreements between the members representing those departments. Some engineers may be too aggressive for the tastes of the less confident hourly personnel. Several strong personalities may clash over control. Language differences can show up in two different ways—nationality or the use of overblown terminology that others do not understand. Of course, a lot of people at all levels in the company are just not team players. A strong team manager will be watching for signs of disharmony and be prepared to reorganize the team quickly at the first sign of personality clashes.
- *No purpose for the team.* Many companies anxious to show their sensitivity and commitment to employee involvement have assembled teams for no clearly defined reason. Absence of purpose makes both operation and evaluation difficult to impossible.
- *Unrealistic expectations.* Teams, like quality circles, are not omnipotent. Starting out in anticipation of improving operations by an order of magnitude over a long period of time guarantees failure. Targets should be established that can reasonably be attained in a short time. Buoyed by the positive experience of having met or exceeded one goal, the team will then be eager to pursue another. Six months is the longest time period to which individuals can realistically be expected to relate.
- *Insufficient imagination.* A company that has always stifled creative thinking in its personnel should not expect team members to break out of the mold. Companies reap what they sow.

12.7.5 Temporary Teams Work Best

One lesson to be learned from the failures of quality circles is the danger of forming permanent teams. A better choice is the formation of a temporary problem-solving team when an issue arises. The team is assembled for the sole purpose of resolving that specific issue. Members are selected on the basis of how closely their individual abilities meet the perceived requirements for solving the issue—and their personal interests in attacking that issue. When the issue has been resolved or it becomes apparent the team has reached a dead end, the team disbands. The members of any former team can be selected for future teams if their knowledge and skills match the perceived requirements for solving the new issue.

12.7.6 The Permanent Team

The system of temporary teams requires the maintenance of one permanent team to manage the teams. The functions of this permanent team include:

- Collecting suggestions for items to be addressed
- Ranking suggested action items in order of perceived importance to the company
- Choosing the members for each temporary team and assigning one problem to each team
- Tracking the progress of each team (generally by reviewing weekly progress reports from the teams)
- Prodding teams that appear to be taking too long
- Providing additional assistance to any team that is at an impasse (or choosing to disband the team and start over with a new team)
- Disbanding a team once it solves the problem
- Maintaining a running account of costs and benefits from the program

The permanent team members are normally drawn from middle management ranks of various departments. But how should the list of issues to investigate be assembled? Who identifies the problem areas and brings them to the permanent team's attention? The answer comes back to education. Only a small number of problems reach management's attention in a normal plant; moreover, the problems that are raised will generally be only superficial. The only way to ensure a steady flow of meaty issues from workers to managers is through education. When the workers, armed with their newly acquired knowledge, can look at their jobs from different perspectives, they will quickly begin to question dubious practices that were previously taken for granted. Even before they leave the classroom, worker-students will begin to realize that many of their existing procedures or standards are inconsistent with reliability or efficiency. The teacher writes down the questions that arise in the class and passes them along to the permanent team for consideration.

12.7.7 Cross-Pollination

Failures in quality circles also demonstrate that team members should be drawn from a variety of departments and responsibilities. Restricting team membership to a single department leads to two problems of consequence:

1. Limited insights into all possible solutions.
2. A tendency to "solve" the problem by transferring the burden to another department

Members should be drawn from all departments that might be affected by the issue at hand. The members' levels of expertise should also vary so that assemblers participate on teams with engineers and supervisors.

12.7.8 Team Failures

The failure of quality circles taught yet another valuable lesson about teams: They can take up a lot of time without producing a solution. Even when all the rules for good team selection, development, and management are followed, failures happen. This is worse than disappointing; in a dynamic industry, time lags can mean the difference between helping the company succeed and ensuring the company fails. There is no way to guarantee success, but the probability of victory can be greatly enhanced by specializing in temporary teams. The temporary team—prodded by a completion deadline, monitored constantly, and provided with whatever resources are needed—offers the best prospect for rapid, consequential change.

12.8 The Challenges of Knowledge Acquisition

Hourly workers are certainly not the only personnel in the company needing more and better information in order to perform their jobs satisfactorily. All members of the organization need to know more—if not about their own jobs then about the activities of other areas and departments that they affect or are affected by.

12.8.1 The Learning Organization

Academics have taken considerable interest in the question of how individuals within the company acquire knowledge. For example, much has been written in recent years about the "learning organization" (what MIT Sloan School of Management professor Peter M. Senge chose to call "the Fifth Discipline").[29] The high degree of executive interest in the need to develop employee knowledge can be seen in the enormous commercial success of those books despite minimal applicable content. Senge's book is the outstanding example: Although primarily an assortment of platitudes and jargon, it has sold hundreds of thousands of copies. While the concept that organizations should always be striving to learn more is solid (albeit essentially tautological), Senge has nothing to offer in the way of implementation. Even his definition of the "learning organization"—an organization that is continually expanding its capacity to create its future[30]—is unsatisfying.

Charles Handy—British economist, management consultant, and visiting professor at the London Business School—published a much more compact and slightly more useful book about knowledge in business a year before release of *The Fifth Discipline.*[31] In Handy's opinion,[32] "The learning organization can mean two things, it can mean an organization which learns and/or an organization which encourages learning in its people. It should mean both."

12.8.2 The Need for Fundamental Knowledge

It is generally assumed that employees must constantly be reeducating themselves because the company is not static and therefore the em-

[29][Senge, 1990]. Senge admitted in a 1994 *Fortune* interview that the fifth discipline designation was pulled out of a marketing hat; evidently there are no prior disciplines.

[30][Senge 1990], p. 14.

[31][Handy 1989].

[32][Handy 1989], p. 179.

ployee with stagnant knowledge will become obsolete. Clearly there is merit in that belief—provided the employee starts out with useful knowledge.

A more fundamental reason for constant educational effort is that most employees arrive at the company lacking critical knowledge of how to perform their jobs. The challenges facing the young design engineer are typical. Most electronics assembly companies hire electrical engineers straight out of school and put them to work designing layouts for circuit assemblies. The engineer turns out designs that (all going well) function suitably in electronic terms. Unfortunately for the production personnel, however, the designer has no experience with design for ease of manufacturability. School did not teach that component alignment can cause too much solder (bridging) or too little solder (skips). The designer rarely knows that a narrower range of component types and values will reduce setup time and minimize the chances of production personnel inserting the wrong components (whether by manual insertion or placing the wrong component magazine in the automatic placement equipment). Hundreds of similar issues—all of them completely unrelated to the electronic performance—determine whether an assembly will help or hinder the production department. Through no fault of their own, design engineers have no experience with any of the issues.

12.8.3 Engineers' Knowledge Problems

The design engineer's dilemma is not unique. On the shop floor, a similar type of knowledge chasm exists. Recently graduated electrical or mechanical engineers join the production department and are assigned to run the soldering machines. Nothing in their education prepared them for the chemistry and metallurgy of this job.

Faced with managing a process for which they have received no preparation, engineers emulate whatever was in place before. However, maintaining the status quo is seldom sufficient to satisfy the engineer's superiors; some improvement in output is expected as a result of the additional "talent." Thus the engineer embarks on desperate ad hoc "experiments" by changing dials and chemicals until finding a combination that provides improved cosmetic results.

The inability of newly graduated scholars to quickly master the production environment should be obvious. Yet many companies believe they can overcome complex process problems, such as machine soldering, by hiring freshly minted chemistry or metallurgy graduates. As previous chapters have stressed, chemistry, physics, and metallurgy certainly determine many of the outcomes in electronics assembly processes; therefore, given time to supplement their knowledge of academic sciences with manufacturing experience, these young scientists can become enormously valuable process designers and managers. Without the first-hand manufacturing experience, however, neither knowledge of chemistry nor knowledge of metallurgy—on their own or combined—will cure the company's process headaches. Sadly, no schools produce graduates with applicable knowledge of the core electronics assembly processes.

12.8.4 Limitations of a University Education

A large part of the problem involves America's unwarranted belief that a university education prepares the individual for immediate participation in the day-to-day operations. Despite decades of evidence that this is not true, American business still believes the myth and hands over greater responsibilities than the new graduate can handle. The sad reality is that the American education system cripples the minds and warps the perspectives of students before they enter the workforce. Starting out with such enormous educational disadvantages, the student can hardly be expected to assume responsibilities in the manufacturing world quickly.

In *Out of the Crisis,*[33] written during the peak years of post-graduate business studies, Deming wrote of the contrast between *tatemae* and *honne* in America's graduate business schools. "Students in schools of business in America are taught that there is a profession of management; that they are ready to step into top jobs," Deming fumed. "This is a cruel hoax. Most students have had no experience in production or in sales. To work on the factory floor with pay equal to half what he hoped to get upon receipt of the MBA, just to get experience, is a horrible thought to an MBA, not the American way of life. As a consequence, he struggles on, unaware of his limitations, or unable to face the need to fill in gaps."[34]

---

[33][Deming 1986], p. 130.

[34]Ironically, despite Deming's importance in Japan and the fact that *Out of the Crisis* is Deming's most successful (in numbers of copies sold) book, by the late 1980s large numbers of Japanese were seeking MBA degrees from American schools. According to *The Wall Street Journal* (June 10, 1991), an estimated 700 Japanese students attended U.S. business schools for the 1990–1991 school year. The *Wall Street Journal* article stated that an MBA degree from an American school had become the most prized academic credential in Japan. However, as would be expected in a country where only a degree from Tokyo University can open the best career doors, the prospective Japanese students restrict their interest to a small number of the most prestigious American business schools.

The choice of schools is not left entirely to chance. Several American business schools send admissions officials to Tokyo to conduct face-to-face interviews with candidates. The *Wall Street Journal* named the University of Chicago, New York University, and Cornell but note other schools are also actively pursuing Japanese students in the same manner.

Deming always singled out American business education for particular scorn. While the value of a North American MBA tends to be overblown (and, in the mid-1990s, in lesser demand), it seems rather unfair to single out business graduates as exceptionally unprepared for business life. The newly graduated engineer (or, for that matter, anthropologist, psychologist, or astrophysicist) also requires some years of professional experience before reaching peak performance.

If anything, the "hoax" (to borrow Deming's term) perpetrated on engineers is even more cruel than the misleading of MBA students. The MBA may have adjustment problems upon running head-on into the wall of real world business, but at least the theoretical underpinnings of the business school education are vaguely related to future job activities. Electrical engineers destined for the electronics assembly shop floor receive virtually no information related to the work they will be doing. Only after years of on-the-job experience is it possible to look back and realize how little of value we really knew upon leaving the academic world.

12.9 The Japanese Worker's Induction

Japanese industry has no such high expectations for its newly hired university graduates. In fact, no one in Japan—notably including the university graduates themselves—believes a university graduate is ready to play a meaningful part in any aspect of the company's operations. This distrust of universities in Japan is well founded.

Takashi Kenjo, an electronics graduate with a degree from Tohoku University and now a professor of vocational education at the Polytechnic University of Japan, uses blunt words to describe the Japanese education system: "Teaching in...universities is mainly one-way lecturing, with no questions from the pupils or students," he notes.[35] "This is a Confucian approach (interpreted in Japan as the teacher being superior to the pupils and not to be questioned). It is probably only in the East that an education system could be tolerated in which there is such a pure divide between the largely academic, non-practical, and non-applied school and university curricula, and the national application of that knowledge in industry and commerce." (Apparently Kenjo's familiarity with schools and universities in the West is rather limited.)

---

[35][Lorriman and Kenjo 1994], pp. 48–49.

On the other hand, Japanese universities may offer certain advantages over their Western counterparts, even though the first 3 years of a Japanese university education are devoted solely to rote learning. Kenjo quotes an unnamed Japanese professor who, when challenged about the memorization aspects of Japanese education, admitted that much of the criticism was true. However, he added, "students must first understand the basics. The first three years of a degree do this, and then in the final year the students work together in a laboratory and learn to apply the theory." These final-year research groups, called *Kōza*, are usually made up of a professor, an associate professor, two assistants, and a number of undergraduate and graduate students. By the time science or engineering students graduate, they probably have more experimental and practical experience than their American counterparts.[36]

12.9.1 Educating the New Japanese Employee

When new graduates arrive for work at a Japanese manufacturing company, they are considered blank canvases on which the company will paint its own pictures. Regardless of the new employee's academic credentials, therefore, the first job will be as an assembly operator followed by gradual movement up the production ladder. Fujitsu, for example, begins a correspondence indoctrination course 3 months before the employee even starts on the job. All new employees at Mitsubishi Motors spend at least 6 months on the shop floor; in the case of new graduates, it is up to 2 years. Toshiba's 1-year induction program involves the normal shop-floor acclimatization but also sends the new employee out to sell Toshiba products.[37] Bear in mind that these are graduates of university engineering programs.

12.9.2 Indoctrination at Hitachi

Hitachi's Musashi Works employs the following induction program for all new university graduates:

1. Orientation by head office (2 weeks)
   a. Lecture by top management
   b. History and outline of Hitachi

---

[36]p. 49.

[37]p. 108.

   c. Discussion of company regulations
   d. Visit to Hitachi's major factories
2. Introduction to Musashi (1 week)
3. Fundamental training (5 months)
   a. On-the-job training in manufacturing section (blue-collar)
4. Assignment to a section (18 months)
   a. Computer training (1 month)
   b. Basic semiconductor course (100 hours, 8 hours per week)
   c. On-the-job training in a specialized field
   d. Preparation of a treatise, with the subject chosen by the trainee's section manager
   e. Oral and written thesis presentation
5. Writing and presenting a thesis on a topic the section manager feels will be of use to the company. This fascinating requirement—totally and unfortunately unknown in Western industry—demonstrates why teams work better in Japan than elsewhere.

   The thesis presentations last between 20 and 30 minutes, and all hires from any given year make their presentations at a single gathering called a "conference." The company places great importance on the conference. Senior management make a point of stopping by, and an engineer's career can be determined on the spot by a particularly good or unusually poor presentation. The poor presentations are apparently minimal because the section manager's reputation also gains or loses according to the performance of the section's new hires at the conference; consequently, each young engineer receives considerable coaching from the section manager during preparation of the thesis.[38]

   Hitachi originated the conference concept in the mid-1950s, and the concept has since been copied by virtually every Japanese electronics company. In the meantime, Hitachi has expanded on the conference concept to the point where every engineer can expect to participate in three or four events during the first 10 years with the company.

12.9.3 The Need for Lifetime Employment

Clearly, the company's investment in each employee is enormous. Under any circumstance other than lifetime employment—actually a mutual lifetime commitment between employer and employee—taking such a financial stake in an employee would not be practical. That

[38]pp. 108–110.

mutual commitment has prevailed in major Japanese corporations for decades.[39] Japanese companies have not raided competitors for employees and employees have not been inclined to leave their original employer for another. It is the knowledge that the employee will be on board for life that allows the major Japanese companies to make massive investments in their people.

In America, where job hopping is a way of life and companies regularly raid the talent banks of their competitors, this system simply will not work. Only if the state, through the education system, provides the kind of applicable skills education in trade schools and universities will America be able to match the Japanese labor force's competency. Alternatively, some form of employment contract could conceivably be drafted between the company and new employees, although American employment law would make such contracts difficult to enforce.

The problems American companies would face trying to hold employees to contracts have already been experienced in the United Kingdom. Companies there in the 1970s and 1980s paid for expensive postgraduate education; MBAs and MMTs (Master of Manufacturing Technology) were popular choices, with the latter costing £20,000 for a 2-year program. Attempts were made to ensure the employees participating in such programs would remain with the company for a reasonable period after completing the program. Generally, the attempts took the form of contracts between employee and employer. The contracts have not prevented beneficiaries of the free tuition from moving to other employers once they have their degrees. Court remedies for the employer are simply too expensive and time-consuming to be feasible. The result has been that the contractual arrangement is seldom used today.

The light at the end of the tunnel for America, unfortunately, does not involve higher standards here or abroad. Rather than American workers being brought up to recent Japanese standards, it seems that Japan will be reduced to the lower American level. Unemployment in Japan in 1996 reached post-war record high levels and was still climbing. So, for the first time in at least 75 years, Japanese companies and employees must experience layoffs or bankruptcies. Without the security of knowing today's new employee will work the entire career in that company,

---

[39]The "lifetime" employment coverage typically extends only to those direct employees of the major corporations; employees of smaller companies—even smaller affiliates of the major corporations—do not enjoy this protection. Consequently, the "lifetime" employment coverage has applied to fewer than 20 percent of Japanese workers, although Western observers have been encouraged to believe the practice is universal. A seldom mentioned liability of "lifetime" employment is that mandatory retirement age has been around age 55!

long-term development of employees is on the verge of extinction in Japan as well as America.

If Japan loses the lifetime bonding between company and employee that makes great investment in every employee feasible, Japanese companies will lose what is probably their greatest manufacturing advantage. The current system provides more than just greater technical knowledge. Equally as important as development of technical know-how, a human networking builds up throughout the company so that there is always someone in the system to whom any employee can turn for assistance. The closest to this kind of networking that we have experienced was found in Edinburgh's Ferranti Defense Systems, now part of GEC Marconi Avionics; new graduates starting at Ferranti were assigned a "mentor" who had many years of experience with the company's systems and personnel.

12.9.4 Apprenticeship

In Western terms, the Japanese system where everyone starts at the bottom might best be described as apprenticeship. Granted, this is not the traditional indentured apprenticeship system known in America and Europe. Perhaps it is best called *pseudo-apprenticeship*. Whatever the appropriate terminology, every Western country could make enormous improvements in quality and efficiency by requiring that every new engineer and technician start as assemblers for at least 4 to 6 months. The remainder of the employee's first year at the company would be spent gaining hands-on experience in all the other departments. Pseudo-apprenticeship is not a cure-all, especially in the current climate of rampant self-interest where neither employer nor employee feels much loyalty to the other. Nonetheless, adopting some variation on the apprenticeship system will take any electronics assembly company to high performance levels previously thought to be completely unattainable.

The pseudo-apprenticeship role for recently graduated new employees destined for upper-level roles in the company has so many obvious advantages that it can be difficult to understand why so little use is made of it. As it happens, almost all parties within the company and among the prospective employees have reasons for opposing an induction period involving all manner of blue-collar jobs. A few of those reasons include:

- Reluctance of companies to invest heavily in employees who may move to another company (possibly a direct competitor) once the induction period ends.

- Shortsighted desire by the company to get engineering work out of newly hired engineers with the shortest possible delays. Inevitably, short-term thinking leads to long-term pain in the form of inferior employee performance throughout their careers with the company. Investing a relatively small amount in an engineer's early development will be repaid many times over by fewer design flaws or more competent process engineering.
- Managers of departments where the new hires will work likely did not experience the pseudo-apprenticeship induction when they joined the company. Not having any personal experience with the concept, they naturally have no way of understanding the benefits. One thing department managers understand very well, however, is that standard cost accounting practices will penalize them while high-priced help is off working at tasks where wages are normally minimal. (The standard cost penalty can be averted by not assigning the new hires, along with their wage penalties, to an engineering department until the necessary experiences have been acquired. Until that time, they would simply be carried on the company's books as fixed costs.)
- The newly hired young engineers will not be enthusiastic supporters of a system that starts them working beside assemblers who never completed high school. Some inexperienced engineers will refuse to join a company that requires pseudo-apprenticeship. The company is better off without the prima donna mentalities; those same individuals would later prove to be poor team players.

Over the past 150 years, the Japanese have shown that copying techniques from other nations and adapting those techniques to local conditions pays impressive dividends in quality and efficiency. Pseudo-apprenticeship is one Japanese concept that Western companies should copy shamelessly. The opportunities for company-wide improvement offered by such modified apprenticeships are too great to be ignored. We cannot overemphasize the value of this practice for any company.

12.10 The Manager's Role

A failing in contemporary quality ideology, we believe, is the imbalance between glorification of workers and denigration of managers. Managers today are caught between workers who want more authority (but often not the accountability that accompanies authority) and executives who believe middle managers are fat to be cut in an age of lean organizations. A good operations manager (especially in quality) requires the following formidable list of talents:

1. The logic of an engineer.
2. The insights of a detective.
3. Knowledge and appreciation of chemistry and physics.
4. Willingness to get hands dirty.
5. Instinctive recognition of inefficiencies.
6. Impatience with the status quo.
7. Rejection of complacency.
8. The business acumen of a CPA.
9. Sensitivity to people coupled with the strength to lead.
10. Courage to challenge the corporate bureaucracy right up to the CEO if necessary.
11. Willingness to put forth unstinting effort on tasks where the credit will flow to others.
12. A logical mind (too often termed "common sense," even though it is quite uncommon) is essential, as are analytical skills. Higher education is not necessary, but an ability to learn is essential.

Aside from those few requirements, anyone can do the job.

12.11 Concurrent Engineering

Much has been made in the last decade of *concurrent engineering*, an organizational approach to overcoming problems of communication among the various departments involved in the product from original concept to production.

Without concurrent engineering systems, the design engineers work more or less in isolation from the downstream departments that will be responsible for turning the design into a real product. Eventually, a "final" design is handed off to the procurement and production personnel, who would then try to overcome manufacturability nightmares that the designers had not foreseen. While production is trying to devise ways of manufacturing a product that may, for example, require component placement tolerances tighter than the automated placement equipment can meet, the quality department is caught in the classic "fix or ship" battles with sales. More often than any company likes to admit, design changes were still taking place weeks or months after the first units had been sent to customers; this is where the cynical term "designing in the field" originated.

Under concurrent engineering, the design engineer works with liaison personnel from the other departments that will inherit the post-design responsibilities.[40] If the design involves lead-to-hole ratios that will cause alignment problems for the placement equipment, the team's manufacturing engineer can spot the problem months before the drawings would have been finalized and passed along to the production department. (Theoretically, CAD software catches such routine problems even if the design engineer doesn't know better. Somehow, though, these theoretically airtight scenarios manage to become dangerously porous in execution.)

Concurrent engineering has been successful almost everywhere it has been tried, although the magnitude of success varies from one company to another. Concurrent engineering has proved particularly effective in reducing:

- Squabbling between production and design personnel
- Product development time
- Product cost by ensuring that information about new component or materials technologies flow into the design department faster

None of those improvements should be particularly surprising. Nor should the inevitable arguments that arise when some decisions will make life easier for one or two departments at the expense of others catch general managers unprepared. The real surprise is the failure by industry leaders to recognize that concurrent engineering is an inadequate "solution" to what may be the greatest single danger to our international competitiveness: departmental inbreeding.

Western industry has too many narrowly focused young technical experts whose companies are not providing the job experience in varied departments that previous generations of manufacturing professionals took for granted. Somehow the quality movement's insistence on personnel development for line workers stopped short of the production

[40]Logically, field service and other facets of after-sales support would also participate in the concurrent engineering exercise. Ferranti's product support department placed their engineers on design teams in the realization that the department would eventually assume responsibility for post-design services—e.g., in-service modifications and updates to equipment capabilities—on each product. That cooperation eliminated long-standing product support problems.

professionals. The last generation of broadly experienced manufacturing professionals are already retiring. When the last veteran generalists put production life behind them, will stopgap measures like concurrent engineering be sufficient to keep our plants competitive with Japan and the other, faster-growing Asian industrial nations? It seems unlikely.

12.12 Summary

- All companies need their employees' minds and hearts in addition to their hands.
- Enlisting those minds and hearts is an important function of the modern quality department.
- Workers have been inadequately educated by the school system; while the workers generally *want* to do the best possible job, they have not been given the knowledge tools that would make such contributions practical.
- Employers must provide the knowledge and experiences that were denied their workers by the education system.
- Western observers have wrongly interpreted the aspects of Japanese management that motivate Japan's workers.
- Quality circles contained the germ of a good idea but overlooked the substantial challenges involved in making line workers effective problem solvers and de facto managers.
- Western companies should study the way university graduates serve a form of apprenticeship upon joining their new employers.
- Some aspects of Japanese personnel development require substantial investment in the employee and would not be practical without lifetime employment.
- In designing quality participation programs, quality departments must take a more pragmatic view of employee development than what is found in mainstream quality literature today.
- The rewards of teams can be enormous—and the workload for the team supervisors (often the quality department) can be even greater.

13 Design for Dependability

13.1 Chapter Objectives

Product cannot have dependability unless its design is inherently dependable. The design engineering department therefore greatly affects product quality. Design issues and some recommendations for maximizing the dependability of designs are presented in this chapter.

13.2 Design and Quality

Previous chapters have shown that quality is determined by many factors, not all of them directly under the quality department's control. Development and supervision of manufacturing processes profoundly affect the dependability of output, but the process engineering personnel are rarely found under the quality department on the company's organization chart. Likewise, purchasing and stores—although generally independent of the quality department—have been seen to affect dependability. But the difficulties of influencing those departments tend to be trivial compared to the challenges of redirecting the department where the ultimate attainable dependability is established: design. Of all departments concerned with development, production, and verification of product dependability, design engineering tends to be the most inbred and resistant to external influence.

13.2.1 Quality Begins with Design

Product design is, of course, the first vital step toward dependability. No matter how much knowledge and effort other departments expend trying to produce acceptable output based on flawed designs, dependability always suffers when design engineering blunders.[1]

A flawed design means an undependable product—together with endless headaches for the production and quality personnel who must try to turn a design sow's ear into a product silk purse. Despite the best efforts of all other departments, quality can never exceed the limits afforded by design. A defective design will result in consistently faulty output. Other departments, including Quality, may succeed in identifying failure modes in designs, but until design is changed the only remedy will be rework—a "solution" that has been seen in previous chapters to ensure increased failures.

The inescapable linkage between design and dependability is well known to quality and production personnel. Time and again, these troops find themselves pitted against design engineers who do not—or will not—understand how certain design practices can create havoc in the downstream departments.

13.2.2 Design Is Not Always the Problem

A word of caution is necessary at this point, however. Many quality problems are improperly attributed to design flaws when the real culprits are process and/or quality evaluation deficiencies. Design personnel are not always innocent, but they typically receive far more criticism than they deserve.

How many quality and production problems can properly be ascribed to design missteps? Obviously the answer varies among companies and even from product to product within plants. However, our experiences have invariably been that the number of perceived design errors greatly exceeds the number of real design errors.

Over the years we have participated in reviews of hundreds of electronics assembly plants around the world. In almost every case, quality

---

[1]The statement that flawed designs lead to undependable products should not be construed as criticism of design engineering in general. It does mean that design engineers, being human like personnel in quality and production, occasionally make mistakes.

and production personnel blamed inferior designs for most assembly problems. But when we examined the "problem" designs, relatively few—typically only about 5 percent—grossly violated appropriate design practices; the other 95 percent could be processed using scientific process management techniques discussed in previous chapters. A substantial portion of the remaining "problem" designs could have been improved—primarily in ways relating to ease of assembly—but would have caused few difficulties for better-informed production and quality personnel.

The quality department is responsible for identifying and, where possible, helping correct all aspects of the plant that reduce product dependability. Design integrity certainly falls into that category of quality department responsibilities. But quality management must always be certain that the identified problems are real rather than illusions. Consequently, before launching offensives against design improprieties, it is best to ensure that the flaws do not lie in quality evaluation or production procedures.

13.3 Some Basic Principles for Achieving Dependable Designs

In a world such as electronics where rapid changes in technology are normal rather than exceptions, design problems will be encountered periodically. Expecting to prevent all design mistakes is probably unrealistic; it is totally realistic, however, to learn from past mistakes and act in ways that will minimize the prospects for failure-prone designs. A few obvious basic guidelines for avoiding troublesome designs include:

- Avoid complexity wherever possible; strive for lower rather than higher component densities and use multifunctional components that reduce the total number of components needed for any given assembly.
- Minimize variety such as the number of component types or values.
- Shun attributes known to have caused problems on previous projects.
- Know the limits of the company's production capabilities—and design in margins for error. It is not always necessary to have an exact knowledge of the process capability; indeed, it is often advantageous to underestimate the process' capabilities (i.e., assume more restric-

tive process tolerances than actually exist), since this will increase the probability of the process being able to accommodate the design.[2]

- Seek innovative ways—such as components configured to make inverted insertions impossible—to prevent assembly errors.
- Document every success and failure as the company's designs change; incorporate these experiences into formal design guidelines.
- Make accurate prototypes of new designs and test them rigorously prior to approving them for production. Harsh environmental stress screening should certainly be employed; the warnings against ESS contained in Chap. 11 apply only to units destined for customers.
- Maintain formal multidepartmental design review boards that can assess all new designs. Restrict turnover of participants in this review process, since the ability to identify potential problems at the design stage can be acquired only through considerable experience.
- Subject all new designs to evaluations such as Failure Mode and Effects Analysis (FMEA) by the design review board as well as the design department. FMEA need not be complex; determine the consequences of failure for each component and invest the most resources in techniques that minimize the possibility of the most serious failures.
- Employ formal CAD libraries that can be—and are—constantly updated in accordance with the plant capabilities rather than arbitrary industry "standards."
- Stay constantly aware that the costs of initial design mistakes have never been greater—and are constantly increasing.

13.4 Design for Dependability

Design has purposes beyond delivering quality. Meeting the marketing department's requirements for certain feature sets producible at a com-

[2]Manufacturing companies often invest considerable funds and effort in measuring the so-called capability index (Cpk), which expresses the relationship between design requirements and process capabilities. While this is a useful concept for processes such as precision machining, very few elements of electronics assembly lend themselves to such analysis. One of the most obvious examples of an electronics assembly operation where calculation of a Cpk would be feasible is automatic placement of components.

petitive price is generally as important to the designer as the quality aspects. In these pages, however, only dependability matters.

Dependability in design consists of four attributes:

1. Functionality
2. Manufacturability
3. Testability
4. Durability

The relative and absolute importances of each attribute varies among companies. The company that has minimized the need for testing by employing strategies outlined in Chap. 11, for example, will value testability less than a company that has high failure rates. Similarly, the relative and absolute importance of each attribute varies among departments. Functionality has considerable importance to sales and marketing (which we will term collectively "Marketing" throughout this chapter) but manufacturability will be more important to production personnel. The quality department—which has as its objective assurance that the product provides its specified functions for the specified period under the specified operating conditions—will therefore find itself negotiating with other departments at the product concept stage, and all parties must make compromises to arrive at the most acceptable combination of dependability and cost.

Negotiating the level of dependability is anathema to the modern quality manager, particularly one who fully supports the core principles of TQM. But the quality professional must recognize that the company is a system, not a collection of independent activities. In a system, it is necessary to find an optimal level for each function. Detailed discussion of quality optimization will be left until Chap. 18. For the purposes of this chapter on interaction between design and dependability, only a general appreciation of the quality optimization concept is necessary. In particular, it is important to note that design-related dependability improvements will affect product costs.

The actual expenses directly incurred by the design department ("Design") in product development account for only a minor portion of total product costs. After taking into account other costs, including presales repairs and post-sales field service, the net effect of design-based dependability improvements can range from substantial reductions in costs to unacceptable increases. Common examples of cost-reducing design actions are those which enhance the ease of manufacture (i.e., manufacturability) while the most costly involve substitution of more expensive parts. Of course, flawed circuit designs requiring modifications of components—including unpopulated circuit boards—represent design at its worst.

13.5 Functionality

The definition of dependability states that the product will provide all specified features under specified conditions for a specified period of time. Collectively, the specified features constitute the product's promised *functions.* The maximum extent to which design permits the product to provide the promised functions can be defined as the design's *functionality.*

While functionality and dependability are closely related, the former is in large measure latent while the latter is achieved. Functionality represents *potential* quality while dependability is the product's *realized* quality. Realized quality can never exceed potential quality. Most often, realized quality falls short of potential quality. In other words, dependability cannot be greater than functionality and will often be less.

Design is the only determinant of functionality. Production, quality assurance, suppliers, and other activities outside Design determine whether dependability lives up to functionality. Accordingly, design for dependability really means setting the upper limits for dependability.

13.5.1 Establishing Features

Features have been discussed in previous chapters, particularly in Chap. 3, where it was observed that features are not synonymous with quality. Features relate to quality only if the product fails to provide any of the promised features for the specified period of time; any product that provides its promised features for the promised time period under promised conditions of use has perfect quality. Marketing specifies for designers the features to be included in the product along with the anticipated selling price. (Whether the customer is attracted to that features set is a separate matter entirely, an issue that is the responsibility of Marketing. Features originate as Marketing instruments.)

The flow of communication is not unilateral from marketing to design. Design personnel not only receive directions about product features from Marketing but should also keep Marketing advised about the feasibility of including requested features and drawing attention to available features that Marketing may have overlooked. This advice from Design to Marketing will be both reactive (evaluating the practicality of features requested by Marketing) and proactive (informing Marketing of new technological availabilities).

13.5.2 Fitness for Use

In Chap. 2, the disagreement between Crosby and Juran regarding the relative merits of their "conformance to requirements" and "fitness for use" concepts of quality was examined. In particular it was suggested that their opinions were much closer than either seemed to appreciate; indeed, we considered them identical for all practical purposes. The reason was that both positions could be reduced to ensuring that the product meets the customer's needs. Needs, in turn, are not necessarily reflected in the product's specifications or in the customer's desires. A product can meet its specifications but neither conform to requirements nor be fit for use.

13.5.3 Optimum Specification Levels

Specifications, of course, first enter the picture in the design stage.[3] There are three possible levels of specification:

1. Specifications are below customer requirements.
2. Specifications equal customer requirements.
3. Specifications exceed requirements.

Which of these three possible relationships between specifications and requirements are acceptable?

Setting specifications below requirements or needs will certainly not help the company prosper. Customers may initially be attracted to the product for reasons including (1) not fully appreciating the limitations of the specifications and (2) buying according to wants without recognizing that their needs are greater than their wants. Many of those customers will return the underspecified products for refunds; others will not buy the product again and may tell acquaintances of the product's shortcomings. In some cases—where use of the product as provided endangers the customer's health, for example—the manufacturer may be forced to retrofit improvements to sold units, recall all units, and/or face legal actions from injured customers (or their estates). All told, insufficiently demanding specifications can easily drive the company out of business.

---

[3]Design rarely sets specifications on its own, but it will normally have a voice in the decision making.

While instances of underspecification are certainly not unknown, few design engineers today are guilty of setting their sights too low. A much more common problem involves setting standards unreasonably high. Designers occasionally feel that the company's interests are best served by designing in features that are not part of the specification. In all too many cases, the company's quality personnel do not appreciate that overdesign does not result in higher quality.[4]

Features that go beyond specification most often relate to variation in performance such as higher signal-to-noise ratio in communication transmission or reception equipment. Less commonly, functions that did not exist in the product specification may be added. If the additional features increase product cost, Design has exceeded its mandate and—assuming Marketing knows its business—harmed the company. Quality management needs to treat designs that exceed specification at increased cost as seriously as it deals with product that falls below specification. This facet of the quality department's duty—ensuring that features do not exceed requirements—is seldom mentioned and even more rarely practiced.

13.5.4 Quality Function Deployment (QFD)

The reader familiar with contemporary quality jargon will doubtless have noted the absence of discussion about Quality Function Deployment (QFD) to this point. Considering the great emphasis placed on QFD by many quality authorities in recent years, this may seem a grievous oversight. However, the absence of earlier reference to QFD (and its omission from the remainder of this book) is deliberate. QFD may have value as a tool for Marketing but cannot properly be considered a tool of Quality.

One of the best definitions of QFD is found in [Shina 1991]: "Quality function deployment is an organized, disciplined process for determining the product or service requirements necessary to meet the stated *or implied* customer *wants* and needs. It requires the horizontal integration of those organizational functions that must Plan, Do, Check and Act in order to successfully achieve customer-perceived expressed or *unexpressed* quality. Quality in this context is defined as the ability to meet *or exceed* customer *expectations* while maintaining a cost-competitive mar-

[4]In this respect, they operate very much like quality inspectors who send acceptable product for touchup and rework.

ket position."[5] In other words, QFD centers on the "customer-driven quality" that was examined and dismissed in Chap. 3 as being outside the scope of the quality department.[6]

Several well-known manufacturing companies—particularly in the automotive industry, where Ford has been the movement leader—have employed QFD in the past decade. They have expressed considerable satisfaction with the results. For readers with an interest in this marketing tool, Shina is an excellent reference.

13.5.5 Engineering Change Orders

Every electronics assembly department's worst nightmare is the "engineering change order" (ECO)—also commonly known as "modifications" (mods)—to parts, subassemblies, and final assemblies. The need for ECOs comes from failure to deliver a dependable design to production the first time.

ECOs mean significant extra work, often excruciatingly complex or precise, for assembly personnel. This then translates into extra work for the quality department. The extra work comes in the form of record keeping, visual and test checking of the materials revisions, higher failure rates, and so on. The quality department's anxiety level should also vary directly with the number of ECOs, reflecting the inverse relationship between dependability and the number of engineering changes.

ECOs are employed to compensate for faults discovered in circuitry after the system has entered production (or, if production has not begun, custom components such as printed circuit boards have been ordered). There is no "good" ECO scenario; however, the magnitude of the problem increases according to the stage in the product's life at which the design flaw is discovered. Often the problem will show up during prototype testing but after the circuit boards have been ordered; the problem at this stage may be limited to opening a trace prior to populating the board, although with modern multilayer circuit boards, the "fix" may involve altering interior tracks, running jumper wires, or hard wiring components such as resistors or capacitors to the leads of com-

[5]pp. 148–149. Italicized emphasis added by the authors.

[6]Our experiences being limited primarily to the electronics industry, we were interested in whether the industry had misinterpreted the true meaning of QFD. Discussions of QFD with Suntory Brewery in Japan made perfectly clear that its aim was to determine the preferred characteristics of beer. In other words, QFD is evidently employed throughout Japan primarily as a marketing research tool.

ponents specified in the original designs. If the problems do not show up until test, it is often necessary to replace components and always necessary to retest; therefore, the workload and costs increase. Of course, by far the most costly point at which design flaws can surface is after units are in the field and must be recalled or serviced in the field. Regardless of the point in the product's evolution where a design flaw surfaces, an ECO costs more than doing the design right in the first place—more for labor, more for test, and (by subjecting the assembly to greater stresses) more for undependability.

Serious as the industry's experience has been with costly ECOs and mods in the past, the future almost guarantees far more dependability-related design problems as companies increasingly rely on new features to create demand for their products. The search for new features requires that designers constantly rush products into production before designs have been proved and include features based on technologies which they have not mastered.

13.5.6 Committing to the Unknown

Undertaking new products incorporating unproven features is a perennial quality problem for electronics assembly companies. Few companies beyond the startup stage have avoided committing to features that they subsequently are unable to deliver. "Unable to deliver" may be absolute, in that the feature cannot be included at any price, or relative in the sense that, while technically possible, the feature proves economically impractical.

In certain industry sectors—avionics, defense contracting, and aerospace, for examples—contracts are few and very large. Sales and executive management regularly bid on contracts that involve unsolved technical problems in the hopes that the problems will be solved in time for product delivery. Design personnel generally find themselves placed under enormous (even, in the opinions of many, unreasonable) pressure to sign off on bids as being capable of full compliance.

In some cases involving highly complex development programs (again typically found in industry sectors such as avionics, defense electronics, and aerospace), suppliers and customers both recognize that early equipment (the development models) will not be fully compliant with the contract terms. Agreement on some contract specifications will not be reached until long after the project is underway.

Additionally, the software for complex applications (for example, the "fly-by-wire" flight controls of new generation commercial aircraft) cannot be fully exercised by simulator techniques. The final evaluations of such systems must wait until the hardware is complete and evaluation units ready.

13.5.6.1 Consequences of Committing to the Unknown. The consequences of unsuccessfully pushing the technological envelope can range from disappointing to embarrassing to calamitous. If the company has never publicly announced plans for the product and it can be abandoned before investments in product-specific materials and equipment are made, the only loss is internal. Despite lost development funds and bruised egos, the net effect may even be positive if the company acquires additional expertise—or just new found humility that prevents future costly acts of hubris.

After contracts have been signed or marketing programs begun, the costs of failure increase substantially. Failure to deliver contracted features in a timely manner may bring about customer lawsuits that could jeopardize the contractor's existence. Bringing the product to market even though the manufacturer knows performance will be less than specified undermines the company's credibility and angers customers.

13.5.6.2 Who Is at Fault? Committing to features that turn out to be undeliverable is widely perceived as the fault of overzealous sales personnel. Unquestionably, as was seen earlier in this section, sales departments do enter into unfulfillable contracts for blue-sky products. Indeed, some companies seem to specialize in promising much more than they can deliver. But not all the blame can be attributed to sales personnel. Naive or overconfident design engineers commonly commit to delivering technological breakthroughs without any experience to back up their self-assurance.

13.5.7 Functionality and the Quality Department

Since designs that lack full functionality limit the maximum attainable product dependability, functionality would be a matter of great concern to the quality department even if no other factors entered the equation. However, errors in design too often translate into conflicts between ethics and career prospects for the quality management. Many—probably most—quality directors at some point have been asked by powerful executives to sign off on product that falls short of requirements. In a few instances, quality personnel have been coerced into forging test results and other documentation. Quality managers who insist on conformance to requirements in the face of relentless executive demands to ship nonconforming product can find their jobs extremely arduous—assuming, of course, that the quality department wields sufficient power to prevent shipment of nonconforming product.

Occasionally, quality managers resign in the face of what they perceive as undue pressure to ship unacceptable units. But the quality manager who resigns to protest company attempts to ship nonconforming product only to be replaced by a more pliable individual who authorizes the shipment has sacrificed in vain and will not be around to influence future company quality decisions. Fortunately, there are usually alternatives that allow quality management to ensure output conforms to requirements without burning career bridges.

13.6 Manufacturability

Manufacturability refers to the relative amount of effort required by the production department to turn out product that meets requirements. A product with high manufacturability can be thought of as "production friendly." Manufacturability is also known as "producability." A highly manufacturable design requires little unusual effort on the part of production personnel.

Manufacturability is often regarded as strictly a production department issue. In fact, it profoundly affects dependability and is very much a quality issue. As should be apparent from earlier chapters, production personnel can tame unfriendly designs through extraordinary process management efforts or camouflage (i.e., touchup, rework, and repair).[7] This extra effort places burdens on already overworked production personnel and diverts attention from other pressing matters. Camouflage, of course, ensures higher failure rates.

13.6.1 Process Capabilities and Designs

Any product designed in such a way that the process specifications—such as tolerances of component placement equipment—exceed the

[7]It is not uncommon to find manufacturing engineers who take great pleasure in attempting to produce the unmanufacturable. This attitude, of course, is entirely consistent with other types of managers who feel most secure when required to resolve endless crises. (That is, individual objectives are not always consistent with company goals.) Therefore, Design cannot always count on production department input to identify and salvage design defects—even when the design engineers provide genuine opportunity for interdepartmental participation in design development.

process capabilities is, of course, destined to fail. Such a product will be uneconomical to build and undependable to use. Accordingly, the design engineer's concerns must go beyond whether the product will work to ensuring that the product can be easily produced.

Process capabilities are determined by the theoretical tolerances of the processes and the actual tolerances dictated by the company's process management skill and discipline. Often, apparent problems in manufacturability can be overcome by:

- Acquiring and applying more advanced process management knowledge to existing equipment and materials
- Purchasing more precise production equipment
- Improving maintenance of existing equipment
- Generating designs that require less rigorous tolerances

As earlier chapters have indicated, we consider acquisition and implementation of additional process knowledge to be the most desirable option. Companies that possess better process knowledge are able to cope most effectively with designs most dramatically affected by process flaws. Moreover, technically knowledgeable companies are less prone to permit unreasonably demanding designs to take hold in the first place.

Design engineers who try to stretch processes to their limits will end up "breaking" those processes. Design engineers who put the company ahead of their personal convenience will always strive to minimize the demands placed on processes, equipment, and personnel. The design engineer of greatest value to the company is the individual who refrains from pushing more complex or tolerance-sensitive assemblies on the production department.

Of course, most designs today are laid out on computers running CAD programs. In theory, this ensures satisfactory layouts, since the CAD software contains data about unacceptable design patterns and intervenes when the designer violates manufacturability guidelines. In practice, the software's capabilities are no better than the knowledge of the programmers. With input from the production department, some designers can—depending on the software's flexibility—modify software parameters to more closely match the shop's needs.

13.6.2 Some Easy Ways to Enhance Manufacturability

Not all techniques for reducing the difficulty of assembly involve great effort on the designer's part, extensive investment in elaborate capital equipment, or intimate knowledge of advances in component technologies. Many of the most effective manufacturability enhancements actually involve little more than taking a few minutes to think through the assembly logistics.

A few examples of important, though obvious, ways of increasing manufacturability include:

- Given the choice between a component that can be automatically placed by normal placement equipment and a functionally identical component that is not compatible with any standard placement equipment, pick the former.
- If assemblies must be washed after soldering, specify sealed components that can be washed rather than open components that must be applied post-cleaning.
- Specify plastic snap fittings rather than screws and rivets.
- Use checkerboard rather than solid ground planes if possible.
- Avoid sockets except where integrated circuits must be changed regularly in the field. If all upgrading of integrated circuits will be conducted at repair depots, dynamic solder pots allow easy removal and replacement of soldered integrated circuits. Sockets often cause connection problems.

The list is by no means comprehensive, but the simple nature of the procedures shows that enhancing manufacturability need not involve great expense, effort, or special knowledge.

13.6.3 Physical Separation of Design and Production

All too often, however, designers and manufacturing personnel occupy parallel worlds that never meet. Manufacturing has increasingly migrated to lower-wage areas, but design engineering has not always followed that geographical movement. It is not unusual today for designers to be located in another state or even another country from the production facilities. To the designer, therefore, the manufacturing engineer who complains about the designs is only a voice on the phone with no ability to affect the designer's career. But the design department manager—who certainly will determine the designer's career path and is in an office down the hall—decides pay and promotions by the number of new product designs that come out of the designer's computer. Forced to choose between requirements within the office and complaints from a disembodied voice, only the most unusual designer will put manufacturability at the top of the list.

13.6.4 Other Obstacles between Designers and Production Personnel

The manufacturing department has additional obstacles preventing it from communicating manufacturability requirements back to the design office. A few such problems include:

- The manufacturing department often cannot articulate what makes one design easier to produce than another. At an instinctive level, production personnel may know there is a problem but lack a scientific or statistical foundation that they can take to the designer as evidence.
- The manufacturing processes may vary from one day to the next. When this happens, the best results are often achieved with a layout that would generate horrendous difficulties if the production processes were consistent and scientifically valid. As an example, the number of solder bridges that will be produced by a properly calibrated wave soldering machine will be minimized when leads of dual in-line processors run parallel to the wave; if the wave profile is not set properly, the least number of bridges may be obtained when the leads are offset 90° from the solder wave. (Obviously, if the production processes are running out of control and the manufacturing engineers do convince the design engineer to change the CAD parameters accordingly, future designs will generate high defect levels if the processes should ever be perfected.)
- The design department may contain several design engineers. To gain compliance with its needs, the manufacturing department must obtain agreement from all designers involved in the project. The difficulty of obtaining changes in the way designers approach their work increases geometrically with the number of individuals involved. The complexity increases even more dramatically when the designers lay out circuit assemblies for different plants; if each plant runs a unique set of processes, there will be disagreement among plants as to the best design attributes.
- The manufacturing department may speak a different language than the designers. Differences between working languages have played

an increasingly important role as more production has been moved to foreign countries.

Manufacturability raises a vast array of production investment questions that are not always easily answered. Most companies are constantly struggling with the choice between adopting new bonding technologies that may allow package size reductions but also require new capital equipment and staying with older technologies that are compatible with existing equipment. The quality management is not always invited to participate in these debates but needs to be involved when changes in manufacturability demand changes in processes or equipment that may affect dependability.

The answer to the question of whether to adopt new component technologies is actually more easily reached than the vast amount of research and numbers of articles published on the topic suggest. Being a leader in adopting new types of component technologies makes sense for very few types of electronics companies. Aerospace, aviation, and mobile communication are among the few industry sectors where significant market advantages can be gained from early use of shrinking components.[8] The need for adopting smaller packaging in other electronics sectors is less obvious.

Under certain circumstances, the only consequences of replacing standard components with smaller packages may actually be negative. One North American telephone company miniaturized the circuitry of its desktop telephones to an unprecedented degree. The resulting unit was so light—and easily pulled off desks—that the manufacturer was forced to put lead weights in the housings. The miniaturized design was more expensive than the design it replaced in several ways:

- The miniaturized components cost more than their larger mature technology counterparts.
- Tighter packaging caused manufacturability problems.
- Expensive new production equipment was required.
- Test of final units was difficult.

The phone sets are now made by a Third World contractor.

[8]Interestingly (and quite properly, in our view), McDonnell-Douglas traditionally expressed grave concern over design proposals where new technologies or techniques represented more than about 30 percent of the proposed equipment. It was felt that exceeding this upper limit of unexplored technology would present a threat to equipment dependability.

13.7 Testability

As Chap. 11 noted, test can constitute a third or more of all non-material costs. Therefore, designing units for ease and speed of testing can greatly reduce total costs.

Taken literally, it can readily be argued that the best form of testability is a design which is so free from problems and easily manufactured that no test is required. In practical terms, however, some test stage(s) will normally be necessary. The design engineer must be aware of what test procedures will be involved in product evaluation and design attributes that will be of most assistance in conducting those tests. This information can properly be obtained only through direct discussion with the test engineering section of the quality department. In other words, the test engineers must be involved in any design evaluation (participating in any concurrent engineering activities that may be undertaken).

Since the test activities vary widely throughout the industry, there are few absolute rules governing the testability aspects of design. One absolute rule where bed-of-nails testing will be employed is to provide readily accessible contact pads across the surface of the circuit board. "Readily accessible" means that no components will interfere with the action of the test pins. Building self-testing circuitry and software routines into the product can be a cost-effective means of enhancing testability in complex systems.

Testability should be an important design concern of the quality department. The costs of creating formal design testability requirements based on the company's test procedures are generally much lower than the savings achieved by reducing the effort involved in test.

13.8 Durability

Durability incorporates two main issues:

1. The environmental extremes in which the product will operate dependably
2. The stability of performance over time

Greater durability usually coincides with higher costs (although, as the discussion of heat damage in Chap. 7 makes clear, some durability enhancements are available through costless refinements to processes). Arriving at the proper level of durability therefore requires balancing the customer's requirements and willingness to pay against product cost.

13.8.1 Upgrading Components

Durability can be enhanced by employing more rugged components. Some elements that could be construed as "ruggedness" might include physical sturdiness, resistance to fungal growth in tropical climates, and ability to withstand rapid changes in thermal conditions.

Most components are available in various grades. Printed circuit boards can be fabricated from a wide range of materials ranging from resin-impregnated fiberboards to complex glass laminates. Components for mounting to printed circuit boards are often available in through-hole or surface-mount and leaded or leadless configurations. Packages may be plastic or ceramic (in some cases, circuitry is not packaged at all). Variations from rated values can be negligible or significant. The possible permutations and combinations of components to perform what are nominally identical functions can be overwhelming. And the design engineer must take account of all these options.

In a relatively small number of applications—particularly those where lives are at risk—the designer's decision is straightforward: Use the most dependable (i.e., least variable and most durable) possible components *regardless of price.*[9] But most of the manufacturing world does not enjoy the luxury of freedom from cost constraints. Even where customers will pay enough that a profit can be achieved even if the most costly components are used, greater product cost that does not translate into tangible market benefits cannot be tolerated.

Compromises must be made between lack of variability (or other significant performance characteristic) and cost. Typically, the tradeoff decisions are made by the design engineer(s), but there are definite advantages to including representatives from other departments—primarily quality and production—in the decisions. Certain component attributes that would seem important to the design engineer will be recognized as very significant by other departments that work with the actual product.

For components commonly used in the company's products, the design department will typically keep a list of specific makes and models of components that have been proved acceptable for use in various grades of product. This approach normally works very well (and, more importantly, is usually the only feasible way of specifying parts). Occasionally, however, a previously unused combination of approved

[9]As was seen in Chaps. 7 and 9, however, inept assembly and quality verification techniques will render even the finest, least variable component undependable and thus eliminate all the benefits anticipated by the design engineer who specified the components.

components will operate with greater than acceptable variability; this design flaw will only be apparent when prototypes are built and tested.

The only rigid guideline for optimizing design according to variability and cost of components is that the upper and lower performance extremes possible with stacked tolerances are within the specification limits. All other choices of components must be based on balancing costs and benefits—choices for which no single set of rules applies.

13.8.2 Redundant Circuitry

Where failure involves unacceptable financial or health consequences, critical functions can be replicated a number of times with a backup circuit available to take over immediately (and automatically) if the primary circuit fails. This design strategy is common in defense electronics. The advantage is reduced risk of failure. The disadvantages include higher cost (because the number of components is greater), larger size, and greater weight.

13.9 Process-Specific Designs

The symbiotic relationship between design and process is poorly understood within the electronics assembly industry. While general comprehension exists that design must be compatible with production processes, there is much less appreciation that process determines design. That is, the "best" design for a given process will be the same as the optimal design for the perfect process if and only if the given process is also the perfect process. Where the given process is not identical to the perfect process, the "best" design will represent only a local optimum. A design created in total conformance to the perfect process would represent the global optimum.

Using the terminology introduced in Sec. 6.4.2.1, the "given" process must be the "real" process (i.e., what is actually done) while the perfect process is the "natural" process (i.e., what natural forces require to be done). We also know from Sec. 9.2 that process consonance—the state in which the real process coincides exactly with the requirements of the natural process—occurs very rarely. Thus the design engineer ends up crafting printed circuit layouts that conform to the unique (and abnormal) requirements of the company's real process. In other words, the design engineer creates local optimum designs, which generally differ markedly from global optimum designs. Indeed, as will be explained in Sec. 14.6, a global optimum design will normally produce high defect levels in the absence of process consonance.

13.10 Factors Impeding Dependable Design

The primary cause of defective design is rarely incompetence or lack of desire to perform well on the part of the design engineer. More frequently, errors creep in because unreasonable demands are placed on the designer. Just a few (increasingly) common unreasonable demands include:

- Pressures to reduce the time spent on any given design
- Insistence by sales, marketing, or general management that unproven functions be added to new designs
- Exhortations to constantly increase the population density of components on a given area of circuit board
- Desire to employ ever-smaller components

All of these demands drive the probability of serious flaws beyond usually acceptable limits.

13.10.1 Shorter Times to Market

Historically, electronic designs have followed a predictable pattern of improvements. First prototype units would be assembled and subjected to rigorous tests under extreme operating conditions; from these tests, the most serious performance deficiencies would be identified. Only then would the design be released for production. The number of design-related problems found in the first units produced would tend to be quite high. Working on the basis of feedback from the quality and production departments, engineers would gradually improve the designs to eliminate the more serious deficiencies. Eventually (and relatively early in the production life of the product), all meaningful design defects would be eliminated. At this point—often referred to as a "mature" product—the amount of effort required to produce an acceptable unit would be substantially less than the effort required for each unit in the early days of production. Moreover, average dependability of mature products typically is markedly higher. The better dependability is attributable in part to design improvements and partly to refined production and quality procedures that reduce product handling.

13.10.1.1 Short Production Lives and Dislocation Effects. Unfortunately, what was true historically is markedly less common today. Moreover, the numbers of companies where production is continued up to the point of product maturity—let alone for any meaningful time after—are shrinking quickly. Products today often have a production life—not to be confused with the "product life" that is the time the unit should work dependably for the customer—of a few months. In some industry sectors, products are discontinued after just a few weeks.

The dislocation effects created by replacing one design with another are not necessarily substantial. In many cases, the "new" design is only a minor modification of the previous design, so the opportunities for design errors are few. It is intuitively obvious that greater changes in layouts and technologies lead to greater probability of meaningful design failure. However, the risk always exists that an "evolutionary" model may in fact be more revolutionary—with substantial dislocation effects on designs and processes—than initially realized. There will be no certain way of knowing performance characteristics until actual working models are produced.

13.10.1.2 Rapid Design Changes and Verification. One of the basic principles for preventing design flaws from reaching production included testing prototypes thoroughly and harshly before the design is approved for production (see Sec. 13.3). Mature designs have high dependability because they have been subjected to many types of test, including actual field performance, which is the most useful test of all. But rushing new designs to market considerably reduces the opportunities for verifying those designs through preproduction testing.

Where production lives are very brief, the time available for verifying design integrity through test may be considered excessive. Increasingly, the time needed to build and test prototypes is greater than the total time allocated for all aspects of the design process. To reduce the verification time, many design departments now conduct "virtual testing" using software to simulate performance of a design that exists only in computer memory.

Virtual testing has been employed most extensively in the microprocessor industry, generally with satisfactory results. However, the characteristics of microprocessors involve fewer types of variables (even though the numbers of elements to be tested may be enormous) than a subassembly or assembly consisting of many varied components. Attributes verification software for microprocessors is also quite advanced relative to the software available to other industries.

As it turns out, the virtual world of software simulation reflects the real world only to the extent that the software programmer understands and can code that real world. Whenever models are substituted for the real thing, the chances are great that critical attributes of the real environment will be overlooked. (Economists, for example, have long employed elaborate software models to predict the performance of the economy over a given period of time; the failure of those models to anticipate shocks to the economic environment has been a primary cause for the general decline of economists' influence.) Above all, software itself is inherently prone to programming errors (the infamous "bugs") that can produce erroneous results.

Since the accuracy of virtual testing will be greatest when the software programmer is most familiar with the attributes of the system being checked, it follows that the most accurate results will be achieved with mature designs. But mature designs are precisely the units for which verification is least necessary. Whenever designs break new technological ground—replacing through-hole components with surface-mounted components, for example—the need for verification of design performance will be greatest at the same time as the software programmer's familiarity with component performance attributes is lowest.

The potential fallibility of virtual testing was demonstrated quite dramatically in February 1996 when a cable linking the space shuttle Columbia with an experimental satellite broke. Failure of the cable, which had been designed and "tested" by computer simulations, caused irreparable damage to a $400 million experiment that had required years of preparation. NASA spokesmen expressed great surprise that the computer simulated testing had not revealed this possible failure mode, although the design should have been recognized as a prototype rather than a proven design.

Computer analysis of designs can be a valuable *additional* check on product reliability and may reduce the time required to produce a prototype for real-world testing. However, it must be recognized that the development of virtual testing software remains in its infancy. For new products in particular, the risk that virtual testing will let defects slip through is very high.

13.10.1.3 Lessons from Automotive Electronics. The emphasis on features rather than reliability was very pronounced in automotive electronics in the mid-1980s. Automotive electronics plants at that time were so accustomed to receiving new product designs almost simultaneously with the production start-up date that they spoke facetiously of "designing in the field" (i.e., the real product development began only when malfunctioning early production units were returned to dealers for retrofitting). Fortunately, conditions have improved meaningfully in recent years.

The auto industry suffered horrendous dependability problems—and spectacularly high needless costs—by rushing unverified new designs into production. The industry learned some valuable lessons from the painful past, with the result today that they generally (but definitely not always) provide designs to production plants well before production begins. Detroit's lessons were not learned by other segments of the electronic industry, however. So non-automotive electronics companies increasingly find themselves "designing in the field."

The auto electronics assembly companies' experiences have as much or more value for quality professionals as for design engineers. Most important of all, quality managers need to be aware that electronic products are evolving so rapidly that several generations of a product may come and go before design problems in the first generations show up—and those problems have likely been institutionalized in subsequent design models.

13.10.1.4 Causes of Shorter Model Lives. The impetus for the ever-faster shuffling of product models was first seen in Sec. 3.5 during the search for a definition of quality. In that section, dependability (i.e., quality) was shown to be unrelated to features, although both variables taken together constituted performance and entered into the demand equation (along with price and value). Despite the considerable talk about quality being the company's primary concern, Marketing increasingly recognizes that features (also known as "bells and whistles") drive sales at least as much as quality in many markets. Though there may not be any conscious realization among marketing personnel that short product life cycles hinder quality, their actions have the same effect as if a deliberate decision had been taken to sacrifice dependability.

Pressure to reduce design cycle time is not found exclusively in the electronics industry. Software designers (i.e., "programmers") have been among the early victims of rush to market. A letter to the editor of a prominent computer industry trade publication echoes refrains often heard in today's electronics industry. "As a programmer, I'd like nothing better than to have the time to tweak and fiddle with the code until it runs as fast as it can, but the reality is that my salary gets paid by the sale of products, and the buying public isn't interested in upgrading its software because it runs 10 percent faster.

"They'll only upgrade when there are enough new features. And every new feature added makes it a little harder to optimize for speed without breaking."[10] Substitute "design engineer" for "programmer" and "electronic circuits" for "code" and the letter captures the realities of modern electronics marketing perfectly.

13.10.1.5 The Quality Department in Time to Market. The inverse (though unpredictable) relationship between time to market and flawed designs reflects an innate conflict between the company's sales and quality departments. The sales department tries to accelerate the development process in the hope that earlier delivery to customers increases the probability that the product will contain the market's most advanced features and thus command premium prices. Provided the rush to market does not unduly compromise dependability (a provision for which, unfortunately, there are no guarantees), the financial benefits of being first with attractive features are generally much too significant to ignore. This is where freedom from the burden of rigid quality orthodoxy gained in earlier chapters proves especially worthwhile. Knowing that higher quality is not always consistent with the goal of maximizing long-term profits, we should also be sufficiently enlightened to recognize that the quality manager's role is not always to blindly push for ever-better dependability. The enlightened quality manager when faced with evidence that faster time to market can generate greater demand cooperates with sales to find the combination of time to market and dependability that moves the company closest to its goal. One important activity in finding that middle ground will be joint preparation of cost-benefit analysis. This interdepartmental collaboration constitutes a superb example of quality optimization at work.

While the quality director may be required in the company's best interests to compromise between time to market and reliability, the necessity of compromise by no means makes time to market a quality initiative. This point may appear to be tautological, since shorter development time undermines dependability. However, not all quality professionals recognize the inherent contradiction between abbreviated design periods and quality improvement. Many total quality proponents contend that shorter times to market should indeed be included among quality functions.

So the quality professional must approach inherently alien concerns

[10]Van Lydegraf, James K., "Why Blame Programmers for Bloat?" *InfoWorld*, Feb. 5, 1996, p. 50. One particularly interesting aspect to Van Lydegraf's letter is that it was written to chastise other *InfoWorld* columnists and correspondents who were defining quality in terms of speed and suggesting that better programming would eliminate some bells and whistles and tighten code to enhance performance.

such as faster turnover of product models with an open mind. It must immediately and forcefully be added, however, that keeping an open mind is not synonymous with spineless acquiescence. Decisions on the relative emphasis to place on dependability and faster product development require careful analysis of all available data. More impassioned argument by sales personnel on behalf of reducing time to market carries no weight without objective evidence. By the same token, a quality director who carries the day with emotional appeals has no reason for satisfaction; the company, after all, has not been shown to benefit from the subjective "victory."

The implications for quality should be clear. Companies will be forced to choose a position somewhere between churning of products and dependability. Some lessons may be drawn from the computer software industry, where rushing bug-ridden products out the door has cost many companies their lives and customers are increasingly inclined to wait out the initial shakedown period before buying any new software. The electronics industry may segment itself into companies known for slower innovation but higher reliability and others who will be pushing technological and dependability limits. The long-term winning companies will likely be those that choose the more conservative strategy of holding back new products until dependability is assured.

13.10.2 Pushing the Technology Envelope

In addition to compressing the normal development time for product evolution, companies increasingly look to revolutionary rather than evolutionary changes in their product technologies. Unlike the inadvertent tumble into revolutionary design change described in Sec. 13.5.6, this venture away from gradual evolution is undertaken deliberately. Market pressures—which it seems are often more perception than reality, as in the early adoption of fine-pitch surface-mounted components by manufacturers of products where size is inconsequential—constantly drive manufacturers to press the technology limits and bring out more fully featured (which typically means more complex) products. The less charted[11] the technological territory into which the company pushes, the greater the likelihood of unforeseeable design problems. Mature technology is synonymous with fewer unpleasant surprises as products move from design to production to field use.

[11]Whether by the industry in general or the company in particular.

13.11 Quality Management When Designs Enter Production

During the design phase, the amount of participation by quality personnel is typically limited to—at most—answering questions posed by design engineers or critiquing submitted designs. The quality department may also carry out the tests that verify integrity of the design. In many companies, however, the design department does not invite participation by other departments, including quality.

Once the design actually reaches the production stage, the quality department's role expands. Finding flaws in circuitry, layout, or other attributes that prevent the product from performing any of its stated functions becomes a primary concern of quality management. Whenever design defects are discovered by quality personnel, they must immediately be fed back to the responsible designers for remedial action.

On occasion, it may not be possible to bring out a revised design in a timely or cost-effective manner. Although many quality departments ignore known design deficiencies until the next generation of product is ready, the ethical course of action is to reduce the product's claimed features set and offer compensation to existing customers. Such decisions normally require consent from Marketing—or even higher executive levels—but their involvement does not mitigate the vital quality aspects of this problem.

13.12 Summary

Quality begins with design. Maximum attainable quality is limited by flaws in the design. No product can ever be more dependable than its design allows. The design engineering department must therefore ensure that all products are designed with dependability in mind.

Attributes of design for dependability can be grouped into four categories:

1. Functionality
2. Manufacturability
3. Testability
4. Durability

New designs are most likely to contain flaws that will harm dependability. As a product matures, its design stabilizes, production processes are improved, and quality evaluation can be relaxed. Therefore, longer production life leads to better final designs and higher dependability.

Competitive pressures are causing electronics manufacturers to reduce design time and introduce new models more frequently. By eliminating some or all design verification (testing of new designs) and reducing the opportunity for all product personnel from design through quality to become familiar with any special characteristics of given models, electronics manufacturing companies are increasing the probability of design-related failures. This problem threatens to become even more significant in the coming years.

Quality managers must recognize that the rapid evolution of electronic products means several generations of a product can be produced and sold before design problems in the first generations show up—and that there is high probability the first-generation problem has been continued in subsequent designs.

14 Statistical Methods

14.1 Chapter Objectives

Deming could have been describing this chapter when he wrote, "This book is not a book on techniques....The reader who wishes to pursue study of techniques is advised to place himself under the guidance of a competent teacher."[1] The intent of this chapter is to explain the underlying assumptions—together with the strengths and weaknesses—of statistical methods as they apply to the electronics assembly industry. Some objectives of this chapter include:

- Examine the nature of variation and its effects on quality.
- Introduce statistical process control (SPC) and the closely related statistical quality control (SQC) that is particularly relevant to quality management.
- Consider the distinction between statistics and science as seen by Deming and Goldratt.
- Discuss the decisions and actions that must be taken for *effective* SPC or SQC.
- Explain the nature and significance of a "normalized" process.

---

[1][Deming 1982], p. 111.

- Review the structure, assumptions, uses, and limitations of control charts.
- Investigate the statistical methods best suited to the electronics assembly industry.
- Study the requirements and usefulness of experimentation, including the popular Taguchi methods.

14.2 Variation and Quality

Variation as a determinant of quality has been a prominent theme for almost seven decades, since 1931 when Shewhart[2] wrote, "A controlled quality will not be a constant quality. Instead, a controlled quality must be a *variable* quality." Half a century later, Deming's new-found popularity in America focused attention once again on the idea that quality equals consistency in output. Largely due to Deming's influence, *Business Week* would write more than 60 years after Shewhart's pioneering book: "It's not that quality is hard to define; it's simply the absence of variation."[3]

14.2.1 Absence of Variation

As explained in Chap. 3 (particularly Sec. 3.3.1) quality cannot be synonymous with absence of variation. Quality is dependability, a concept that entails much more than mere consistency. Output may be exactly the same from one unit to the next but be unable to perform the specified functions for the specified period of time under the specified conditions of use. In other words, lack of variation is not a *sufficient* condition for quality.

At the same time, if output characteristics vary excessively from unit to unit, not all units can be equally dependable. Therefore, absence of variation may be *necessary* for the existence of quality. Variation must be managed, as Shewhart contended, because quality and variation are antagonistic states. But it cannot be said that absence of variation alone means quality has been achieved.

---

[2][Shewhart 1931], p. 6.

[3][*Business Week* Editors and Green 1994], p. 3. The strengths and weaknesses of this statement were examined here in Chap. 3.

14.2.2 Attributes and Variables

Data collection can involve integers (e.g., the number of units that fail) or continuous values (e.g., measured output of power supplies). The integer form of data are said to describe "attributes" while those that generate continuous values relate to "variables." Attributes and variables are the two subgroups of a generic category known as "characteristics."

The nature of the characteristic being measured can make a difference to how data should be collected and analyzed. In this chapter, the precise term—"attribute" or "variable," as appropriate—is used where it is important to know the specific nature of the characteristic. Otherwise the generic term "characteristic" is used.

14.2.3 The Nature of Variation

Variation exists in a characteristic whenever the value of that characteristic is not identical for all units of output. The amount of variation can be expressed as the extent to which the value of the characteristic changes among units. In practice, several characteristics normally make up the product's "quality," and an element of variation is associated with each characteristic. The need to consider several characteristics simultaneously greatly complicates management of variation.

The values of any characteristic for an output population greater than one unit will always contain some variation except by coincidence. The presence of variation is inevitable because fluctuations exist in the raw materials, components, machine operations, labor, and other constituents of the output. Every one of those constituent elements—the process "inputs"—fluctuates during production and over time.

As we have seen several times throughout this book, Shewhart concluded that process variation consists of two components that he termed "assignable" and "constant" variations. The assignable variation results from forces that can be traced to specific operations, personnel, or equipment. Assignable cause variations tend to be predictable in their effects on output. An improperly calibrated placement machine which periodically misaligns a component would constitute an assignable cause of variation. The constant variations—generally termed "random" variation today—are unpredictable and may be positive or negative (in contrast to assignable variation which consistently will be either positive or negative). Fluctuations in real values of resistors all rated at the same nominal value would constitute a constant cause variation.

Shewhart placed considerable emphasis on distinguishing between assignable and constant variations. While excessive variation of either type would cause product failures, Shewhart contended that only assignable cause variation could *economically* be identified and eliminated by the general workforce. Deming argued that reduction of constant cause variation (which he termed "special cause" variation) is the function of engineers and management. A "stable" process in the context of SPC/SQC is free of assignable cause variations but still contains constant cause variation.

However, if Deming's beliefs accurately reflect conditions in the electronics assembly industry, freedom from intermittent variation may not greatly improve the total variability of output. Deming attributed up to 94 percent of all variation to inherent variation (which he called "common causes") that workers cannot affect.[4] So, if Deming is correct, a process enters the state of statistical control when only 6 percent of its total variation has been eliminated.

14.2.4 The Target Value and Actual Values

Design begins with a desired or "target" value for each characteristic. However, variation means that the actual value of *every* unit cannot be the same as the target value. Consequently, it is also necessary to determine the extent to which the actual value can deviate from the target value without requiring that the unit be rejected. That is, the designers must calculate a range of performance values that will be acceptable. The greater the difference (larger or smaller) between the actual value and the target value, the smaller the probability that the measured unit will perform dependably. The amount of difference between actual and target values that is acceptable—i.e., the degree of precision required in output—is known as the "tolerance" and will depend on the nature of the product.

In most cases, the position of the cutoff points between acceptable and unacceptable values is arbitrary. If $T \pm 10$ is the tolerance range (where T = the target value), why is $T \pm 10.01$ unacceptable? Often the tolerances have no scientific basis; the difference between the extreme acceptable value and the value lying immediately beyond the acceptance limit may be insignificant in the unit's performance. Where the

[4]Deming, W. Edwards. "On Some Statistical Aids toward Economic Production," *Interfaces*, August 1975, p. 2

tolerances are established arbitrarily rather than scientifically, they will often be too high (allowing acceptance of undependable product) or too low (causing rejection of dependable product).

The maximum and minimum acceptable values should be determined by selecting a "worst case" performance that would exist if the actual value of every characteristic was the maximum allowable (or, conversely, the actual value of every characteristic was the minimum allowable). Generally, the "worst case" performance is not realized when some of the values are at the maximum allowable and others are at the minimum allowable, but theoretically at least it is conceivable that a combination of maximum and minimum acceptable values would result in unacceptable overall performance.

Even when substantial variation exists in output, some units will have values exactly equal to the target value (provided the output population is sufficiently large) while the values of others will deviate from the target value. The magnitudes and direction by which the measured values deviate from the target value will not be the same for all units.[5]

Thus, in addition to the target value, the design specifications for any product include the maximum and minimum acceptable actual values. Accordingly, the performance specifications will include a minimum acceptable value, a maximum acceptable value, and a target value that lies between the minimum and maximum acceptable values.

The distance between the minimum acceptable limit and the maximum acceptable limit is the tolerance zone. Any unit whose value for that characteristic lies within the tolerance zone will be acceptable for that characteristic. Values that lie outside the tolerance zone are unacceptable, and the unit fails. Adding the deviations from target values of the many components in a given unit provides the value called "stacked tolerances."

14.2.5 Production and the Tolerance Zone

Although the probability of product failure increases with the amount of variation in the product, the objective in production is not to achieve exactly identical values for all units. That objective would be impossible to achieve, and the costs of trying would be crippling. Rather, the objec-

---

[5]Variation also exists in the performance of test equipment and other measuring devices as well as human interpretations, so that measurements themselves do not generally constitute exact representations of the measured values.

tive is to ensure that the achieved values fall between the specified minimum and maximum acceptable levels. Statistical methods prove very valuable in achieving this objective.

14.3 Statistics in Modern Electronics Assembly

Every process generates numerical information ("data") about its strengths and weaknesses. That information is the amount of variation in the output. If that data can be accurately measured and understood, the findings can help with design and management of processes that are both more effective and deliver higher-quality output (i.e., output which has fewer units whose values lie outside the tolerance zone). The techniques for measuring and interpreting numerical information are collectively known as "statistics."

14.3.1 The Meaning of Statistics

The word statistics has two meanings. When used as a plural (as in "statistics indicate that..."), statistics normally means the body of numerical data itself. Used in the singular ("statistics is..."), "statistics" actually means "statistical methods" or methodology. Statistical methodology can be defined as "the body of principles and methods that has been developed for collecting, analyzing, presenting, and interpreting large masses of numerical data."[6]

Statistical methods offer many different techniques for gathering, interpreting, and applying information about all aspects of life, not just about manufacturing and processes. However, not all of those techniques are appropriate for use in electronics assembly environments. The subsets of statistical methods most directly relevant to electronics assembly quality are those generally used for process management (i.e., statistical process control, or SPC) and the closely related statistical quality control (SQC), which is the use of statistical methods to manage quality. The emphasis in both cases is on "control"—or management—of the processes and quality; simply compiling raw numbers that are never used for control or management constitutes neither SPC nor SQC.

[6]*Compton's Reference Collection 1996*. Compton's NewMedia, Inc.

14.3.2 Knowing *Why* vs. Knowing *What*

In other words, simply knowing *what* happened (i.e., the results) is not enough. We must also understand *why* it is happening (the causes) and *how* we can change it (the corrective actions). Mainstream statistical methodology—the gathering and display of outcomes—tells us only what has happened and, possibly, suggests hypotheses for causes that can be tested by experimentation. (Note that statistical methods rarely act in real time. Therefore, the information will always concern what has occurred rather than what is occurring.) Only through experimentation, itself a branch of statistical methods (see Sec. 14.9), can the "why" be determined—and that desired outcome is by no means certain.

The steps involved in SPC/SQC consist of:

1. Gathering data.
2. Organizing the data in ways that permit identification of changes in the values of process.
3. Using the information to manage the processes.

14.3.3 Statistical Activities Involve Substantial Costs

SPC and SQC are employed today throughout the electronics assembly industry. The trust placed in those statistical programs is immediately apparent upon entering production areas. Walls are papered with printouts of control charts, Pareto analyses, defect trends, performance against AQL, yields at various points in the assembly cycle, and seemingly endless other graphical representations of the company's quality measurements. Each of these charts is based on statistical data collection and is intended to assist in interpretation of process behavior. In addition, each department typically maintains sets of raw data for use whenever questions arise that are not readily answered by reference to the analyses already carried out as regular practices.

None of these statistical activities is free. Indeed, the expenses and effort required in collecting, interpreting, and reporting statistics can be substantial. Despite the costs involved, however, there is almost universal belief that statistical approaches must inevitably lead to identification and extermination of quality problems. In some minds, quality management and statistical controls are inseparable.

14.3.4 The Benefits Question

Unfortunately, the realized benefits of statistics rarely live up to expectations. Most SPC / SQC endeavors end in failure, even though the mathematics of statistical tools needed in electronics assembly are neither terribly complex nor difficult. Few companies are aware that their statistical programs are useless or possibly even harmful. The lack of value in these programs is apparent only to persons with broad *and* deep experience in establishing and operating statistical programs for identical processes. In most cases, a plant will be fortunate to have even one individual who meets those requirements. Statistical methods on their own have no value. They acquire value only when used in tandem with process knowledge to refine the processes and improve quality.

14.3.4.1 Deming's Views on the Usefulness of Statistics. Deming contended that difficulties should be anticipated when applying statistical methods in an industrial setting. "No one should teach the theory and use of control charts without knowledge of statistical theory...supplemented by experience under a master," he argued.[7] The reference to a "master" in that statement is often misinterpreted to mean an authority on statistical methods. In fact, Deming's "master" was intended to be an expert on industrial processes. The master's guidance is necessary because numbers alone do not help solve analytical problems.[8]

Deming's position can be paraphrased as "theory without experience is inadequate."[9] In the terminology developed in Sec. 6.2, theory was referred to as "*academic* knowledge" and experience as "*working* knowledge." Using that terminology, we can say that effective management using statistical methods is possible only when the decision-makers possess both working knowledge of the plant and processes *and* academic knowledge of statistical methods.

It is interesting to note that Deming, generally regarded as a founding father of statistical process control, held so many reservations about the application of statistical methods in industry. Of course, Deming was not condemning all use of statistical methods in management; he was only cautioning against thinking that statistical methods are easily and

[7][Deming 1986], p. 131.

[8]Although we believe Deming's generic quality philosophy contains many serious failings, his competence in statistical matters is beyond question.

[9]That statement is not to be confused with an actual quote of Deming's that "Experience without theory teaches nothing" [Deming 1986], p. 317. The need for theory to make experience meaningful refers to managers who base decisions on experiences but have no testable hypothesis as to why those experiences may be valid for policy purposes.

universally applicable. Rational, insightful use of statistical methods *will* benefit most operations; the challenge lies in trying to determine what applications of statistics constitute "rational" and "insightful."

In Deming's mind, statistics could most usefully be applied to enumerative (i.e., counting) problems. "Application to analytic problems—planning for improvement of tomorrow's run, next year's crop—is unfortunately...in many textbooks deceptive and misleading," he wrote.[10]

Ironically, the rise of Demingism brought with it a surge in demand for statistics training without any corresponding increase in mastering process knowledge. Almost every post-secondary course of education—and definitely those involving the hard sciences and engineering—now devotes considerable attention to statistical techniques. Employees who had not been exposed to statistics classes earlier in life were sent en masse to seminars. As a result of the extensive training in statistics, modern quality and production personnel possess substantial theoretical statistical knowledge. But this knowledge is a very mixed blessing because it has not been accompanied by greater understanding of the processes and variables. Consequently, manufacturing companies are now plagued by a surfeit of employees with academic knowledge of statistics but little working knowledge. These classroom statisticians lack the practical process knowledge required to select suitable statistical tools—but are not aware of their limitations.

14.3.4.2 Goldratt's Contribution. Statistics are dangerous when improperly used because they convey a false sense of scientific significance. Goldratt, the physicist turned management scholar, explains the difference between statistics and science extremely well.[11] The evolution of scientific knowledge, Goldratt points out, involves three stages:

Stage 1: Classification. At this earliest stage, the existence of a phenomenon is noted, but no one understands the phenomenon or has the vaguest idea of what it may represent. Knowing the phenomenon exists, terminology can be developed to describe it. The terminology enhances communication and draws attention to the phenomenon but has little more practical value. In electronics assembly, we can think of the list of workmanship defects as an example of classification. When solder creates a short between two conductive points, the terminology "solder bridge" is

---

[10][Deming 1986], p. 132; Chap. 7 of Deming's book *Some Theory of Sampling* (Wiley, 1950; Dover, 1984) explores the differences between enumerative and analytical applications of statistics.

[11][Goldratt 1990], pp. 22–35.

used. Similarly, it is common to speak of dark, irregular solder flow as a "cold solder joint."

Thus, beyond noting some phenomena that interest us and developing the terminology to describe those phenomena, classification represents only a tentative step forward. Though the terminology exists, there is no reason to believe that the causes—or even the consequences—of the phenomena are understood.

Stage 2: Correlation. The intermediate step toward science is taken when two or more phenomena are found to move simultaneously. This commonality of movement is known as correlation.

The discovery of correlation can be explained by considering the cause(s) of rain. In the beginning, it was not noticed that other phenomena occurred at the same time as rain. Eventually, some observers noticed that clouds generally accompany rain. A relationship was therefore recognized to exist between clouds and rain. But the exact nature of that correlation was not yet known. Some possible relationships between clouds and rain included:

- Clouds cause rain.
- Rain causes clouds.
- Other factors are responsible for both clouds and rain.
- Clouds can exist without rain.
- Rain can exist without clouds.

Therefore, the relationship between clouds and rain may be entirely coincidental. Or there may be a closer connection. The exact relationship—if any—between clouds and rain is unknown.

Establishing correlation—finding the linkages between two elements without knowing why or how those links exist—is the most basic aspect of statistics. Correlation is interesting and statistics can show if it exists, but an enormously important question—"Why does this work?"—has not yet been asked let alone answered. Thus, it is impossible to be certain that the "correlation" is not, in fact, simple coincidence rather than a genuine consistent natural linkage.[12]

It might be established, for example, that the number of solder bridges during wave soldering varies inversely with the level of flux solids or directly with the speed of the conveyor. Or the number of sol-

---

[12]Goldratt provides a fascinating real-life example of a phenomenon based on correlation that never matured beyond that state: just-in-time inventory management. The Japanese architects of JIT know only that it does work (correlation), not why it works. Thus, they cannot be certain that the relationship is not purely coincidental. [Goldratt 1990], pp. 27–28.

der balls from reflow of solder paste may vary directly with the temperature gradient in the reflow chamber and inversely with the number of solvents used in the paste. Are these observations indicative of true correlation or coincidence? And if they are real, what do the observations tell us about *why* the correlations exist? In all cases, the answer is the same: We don't know.

Stage 3: Effect-Cause-Effect. The evolution moves from correlation to true science when a cause for an observed effect is hypothesized and used to predict another effect that also turns out to exist. A possible hypothesis to explain the inverse correlation between the level of flux solids and the number of bridges in wave soldering could be that the flux solids exclude oxygen, thus preventing the formation of oxides on the solder and reducing the surface tension. The validity of that hypothesis could then be tested by seeing whether excluding oxygen in other ways—an inert gas blanket, for example—would also reduce the number of solder bridges in wave soldering.[13]

Effect-cause-effect[14] does not necessarily lead to a universal truth. The cause is "true" only so long as no other cause can be presented that proves to be more broadly applicable. Science is only the best answer available at the moment; today's best available answer may be tomorrow's proven failure.

14.3.4.3 Implications of Goldratt's Model for Statistics. Several important implications for statistical methods emerge from Goldratt's three-stage evolution of science:

1. Data collection cannot occur until observations of phenomena and classification have been made. Without knowing the phenomena to be studied and having a vocabulary to describe those phenomena, it is impossible to know what data should be collected.

2. The most common statistical operations involve the second stage, Correlation. Measurements are compiled and arranged in various ways that the statistician hopes will reveal patterns of interaction among variables.

3. More elaborate exercises in the broad family of statistical tools—i.e., experiments—are required before any knowledge is gained about how to manage the process. The term "experiment" has a very specific scientific meaning: an exercise to validate or invalidate a hypothesis. Often, activities intended only to look for patterns in relationships among variables (in

---

[13]It does, and for precisely the postulated reason.

[14]Which, incidentally, is Goldratt's personal terminology.

other words, Correlation) are erroneously called experiments. Taguchi's orthogonal array Design of Experiments "cookbook" methodology is a leading example of correlation masquerading as experimentation.

14.3.4.4 Common Elements of Deming and Goldratt. Clearly, Deming and Goldratt both had the same issue in mind: the distinction between compiling numbers (enumerating) and knowing how to use those numbers in management strategies. Data compilation without support of a valid scientific hypothesis is only an exercise in tabulation. The data itself in the absence of a validated scientific hypothesis does not provide any guidance for tactics to improve the process or predict outcomes when variables change.

What does this knowledge of scientific evolution tell us about SPC and SQC? We know that data collection makes up an integral part of both SPC and SQC. And experience has shown that the net benefits of SPC and SQC can be very high. But in most plants the level of scientific evolution has not progressed beyond the Correlation stage; the plant personnel know *what* correlations seem to be important but have no reason to believe they know *why* those correlations are important.

How, therefore, can either SPC or SQC be described as tools of scientific process management? The answer is that SPC and SQC are scientifically meaningful only when carried out after effect-cause-effect has been employed to establish why certain variables affect the outcomes in specific ways. When the exact nature of the interaction between variables and outcomes—the *whys* and *hows,* not just an observed correlation—is not known, the data collection has no predictive value. In other words, there is an enormous difference between keeping score and carrying out SPC/SQC. Most of what passes for SPC/SQC in industry today is not scientific and has no useful control function.

14.3.4.5 Ability to Measure vs. Ability to Manage. Students of statistical management are told: "That which cannot be measured cannot be managed." A corollary—"That which can be measured can be managed"—is also widely believed to be valid. While the basic thesis stands up under close scrutiny, it should be apparent from the preceding discussions about the distinction between counting and scientific statistics that the corollary is false. There is no assurance that we will be able to manage any operation just because it can be measured. We must have knowledge of processes to accompany the measurements before management is possible.[15]

[15]Deming did not agree. It was his contention that the most important features of a firm's performance cannot be measured. Essentially Deming believed that "The role of managers is to manage what can't be measured." [Dobyns and Crawford-Mason 1994], p. 139.

14.4 Decisions in Statistical Methodology

The world offers an almost endless supply of ways in which to learn theoretical statistical methods: books, seminars, even full university programs (post-graduate as well as undergraduate). But, as Deming, Goldratt, and other writers—not to mention real-life experiences—tell us, effective application of statistical methods involves considerably more effort than simply learning the various tools.

Applied statistical management of quality and production is a system rather than a group of loosely connected individual tools. The system involves several events; if any event is overlooked or carried out incorrectly, the system will be compromised, perhaps fatally. The critical events consist of:

1. *Why* to measure
2. *What* to measure
3. *Where* to measure
4. *When* to measure
5. *Who* measures
6. *How* to present the data
7. *Who* interprets data
8. *What* to do with findings

No "cookbook" recipe can be used to specify the proper structure of the SPC/SQC events for any particular company. Since each company has an individual personality and unique competencies, the SPC/SQC system applied to a specific company must be tailored to its specific needs and abilities. The following sections provide some general guidelines for developing and operating an effective SPC/SQC system but must not be construed as immutable rules.

14.4.1 Why to Measure

The first mistake made by most companies embarking on statistical exercises is the implicit assumption that all personnel share a common understanding of why measurements are vital. In fact, a wide range of beliefs about the reason(s) for data collection can normally be found in any plant. A few of the more common beliefs we have encountered include:

- To tell how well the plant is running
- To satisfy the customer(s)' requirements
- To identify worker incompetence
- To understand the processes
- To find weak points in processes
- To qualify for quality awards (such as the Malcolm Baldrige National Quality Award)
- To learn how variables affect output
- To determine number and costs of defects
- To involve workers in plant management

None of these reasons is entirely acceptable, though some have more merit than others. Indeed, most are completely untenable. The inadequacies of the reasons can be seen from previous chapters that have emphasized the necessity of orienting all activities toward scientific management of quality and processes. Only two reasons for data collection are consistent with scientific quality and process management:

1. To predict and prevent impending deterioration of processes
2. To discover the existence of conditions previously unidentified and not understood

The first reason—to manage the process(es)—requires that the company already possess thorough comprehension of the process(es)' scientific foundations. That is, the personnel (1) can distinguish between the causes and effects, (2) understand the messages that changes in effects are sending about changes in the causative elements of the process(es), and (3) know how to adjust the causative elements in the process(es) in response to indications of loss of process stability. If the measured variables remain constant, we know that the processes are stable.[16] Excessive variation[17] demonstrates loss of control.

The second reason corresponds to Goldratt's classification and correlation stages of scientific evolution. No problem is actually solved by gaining this particular information. At best, some previously unnoticed correlations may become apparent. If new patterns are observed in the variables, opportunities exist to develop better understanding of

[16]"Constant" is a relative concept. As Shewhart and Deming both emphasized, some variation will be found in even the best known and most rigorously managed processes.

[17]For discussion of how much variation is "excessive," see Sec. 14.2.

processes—i.e., forming hypotheses and testing the hypotheses through experimentation. But at this stage they are only opportunities, not realities. The step between recognizing an opportunity and finding a valid hypothesis can be extremely—even impossibly—large.

It should also be apparent that the reasons for collecting and analyzing data are consistent with sampling rather than universal measurement of output or process parameters. Sampling is much less costly and, when properly applied, as accurate as universal measurements.

14.4.2 What to Measure

Measurements cannot be taken of the unknown. Therefore, if we can't manage what we can't measure, neither can we manage what we can't identify. In many instances, the identities of critical characteristics are not readily apparent. Accordingly, before data is collected, it must be determined what characteristics require managing.

Generic statistical data are typically obtained in one or more ways: (1) by consulting publications such as magazines, newspapers, industry studies, reports of government agencies, or work by research bureaus; (2) by collecting data directly from the operation being studied; and (3) by conducting scientific experiments and measuring or counting under controlled conditions. The first means is irrelevant to our purposes; there will be no published studies of our in-plant operations. Experimentation will only be undertaken as the final stage in effect-cause-effect analysis. Most SPC/SQC work involves collecting data directly from the process(es)' output.

Most "statistical" quality efforts in electronics assembly consist of counting the numbers of various defects by assembly type produced over a specified period of time. This defects-oriented approach to quality typically contains several serious flaws, among them:

1. *Improper specification of defects.* As earlier chapters, especially Chap. 9, explained, specified "defects" in the electronics industry frequently have little or no connection to either quality or process management. For example, surface-mounted components that are not perfectly centered on their pads (but are not so badly misaligned that a lead contacts more than one pad) are often ruled defective although they are perfectly dependable. Dull solder joints may be called "cold solder" and rejected. Tilted components, through-hole leads that are not visible on the "solder side" of the board although solder flow up the lead on the component side is clearly visible, visible flux residues, spots on PCB laminates...a very long list of meaningless "defects" exists in almost every electronics assembly plant. Clearly there is no value in tracking "defects" that are perfectly good to begin with.

2. *Induced failures.* Forcing a healthy component to fail by stress "testing" does not necessarily mean a dependability problem was prevented. The failure demonstrates only that the component can be induced to fail under artificial conditions that exceed the specified operating environment. Far from facilitating quality management, recording such data makes identification of genuine reliability problems more difficult. Too much "data," by obscuring patterns, prevents useful analysis.

3. *Lack of awareness about real problems.* Internal heat damage to components, ionic contamination on PCB surfaces, and ESD damage are the three most serious types of damage during assembly. They cannot be seen and generally cannot even be measured. This information is not recorded—or, in the cases where failures occur at test, is attributed to the wrong causes such as poor workmanship by the component manufacturer.

Basic information must be collected in such a way that it is accurate, representative, and as comprehensive as possible. Statistical treatment cannot in any way improve the level of inherent integrity or accuracy of the raw data.

14.4.3 Where to Measure

The point in the process at which measurements are carried out will affect the numbers and types of anomalies recorded. For example, the number of machine solder defects found will be much greater if the audit point is located immediately after the assemblies emerge from the soldering machine (or cleaner, if the production flow is organized with the soldering machine feeding in-line cleaning equipment) than after touchup. The pre-touchup number is more accurate than post-touchup numbers, although measurement errors typically prevent a true accounting at any point.

All recording of data must therefore take place *prior* to activities that can disguise the real nature of the output. Although this requirement seems obvious and possibly even superficial, a surprisingly large number of companies that collect defect data after touchup and rework take undeserved pride in their low *measured* defect and failure rates. Their contentment is based entirely on lack of relevant information. Measuring too late in the production flow gives rise to unwarranted feelings of accomplishment that undermine improvement initiatives.[18]

---

[18]This is one reason why "comparing" defect rates to those reported by other companies is a futile activity.

The point at which to measure the process is not always readily apparent. For example, companies often measure machine-soldering defects after hand soldering in the belief that it is logical to evaluate all soldering at once. These companies have overlooked an important detail: Most hand-soldering operators inspect and touch up previously completed work (informally if these activities are not part of their formal job descriptions) in addition to performing their own hand soldering.

14.4.4 When to Measure

Greater numbers of failures are likely to be encountered early in a product's production life. As the product matures and the production department acquires greater expertise in assembly of that product, the numbers of failures are likely to decline substantially. Since there will be more failure data to capture during the product's early production life, logic suggests putting more effort into tracking and interpreting data then.

It is important to realize that the nature of the processes—and thus the usefulness of the data—will change as the product matures. Early in production life, the processes will tend to be out of control and the data (still heavily influenced by assignable variation) will reflect that state. As the product and processes mature, plant and design personnel advance along the relevant knowledge curves. Using the additional knowledge, they will bring the processes under control, and the amount of assignable variation will fall markedly. Most of the failure data collected from mature products and processes will therefore relate to Shewhart's constant variation.

Meaningful statistical experimentation, discussed later in this chapter, can be conducted only with mature, steady-state processes (which, of course, means mature products). Data collection early in a product's production life will have little statistical value beyond telling the quality and production departments that serious but unknown problems exist. It is often felt that gaining this information, flawed though it may be, is better than having no information at all. This belief is not necessarily true. Since the process is not yet under statistical control, actions taken on the basis of that data are apt to have negative consequences. The data will be useful at this stage only to individuals who are intimately familiar with the process causes and effects.

The importance of collecting data on product over its field life cannot be overstated. This data is generally collected only during the warranty period, if at all. Yet some of the most useful information about product performance shows up after the warranty has ended.[19] The proactive quality department will always track field performance throughout the product's field life—even after the product goes out of production.

14.4.5 Who Measures

It is generally agreed that data has the most value when made available to everyone in the operation. For that reason, companies post statistical charts in prominent locations throughout the plant. But while widespread circulation of findings may be desirable, not all—or even most—personnel should be involved in data collection.

Most actual data recording is conducted by plant personnel closest to the operation generating the information. In the quality department, this would mean that inspection personnel and test operators record their findings. Various ways of recording data can be found throughout the industry.

The most common data collection protocol has the inspectors and test operators record their findings. Indeed, this is the only feasible approach where inspectors or test operators carry out rework and/or repair along with their inspection and test activities. It is not possible for anyone else to record the pertinent information after the product leaves rework and repair. Often, however, data collection can be performed more accurately and economically if inspectors and test operators mark their findings on the assembly and pass the unit along to dedicated recording stations on its way to rework or repair.[20] The need for test operators to record observations can be eliminated by data recording functions built into the test programs and hardware.

"Data" that does not accurately and fully describe the condition has zero or negative value. Therefore, the individuals assigned to data collection must:

1. Be fully knowledgeable about the conditions they are to record.
2. Be located at sites where they will observe every occurrence of the condition being tracked.
3. Have time and inclination to record observations.

---

[19]For that matter, it is impossible to know the potential costs of extended warranty coverage without such information.

[20]Attaching color-coded gummed paper arrows to the assemblies at the location of the problems is one way for inspectors and operators to easily identify their findings for downstream data recorders.

Normally—and logically—data about production processes would be gathered by individuals within the production department while quality-related data will be collected by members of the quality department.

14.4.6 How to Present the Data

Statistical tools—which are essentially methods for presenting data rather than proving scientific hypotheses—come in many forms. Some of these tools are so complex they can be understood in a theoretical sense only by a handful of thoroughly educated practitioners—and in a practical sense by virtually no one. Those complex tools are neither necessary nor desirable for *effective* SPC/SQC in electronics assembly.

How to present the data depends to a large extent on the nature of the data. Where the data represents attributes, the values must be integers and the "distribution" of data will be a series of discrete points. Where data is obtained by measuring variables, the results will be distributed along a continuous scale without the discrete steps found when counting.

Determining *how* to measure performance is normally much easier than identifying *what* to measure, especially where performance is completely binary (i.e., "go/no-go" situations that do not involve tolerance ranges). The most useful statistical tools for the electronics assembly industry are:

1. *Pareto analysis.* Failures are categorized by type and ranked according to frequency of occurrence. From that data, a vertical bar chart can be constructed in which the y axis shows the number of defects found and the defect categories are distributed along the x axis (see Fig. 14.1). It is customary to display the defect categories in decreasing order

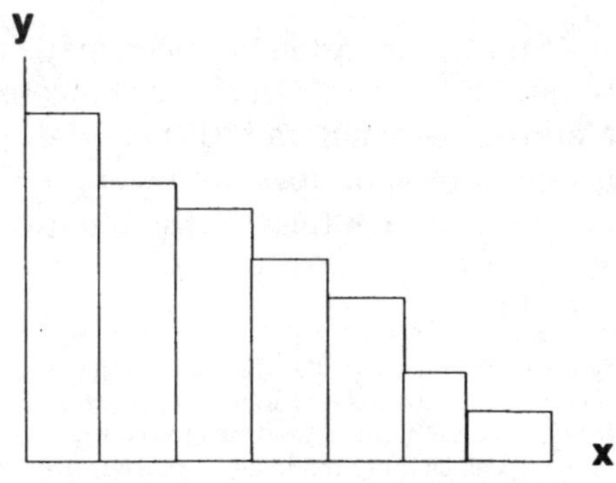

Figure 14.1. A Pareto analysis chart.

of occurrence beginning at the junction of the x and y axes on the left and moving along the x axis to the right.

The implicit assumption of Pareto analysis is that the best return is achieved by attacking the most common problems first. This assumption will be valid if and only if each fault is equally serious. However, where "faults" are measured according to defects rather than failures, Pareto analysis is not the same as saying it ranks the defects by severity. A relatively uncommon defect which invariably leads to hard failure may fairly be described as more serious than a frequent defect which has only cosmetic overtones. However, if the data collection is restricted to dependability issues—the unit either works or doesn't work—close correlation between Pareto ranking and severity of the problem can be anticipated.

2. *Mapping diagram.* One of the most effective techniques for quickly seeing patterns in failures is to map failures by type on diagrams of a printed circuit assembly. The failures will immediately be seen to be random or predictable (systematic). Random failures will be distributed across the printed circuit assembly diagram without any discernible pattern. Predictable failure modes will cluster around a small number of components. Since the causes of and solutions to random failures are quite different from those associated with predictable failures, mapping diagrams (also known as "measles charts") allow fast determination of the sort of problems being faced.

3. *The failure rate.* Having defined quality as dependability, it is logical to track failure rates by product. To obtain a good perspective of dependability at various points in the product's life, the failure rate should be recorded at each test point and time in the field. To know whether any progress is being made in quality improvement, separate time-based charts should be kept for each product at each point in its production life.

The reader will note that no mention has been made of tracking defect rates. The reason should be clear: "defects" are inconsequential unless they are failures. Nothing else matters.[21] (Where the customer specifies cosmetic standards, anything that does not meet the cosmetic standard can be considered a failure. Track these items separately from depend-

[21]The process master who also possesses statistical knowledge will recognize some classes of "defects" as indicators that the process is drifting out of control. For these masters, collection of that indicator data has value because it shows that corrective action of a particular type is required. Unfortunately, as Deming made clear, masters are not common.

ability-related failures, however; in many cases, the cosmetic standards have no bearing on dependability.)

Mixing cosmetic defects with hard failures (e.g., the subjective "not enough solder" mixed with the objective "solder shorts") confuses the picture and thwarts the entire exercise. Nonetheless, most plants today waste resources recording numbers of defects and agonizing over how to translate those absolute numbers into parts per thousand or parts per million. Compounding the problems raised by this exercise, they then try to compare their "defect rate" with the defect rates reported by other plants and companies. The defect rates recorded in other facilities are unlikely to have been determined in the same manner, and comparison is therefore meaningless.

Of course, not every outcome is a binary pass/fail. Many electronic systems can operate over a *range* of values rather than simply working or not working (i.e., the quality characteristic is a variable rather than an attribute). The obvious example of an electronic system that can have a range of output values is a radio transmitter. Although targeted at a nominal frequency, each transmitter will normally have a real output frequency different from its targeted value. A *histogram* (see Fig. 14.2) is an ideal means of identifying the degree of consistency of such performance variables.

Control charts are another—but considerably more complicated—technique for viewing and managing variables data. Discussion of control charts is found in Sec. 14.8.

Although many other statistical tools exist, the handful presented here should normally be more than adequate to maintain scientific process control in any electronics assembly environment.

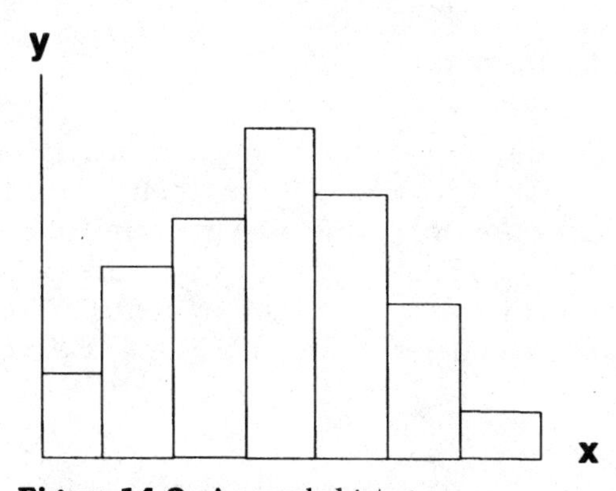

Figure 14.2. A sample histogram.

14.4.7 Who Interprets Data

Data collection is, of course, only the preliminary stage in the application of statistical methods. Even the ways in which those data are compiled and presented will be of little—or no—use to anyone who lacks the knowledge to recognize whatever messages the data may be sending. Consequently, data interpretation should be left to the few employees who are both statistically literate and process knowledgeable. These will seldom be the same individuals who collect the data, though they tend to be the personnel who specified the elements to measure and the analytical tools.

Since intimate familiarity with the company's processes normally comes only from long periods of hands-on experience with those processes, it is unlikely that recent arrivals to the quality or production departments will meet the criteria for competent data interpretation. Nonetheless, it is common to find relatively inexperienced members of the quality department conducting statistical "analyses" using generic computer software packages. Not surprisingly, most of the conclusions reached in this manner are erroneous and can be hazardous to the company's quality and financial health.

Computers unquestionably facilitate the "crunching" of numbers and allow complex mathematical equations to be solved much faster than was possible in the past. At the same time, a computer-compiled report is nothing more than meaningless digits until interpreted by someone who knows the interactions of parameters in the process. That knowledgeable person will either have personally conducted effect-cause-effect iterations in the Goldratt manner or have carefully studied and used the results of effect-cause-effect work by others working on similar processes.

14.4.8 What to Do with Findings

Assembling, processing, and analyzing the data has no value unless corrective actions are taken (provided, of course, the data analysis reveals problems requiring solutions). This is true even if the analysis yields correct answers. Yet in a majority of plants the data is left in files never to be seen again. Posting data charts throughout the company is more valuable as a public relations exercise than as a way of actually improving quality.[22]

[22]It must always be remembered that the meaning of those charts is far from clear to the average person on the shop floor. Far too often, the wrong implications are drawn from the charts and undesirable actions undertaken. For example, operators who see defects recorded will spend more time studying and reworking product before it can reach inspectors.

The only valid reason for employing SPC/SQC is to move the company closer to its ultimate goal of long-term profit maximization. This reason leads, in turn, to the conclusion that only three courses of action can be justified by SPC/SQC:

1. When the data shows a clear-cut course of action, take that action.
2. When the data tells us there is nothing that we are capable of improving (including problems that we can identify but have been unable to solve), do nothing; taking any action under these conditions is as likely to cause harm as to provide benefit.
3. When the data reveals new problems or opportunities, undertake effect-cause-effect actions to see if a clear course of improvement action can be found.

Whenever data has been compiled and assessed with respect to the problem of immediate interest, it should also be reviewed to see if any implications exist for other issues. Data collection is costly, and investments in its collection should not be constrained to a single problem.

14.5 A Caution about Measurement Accuracy

All too often, the measurements are inaccurate. Reasons for the lack of accuracy range from defective measurement equipment to lack of user competence. If the measurement errors are purely random and data are collected in sufficiently large numbers, the consequences will be negligible; one set of inaccurate measurements will simply offset another. But if the measurement error is constant—as might be expected when a measurement device is improperly calibrated—the results will be biased.

What will be the consequences of acting on flawed measurements? It is impossible to predict without knowing the exact distortion of the data. It should be presumed that the courses of action suggested by the analysis will be less than ideal and may well cause deterioration in the processes and product dependability. If the consistent bias in the data is recognized, the data can be adjusted to allow for accurate interpretation.

Errors in measurement are difficult for anyone—whether veteran statistician or master process engineer—to identify. Neither type of individual has an advantage over the other. The only way to compensate for possible consistent bias is to first deliberately change the measuring devices and personnel periodically and then factor those changes into the data analysis. If a consistent bias is present in the measurements of one instrument, there will be a noticeable change of course in the data values when a replacement instrument is substituted for the original. Where two instruments show differing means, checks with one or more additional instruments will be needed to identify which of the first two instruments is wrongly calibrated.

14.6 Global and Local Optimums

The success of any process optimization effort depends on the process managers being aware of all variables that determine the outcome of the process being analyzed. If every variable is known, on-line statistical analysis should theoretically allow the engineers to discover the unique combination of settings for each variable that will minimize defects and failures. That unique combination of settings is the *global optimum* and has the property that it is impossible to achieve fewer defects or failures by changing any or all of the variable settings. But if any variable escapes the attention of the process engineers, experimentation can only result in a *local optimum*.[23] The local optimum is achieved when the defect and failure rates cannot be improved by changing settings for any or all of the known variables. The local optimum will be different each time the unrecognized process variables change value. From this knowledge of the global and local optimums, the following statements must always be true:

- The defect and failure rates at a local optimum can never be better than the rates at the global optimum.
- The defect and failure rates at the local optimum will be equal to the defect and failure rates at the global optimum only if the settings for each of the unknown variables equal the settings for those variables when the global optimum is reached *and* the known variables have been set at their local optimum values when the unknown variables are at their global optimum values. Obviously, whenever one or more crucial process variable[24] is unknown to the process engineers, the local optimum can equal the global optimum only by accident.

[23]Often, the study will include many variables that have no effect on the process. Although enlarging the number of included variables may decrease the risk of overlooking a critical factor, the complexity of the study increases significantly with each new variable. Further, there is no assurance that important variables are not among those which the study omits.

[24]Laying out components along the wrong axis with respect to the circuit assembly's passage through the solder wave is just one example of common oversights.

- Except when the local optimum accidentally equals the global optimum, the defect and failure rates at the local optimum will be greater than the defect and failure rates at the global optimum.

Whenever a process variable is operating at a level different from its global optimum setting, the local optimum setting for at least one of the other variables will probably be different from the global optimum setting for that variable or variables. Depending on the importance of the uncontrolled variable and the magnitude of the deviation between its current value and its global optimum value, the local optimum setting for controlled variables can be vastly different from their global optimum values.

A lesson of no little consequence emerges from this contemplation of the many constraints hampering the effectiveness of statistical tools in the production environment. Specifically, the technical personnel responsible for tuning the processes to maximize product quality too often overestimate the capabilities of statistical methods, whether those methods are termed SQC or SPC. Statistical controls can be of great value when used to provide feedback on fluctuations in a stable, well-understood process; when it comes to taming processes running amok, however, studying what has been published on those processes will usually prove more helpful and considerably less frustrating than resorting to the uncertain world of statistics.

14.7 The Normalized Process

The dispersion of values for the characteristic being studied—known as the data distribution—can take any of many shapes. The shape found most often in explanations of statistical process control, however, is the bell-shaped "normal" distribution. Indeed, so little attention is paid to non-normal distributions that many casual students of industrial statistics assume bell curves are the natural state in manufacturing processes. In fact, "normal" distributions, being dependent on elimination of all assignable cause variation, are rather uncommon. But, whether common or rare in the real world, normal distributions form the core of SPC/SQC theory and practice.

As Shewhart stated, a stable process is not completely free of variation in output. Constant variations remain. But, being random, the constant variations will cause an even distribution of output values around a central mean. The shape of that distribution will be the normal curve

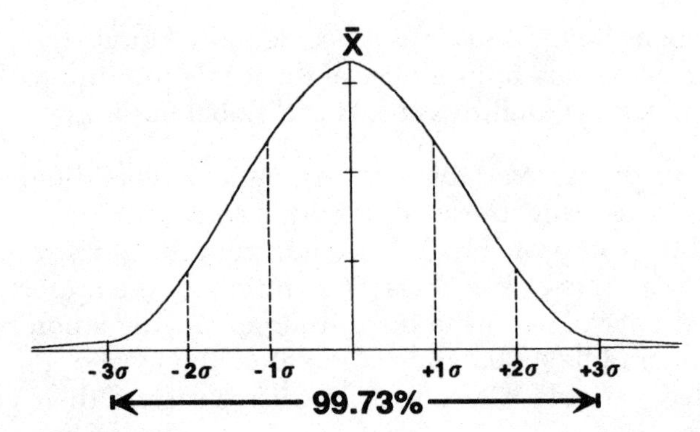

Figure 14.3. Normal distribution curve.

(see Fig. 14.3). The only way to determine whether the process is under control is by examining the output's frequency distribution to see whether it follows the bell curve pattern. If the distribution curve has the required bell shape, the process can be said to be "normalized."

A different type of frequency distribution must be constructed for binary ("go/no-go") outcomes. Data must be sampled in groups of sufficient size that several failures will be found in the mean sample group. The number of failures is recorded for each group, a histogram is created to show the dispersion of values, and a curve is fitted. The histogram of a normalized process producing "go/no-go" output will generate a normal curve.

It must be noted that the distribution "curve" for binary (i.e., attributes) data is actually a convenient fiction. It exists only as an artificial representation of the shape that would be taken by a line drawn through the center top point of each bar of a histogram (see Fig. 14.4).

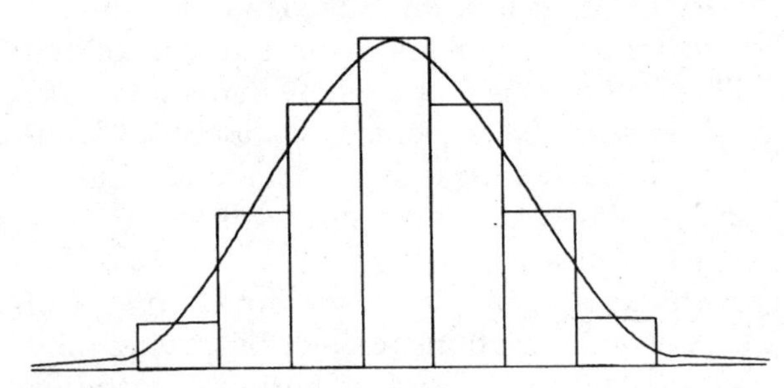

Figure 14.4. Derivation of normal curve from histogram.

14.8 Control Charts

During the 1920s, Shewhart devised a visual method for evaluating the amount of variation in a process and identifying when the process wanders out of control. The tool, known as a "control chart," figured prominently in [Shewhart 1931] and remained popular for some two decades. After a period of neglect in the postwar years, the use of control charts experienced a renaissance with the rise of Demingism. Today, control charts can be found throughout most electronics assembly plants. The following discussion presents some key aspects of control charts in the hope of shedding light on some often overlooked assumptions and realities underlying control chart theory; more complete explanations of the mathematical structures are available from many statistics books or Chap. 24 of [Juran 1988].

Control charts are intended primarily as evidence to support the use of random sampling rather than universal inspection of output for quality control purposes. If the control charts point to a process in statistical control, it is assumed that the output is homogeneous within acceptable limits and verification of quality can be obtained from small sample inspection. Where the control charts indicate that statistical control has been lost, universal inspection is typically resumed until the assignable causes of variation can be found and corrected.

14.8.1 Assumption of Normalization

Use of control charts by operators and similar shop floor personnel assumes that the process has been normalized (i.e., all assignable causes of variation have been identified and eliminated so that only constant causes remain) during construction of the control chart by the statistical process management personnel. This assumption may be ill-founded. Shewhart himself admitted the high probability that the data would not be normally distributed. However, he justified the assumption of normalcy by the "law of large numbers" that says sizable populations will organize in a normal distribution even where assignable causes of variation exist.[25] In fact, if an assignable cause of variation exists, the distribution can be bell-shaped but its mean will not be the same as the mean of the natural process.

Figure 14.5 shows three very similar bell-shaped curves, but only curve A is truly normal. Curves B and C are symmetrical as required of a nor-

---

[25][Shewhart 1931], pp. 121ff.

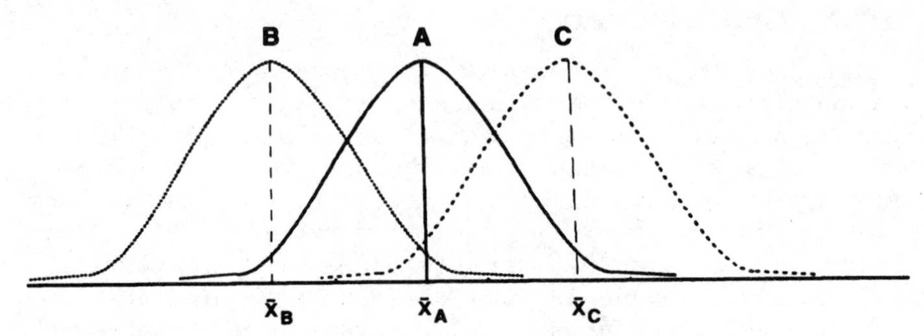

Figure 14.5. Wandering means.

mal distribution, but their means diverge from the "natural process" mean of distribution A. The mean of distribution B is lower than the natural process mean while the mean of distribution C is higher than the natural process mean. The difference in mean values shown in distributions B and C can be attributed to Shewhart's assignable cause variation.

If all three distributions—A, B, and C—were simultaneously visible, the variance problem would be readily apparent. However, only one distribution results over any period. Under such conditions, how is it possible to know whether the bell-shaped distribution is also normal? There is no way to know this vital information.

From the shape of the plotted data curve, it can be seen whether the data distribution is bell-shaped. But it is not possible to know whether the data mean corresponds to the natural process mean. In other words, no way exists for distinguishing between a truly normal distribution and a merely cosmetic "normal" distribution.

In electronics assembly, mass-produced items do reach quantities consistent with the requirements of the law of large numbers. However, many more electronics assembly companies in this country specialize in relatively small volumes of a range of products. For these companies, which might best be considered "job shops," the amount of output does not begin to approach the least restrictive requirements of the law of large numbers and there is no reason to anticipate that data will be normally distributed—or even *approximate* a normal distribution.

14.8.2 Construction of Control Charts

The assumption of normalcy in the data distribution is necessary to support yet another assumption: that the state of control in the process can be visually ascertained by plotting data on a time-based run chart. The chart itself (before adding data) is constructed using the mean and standard deviations of the data distribution derived from the same data that generated the normal distribution (see Figs. 14.6 and 14.7).

The middle line—$\bar{x}$—corresponds to the mean of the data distribution. The standard deviation—σ—is also derived from the data distri-

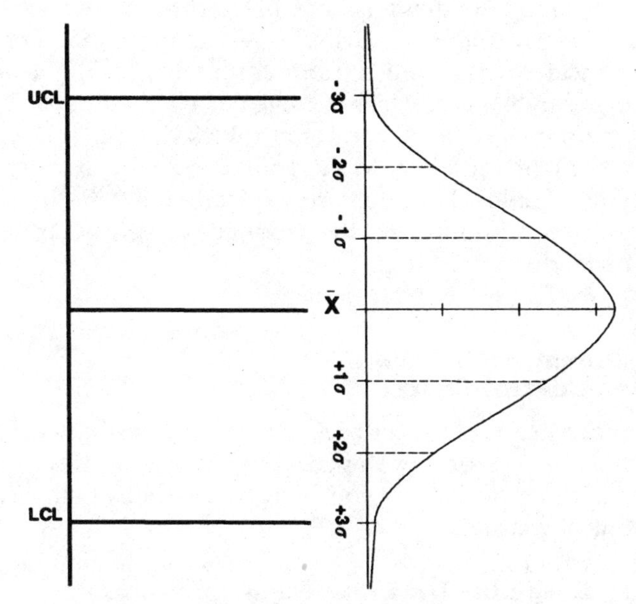

Figure 14.6. Derivation of the control chart.

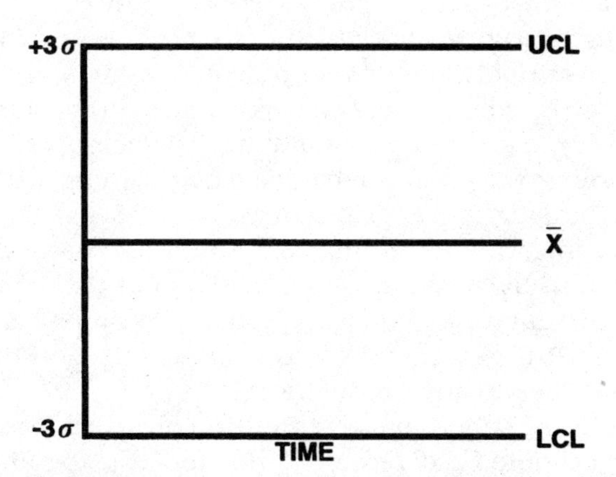

Figure 14.7. Basic control chart structure.

bution. If the distribution is truly normal, 99.73 percent of the population lies between $\pm 3\sigma$. Shewhart and his followers believe that the population falling outside this range—2.7 units in a population of 1000 (or 2700 in a population of a million)—is so small as to be insignificant. If a substantially larger number of data points do fall outside the control limits, the process likely is not in control, in which case efforts to find and eliminate assignable causes should be undertaken.

It is essential to note that any control chart has relevance to only one characteristic—a variable or attribute. Where more than one characteristic is of concern in assessing a product's quality, it will be necessary to operate a separate control chart for each characteristic. Clearly, the ongoing maintenance of several control charts can place enormous demands on plant personnel to the point where the exercise is not economically justifiable. Control charts typically have merit only where a single characteristic dominates the product's quality and that characteristic can be readily quantified.

14.8.3 Constructing the Base-Line Control Chart

Over the years, certain procedures for creating the base-line (i.e., first) control chart have been developed. They are by no means immutable laws but can be helpful to any company that actually believes in the efficacy of control charts.

14.8.3.1 Base-Line Data Collection. Construction of the base-line control chart requires data. The first step in data compilation involves gathering a cluster of output units produced in uninterrupted sequence. The cluster is known as a "subgroup." If the quality characteristic is a variable, the subgroup normally consists of four or five units; where the characteristic is an attribute, the subgroup is much larger and depends on how large a sample is required to reveal occurrences of the attribute. The values of the subgroup units are found and a subgroup mean value computed. The subgroup mean is recorded along with the date and time that the units in the subgroup were produced.

At regular intervals, additional subgroups are assembled, measured, and recorded in the same manner. The length of the interval between collections of subgroups is normally not very important at this early stage, though it is generally best to restrict collection of subgroups to not more than two in any production shift.

The number of subgroups required in construction of the base-line control chart should be 20 or more. Therefore, the total number of units sampled will range upward from 80 or 100. If any changes to the process occur while the data is being collected, the time and nature of the changes must be recorded.

14.8.3.2 Base-Line Data Interpretation. From the subgroups, a universal mean and standard deviation for the process can be calculated.[26] The universal mean is the line labeled $\overline{x}$ in Fig. 14.7. An upper control limit (UCL) equal to $\overline{x} + 3\sigma$ and lower control limit (LCL) equal to $\overline{x} - 3\sigma$ are also calculated and used to complete an empty control chart such as that shown in Fig. 14.7. The base-line control chart is now ready for data plotting.

14.8.3.3 Plotting the Base-Line Data. The means for each of the subgroups are plotted in chronological sequence on the base-line control chart. If all the data points fall between the upper and lower control limits, the process is considered to be in statistical control. If any of the points falls outside the control limit range, the process is considered to be out of statistical control and the search is begun for assignable cause variation responsible for the non-random variance. Once found, a solution to the cause(s) must be found and the problem corrected.

The base-line control chart is more likely to show that the process is out of statistical control than in control. If the base-line control chart shows the process to be out of control, the control chart is not used for tracking process performance. Rather, no work is done with control charts until assignable cause variation has been identified and eliminated. A new control chart is then constructed in the same manner as the base-line control chart using new data. Plotting the subgroup means of the new data will reveal whether the process changes eliminated all assignable cause variation and brought the process under statistical control. If any points in the new data lie outside the control range of the revised control chart, assignable cause variation still exists, and the exercise of finding causes of variation, repairing the process, and constructing a control chart is repeated. The cycle continues until the data points all fall within the control limits, at which point the control chart is ready to be provided to the process operators for shop floor use.

14.8.3.4 A Variables Control Chart. A sample control chart for monitoring a variable quality characteristic is shown in Fig. 14.8. For illustrative purposes, assume that the variable being monitored is output of a 100-V power supply. If the process is in statistical control, the universal

---

[26]The formulas for computing the mean and standard deviation are contained in most statistics manuals. A good discussion may be found in [Juran 1988], pp. 24.8ff.

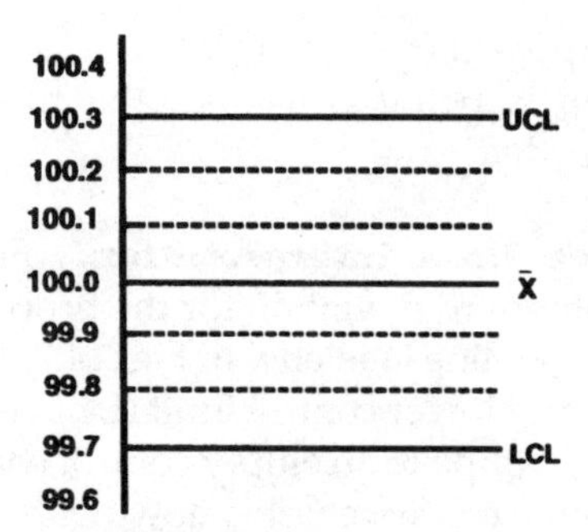

Figure 14.8. A variables control chart.

mean of the subgroups ($\bar{x}$) will be the same as the nominal value of the power supply, 100 V. If the standard deviation is 0.1 V, the upper control limit (UCL) will be 100.3 (i.e., $\bar{x} + 3\sigma = 100 + 0.3$) and the lower control limit (LCL) will be 99.7 (i.e., $\bar{x} - 3\sigma = 100 - 0.3$). The range of values is shown along the y axis; since not all data will fall in the range of $\bar{x} \pm 3\sigma$, the y axis must be labeled for a range above 100.3 V and below 99.7 V sufficient to contain any stray data that falls outside the $\bar{x} \pm 3\sigma$ range. The x axis, though unlabeled, is a time line.

When the control chart is used to monitor production, data is collected in subgroups, compiled, and plotted in the same manner that was used to develop the base-line control chart. The result, if the process is in statistical control, would be similar to Fig. 14.9. All of the subgroup means lie within the control limits.

14.8.3.5 An Attributes Control Chart. As noted previously, the control chart may also be used to monitor attributes (e.g., the numbers of failures) rather than the values of a variable. In such a case, the y axis will be divided into discrete integers, the size of the integers being determined by the sample size. Where attributes are being tracked, the y axis begins at the value 0, as shown in Fig. 14.10. Alternatively, the attributes can be measured according to their percentage representation in the sample population. For example, the percentage of units that are defective could be calculated, and the y axis would be labeled accordingly. Such an attributes control chart, often called a "p chart" because it graphs according to percentages, does have a continuous distribution of values along the y axis rather than the integer values found in a standard attributes control chart. The same basic principles of interpretation apply to all control charts, however.

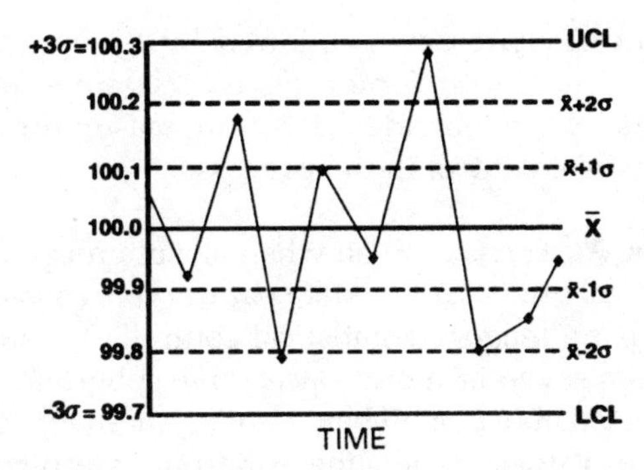

Figure 14.9. Variables control chart with data.

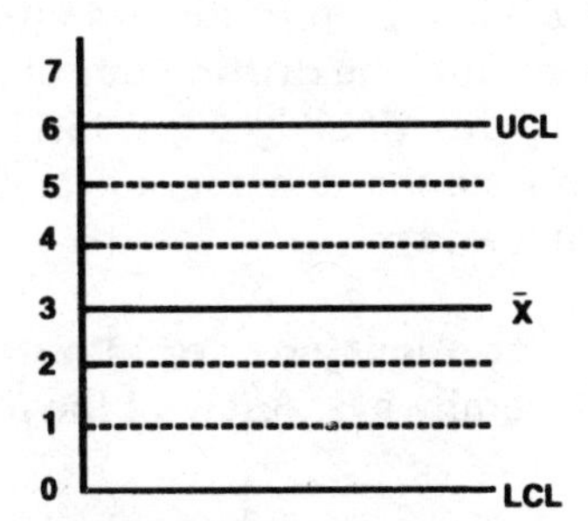

Figure 14.10. An attributes control chart.

14.8.4 Other Criteria for Interpreting Control Chart Data

Other criteria are often employed to determine whether the process is actually in statistical control. They include, but are not limited to, the range chart and identification of non-random patterns in the data distribution.

14.8.4.1 The Range Chart. A second set of control charts, derived from the same data used to construct the $\bar{x}$ control chart, is the *range* chart. Using the set of subgroups from which the $\bar{x}$ control chart was constructed, subtract the mean of the subgroup with the smallest value from the mean of the subgroup with the highest value. This is the range of mean values. Dividing the range by the total number of subgroups provides a mean range $\bar{r}$. Standard deviation for the range of means (σ_r) can also be calculated and a range control chart constructed with a control range of $\bar{r} \pm 3\sigma_r$.

The range control chart data is plotted by finding the mean range value for each subgroup and plotting those values in chronological order. Any points that lie outside $\bar{r} \pm 3\sigma_r$ are reason for questioning the existence of statistical control in the process.

14.8.4.2 Data Patterns. Even when all subgroup means and range lie within the control limits, certain data patterns are considered indicators that the process is no longer in statistical control. One such pattern (Fig. 14.11) is seen when seven or more consecutive subgroup means all lie on the same side (i.e., all above or all below) of $\bar{x}$. Another pattern (Fig. 14.12) that causes concern arises when the subgroup mean values display a repeating symmetrical pattern above and below $\bar{x}$.

Finally, if a series of six or more subgroup means trend consistently farther from $\bar{x}$ (Fig. 14.13) in a pattern toward either the upper or lower control limits, the process may be drifting out of statistical control.

In any of the situations described by Figs. 14.11 to 14.13, recalculating the universal mean and standard deviation of the data would likely result in new values for $\bar{x}$ and σ.

14.8.4.3 Improper Applications of Control Charts. Control charts are misused in several ways. A few of the more common mistakes include:

1. Tracking values of individual units rather than subgroups. An individual unit cannot possess a mean, and the very nature of the control chart structure dictates that on average 27 out of every 10,000 units should have a value outside the control limits even if the data distribution is truly normal. Using the means of subgroups, however, no point should lie outside the range $\bar{x} \pm 3\sigma$ unless the process is no longer in statistical control.

2. Forming subgroups by taking units randomly from pooled output. Trends can be noted only when the subgroup data is gathered in immediate sequence.

3. Monitoring attributes with control chart and subgroup sampling designed for use with variables. This will cause particular problems when the attributes occur sufficiently seldom that they can be tracked only with very large subgroup population sizes.

4. Assigning responsibility to operators for constructing the base-line control chart and all subsequent recalculated charts. In many cases it is feasible and desirable for the operator to collect the data, plot it, and monitor for indications that the process is drifting out of control.

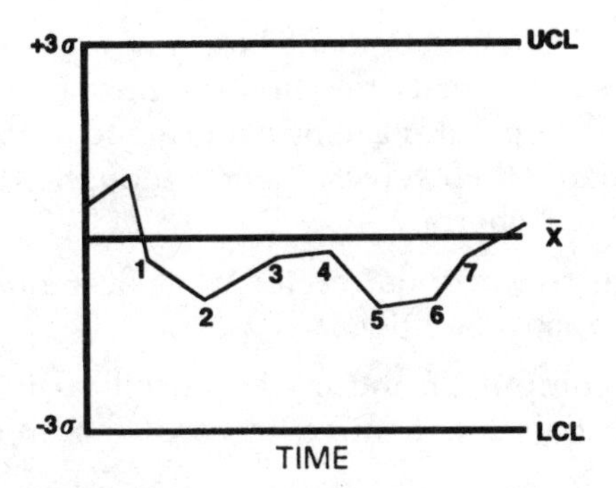

Figure 14.11. Seven sequential data points on same side of $\bar{x}$.

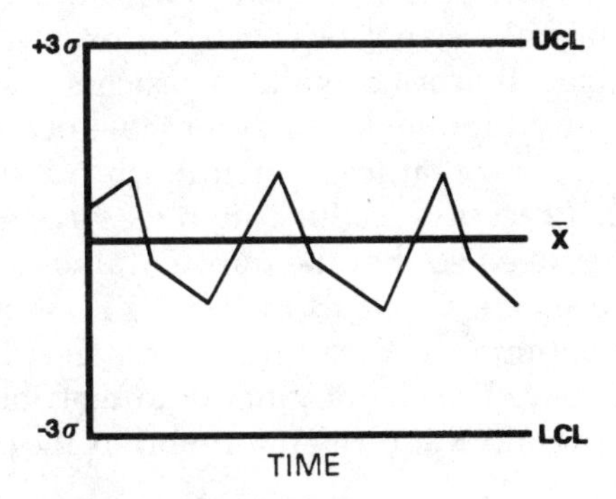

Figure 14.12. Symmetrical data pattern.

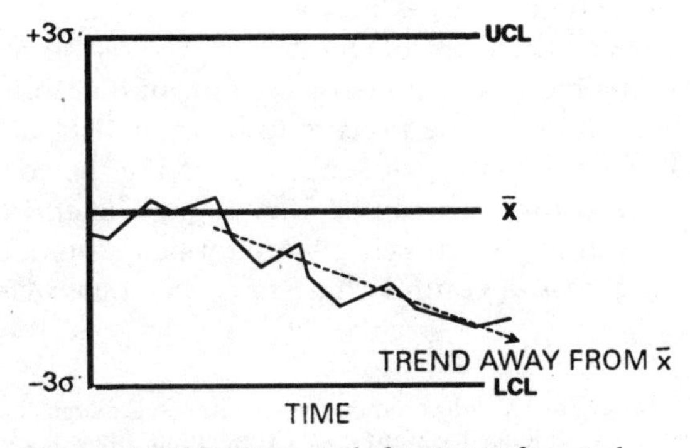

Figure 14.13. Process drifting out of control.

However, the actual construction of a base-line control chart and refinement until the process is ready for shop floor use requires more statistical expertise than the operator can be expected to possess. Educated statistical practitioners themselves have enough trouble preparing representative control charts.

5. Taking "corrective actions" on the basis of trends in control charts without further statistical evidence.

6. Failing to recompute $\bar{x}$ and σ periodically using the numbers of subgroups (more than 20) required to construct an original base-line control chart.

Additionally, a control chart has no value in an operation where the product tolerances (maximum and minimum values) are less than the values of the upper and/or lower control limits. In other words, nothing is gained from a control chart which shows all data lying within $\bar{x} \pm 3\sigma$ if the product usefulness lies between, for example, $\bar{x} \pm 2\sigma$. It must always be recognized that a normalized process may be incapable of producing sufficiently high levels of acceptable output; the shop floor personnel can manage—or, at least, identify the existence of—assignable cause variation. Process capability often requires that common cause variation also be reduced so that the output values are dispersed more tightly around the mean. Curve B in Fig. 14.14 shows how reducing common cause variation results in a narrower yet still normal distribution of output compared to the original data distribution A. Common cause variation generally requires intervention by personnel at more senior levels.[27]

14.8.4.4 Final Cautions on Control Charts. Control charts can make valuable contributions to scientific process management, especially under conditions of high-volume homogeneous output. At the same time, it is vital to recognize that control charts also give rise to numerous troubling concerns. The fact that a non-normal output distribution can result in a control chart showing the process to be in statistical control is one important issue. The probability of misinterpreting the courses of action indicated by the control chart can be very high. The effort and expense required to maintain the control charts cannot be ignored; compounding the workload and cost, a control chart must be constructed and main-

---

[27]Deming, of course, argues that all common cause variation is management's responsibility, but there is no hard evidence to support that contention. Indeed, we have frequently found that operators and other personnel at similar responsibility levels can reduce many forms of variation that traditionally would have been considered common cause variation.

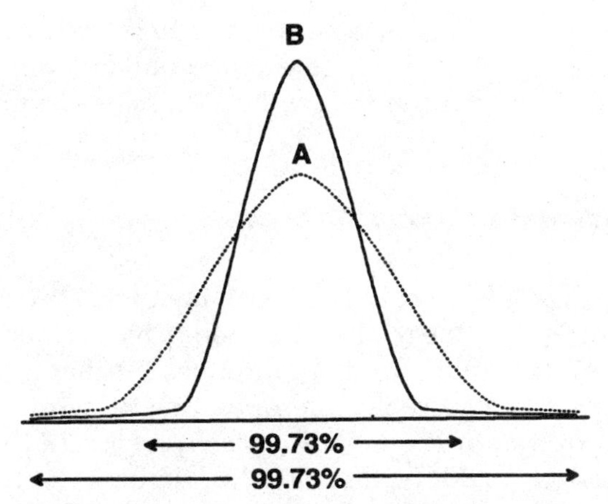

Figure 14.14. Effect on normal distribution of reducing common cause variation.

tained for each significant quality characteristic. And, perhaps most disturbing of all, substantial quantities of nonconforming output can be produced before the control chart reveals that the process is out of control.

14.9 Experimentation

SPC/SQC involves the collection and application of data to processes where cause-and-effect is known. If the cause-and-effect is not known, SPC/SQC is impossible. Therefore, the cause-and-effect links must be established before implementing any SPC/SQC program. The most common technique for establishing the cause-and-effect relationships (and proving the validity of findings by effect-cause-effect analysis) is experimentation, a special branch of applied statistics.

True experimentation is by no means a trivial undertaking. As Goldratt explained in his *Theory of Constraints*, experiments are undertaken to validate hypotheses arrived at by the earlier statistical stage of correlation. Without any hypothesis to test, "experimentation" is only a more complex statistical tool for finding correlations that may lead to hypotheses. Very little true experimentation is conducted in the electronics assembly world, although much activity to investigate possible correlations takes place under the erroneous designation of experiments.

The mathematics of Design of Experiments are so complex that satisfactory treatment requires a substantial book all its own. The following brief explanation of how Design of Experiments works is not intended (or adequate) to equip the reader to actually set up and run statistically valid experiments. Rather, the objective is only to point out a few of the challenges and frustrations that will certainly be encountered.

14.9.1 Identification of Process Variables

As Sec. 14.6 explained, processes involving one or more unknown variables can produce global optimums only by accident. At best, such processes will result in local optimums. Furthermore, processes in which one or more variables are operating at settings different from their global optimum levels will typically generate the fewest defects and failures—i.e., achieve their local optimums—only if known variables are set at levels different from those they would have under a global optimum.

An example of how the search for a global optimum can be constrained to a local optimum without the experimenter's knowledge is found in most "experiments" to determine the effects of certain variables on automated soldering operations such as wave soldering or reflow of solder paste. In automated soldering, several of the most critical process variables involve the layout of the circuit board—including such characteristics as component alignment, pad geometry, and proximity of leads. Therefore, results of any experiment to discover the "optimum" settings for process variables will depend on whether the circuit boards used in the experiment meet the design requirements for global optimization. If the circuit boards deviate in any way from the global optimum settings, the experiment can yield nothing better than local optimum process settings that will be valid for use only with circuit boards having the same design flaws as those used in the experiment. Use of imperfectly designed printed circuit boards explains why experiments over the years to find the forces that determine the outcome of wave soldering have reached contradictory conclusions.

14.9.2 A Sample Experiment

The key to any successful experiment lies in planning. Before conducting any trial runs, the experimenter must first identify those variables which can affect the process. Then sample production runs are conducted, with the variables being fixed at selected values across a limited range of settings. The following example, based on a real experimental study conducted by a noted defense contractor to determine the effects of various known variables on solder bridging, illustrates the technique.

The engineers began by identifying three different variables: conveyor speed, solder temperature, and flux activity. They decided to allow two possible settings for each of the variables. Conveyor speed was allowed to be either 4 ($CONV^s$) or 6 ($CONV^f$) feet per minute. Solder temperature could be 450°F ($TEMP^1$) or 500°F ($TEMP^2$). And flux activity could be either RMA ($FLUX^m$) or RA ($FLUX^a$). Thus the engineers had three variables, each of which could be set at either of two levels.

The factorial table looked like this:

Run #	Combination
1.	$CONV^s/TEMP^l/FLUX^m$
2.	$CONV^s/TEMP^l/FLUX^a$
3.	$CONV^s/TEMP^2/FLUX^m$
4.	$CONV^s/TEMP^2/FLUX^a$
5.	$CONV^f/TEMP^l/FLUX^m$
6.	$CONV^f/TEMP^l/FLUX^a$
7.	$CONV^f/TEMP^2/FLUX^m$
8.	$CONV^f/TEMP^2/FLUX^a$

Thus 8 trial runs representing the eight possible combinations of variables (2^3) were run. The outcome of each test run was as follows:

Run #	Combination	# of bridges
1.	$CONV^s/TEMP^l/FLUX^m$	6
2.	$CONV^s/TEMP^l/FLUX^a$	5
3.	$CONV^s/TEMP^2/FLUX^m$	3
4.	$CONV^s/TEMP^2/FLUX^a$	4
5.	$CONV^f/TEMP^l/FLUX^m$	8
6.	$CONV^f/TEMP^l/FLUX^a$	6
7.	$CONV^f/TEMP^2/FLUX^m$	6
8.	$CONV^f/TEMP^2/FLUX^a$	5

To determine the importance of conveyor speed, the engineers compared the effects of changing conveyor speed when the other variables were held constant. From the factorial table above, we can see that four pairs of combinations meet the condition of varying conveyor speed while holding the other variables constant. The importance of conveyor speed can be determined by subtracting the number of solder bridges produced at higher conveyor speed from the number of bridges produced at slower conveyor speed. That is:

Run #1 − Run #5 = 6 − 8 = −2
Run #2 − Run #6 = 5 − 6 = −1
Run #3 − Run #7 = 3 − 6 = −3
Run #4 − Run #8 = 4 − 5 = −1
Total = −7

From this, the engineers concluded that increasing conveyor speed increases the incidence of bridging.

They examined the effect of flux activity on bridging in the same way:

Run #1 − Run #2 = 6 − 5 = 1
Run #3 − Run #4 = 3 − 4 = −1
Run #5 − Run #6 = 8 − 6 = 2
Run #7 − Run #8 = 6 − 5 = 1
Total = 3

This led them to believe that increasing the flux activity reduces bridging. Those conclusions were used to specify shop floor process settings.

14.9.3 Errors in the Experiment

Unfortunately, the experiment was seriously flawed. Flux activity does not affect bridging; variance in the results led to the wrong conclusion. Among the reasons for the faulty outcome, three in particular stand out:

1. The "experiment" was looking for correlations, not attempting to test a hypothesis; as Goldratt warned, there is always the very real possibility that the "correlation" is coincidental.
2. The number of experimental runs was too small for statistical validity.
3. Several very important forces were overlooked. Some of those variables include:
   a. Flux solids content.
   b. Flux solids type.
   c. Component alignment.

d. Solder wave profile.
e. Solder mask type.
f. Preheat temperature.

If these unknown variables are left out of the experiment, it is still possible to discover that conveyor speed, for example, affects the number of bridges—provided *all* of the omitted variables remain constant from one experimental run to the next. However, there is no reason to expect that any of these variables will remain constant. Moreover, it is impossible to learn whether any of the omitted variables play important roles in the process. Accordingly, the success[28] of factorial experimentation depends in large measure on the quality of the educated guesses made in choosing the variables to be tested.

14.9.4 More Variables Means More Effort

The possibly significant effects of omitted variables introduce new complexities, of course. To be safe, it is necessary to test for all possible variables—and even then some obscure but critical variable is likely to be overlooked. Suppose, however, that all variables are known. A formidable challenge still looms.

Increasing the number of variables or the number of settings for any variable will increase the number of experimental combinations. For example, to test for the effect of all nine variables (conveyor speed, solder temperature, flux activity, flux solids content, flux solids type, component alignment, solder wave profile, solder mask type, and preheat temperature), requires *512* different experimental combinations (2^9) if each variable is only permitted two settings. Obviously, repeating the experiment hundreds of times poses a far greater challenge than the experiment with only eight runs performed described above. The experiment's difficulty increases by several magnitudes when more variables or settings are added. Thousands of experimental combinations may be required. Moreover, each combination should be run several times to reduce the possibility of hidden non-random variation.

Many other complicating factors must also be taken into account. For example, the number of experimental runs may be too large to be completed on the same day using the same personnel and the same equipment. The runs carried out on different days are not strictly comparable

[28]Where "success" equates to finding or not finding previously unnoticed correlation(s) between variables.

because environmental conditions such as air temperature and humidity can vary greatly. For such cases, statisticians have devised a technique called blocking, which limits the significance of the potential day-to-day variations. Again, however, this involves considerable numbers of possibly invalid assumptions along with guesswork.

Therefore, the feasibility of factorial experimentation is limited to the simplest of processes. But a process that *appears* simple often involves unrecognized variables having critical effects.

14.10 Fractional (Partial) Factorial Experiments

Seasoned statisticians often reduce the number of experimental runs to more manageable proportions by discarding factors they consider to have negligible effect on the process outcomes. The usual terminology for such experiments is fractional (or "partial") factorial designs. Fractional factorial designs rely on guessing at probable outcomes for certain variables and eliminating those variables from the experiment. This arbitrary exclusion of variables reduces the science to a pseudo-science.

Generally, engineers who know enough about a process to design and conduct meaningful experiments will not benefit from actually conducting the experiments. Those process masters already possess all the required knowledge to derive a meaningful hypothesis. They can then validate their hypotheses through effect-cause-effect reasoning rather than submitting to the dubious science of statistical experimentation.

14.10.1 Taguchi Methods

In recent years, fractional factorial experimentation has become virtually synonymous—albeit implicitly—with Dr. Genichi Taguchi, a Japanese statistician.[29] In many companies, Taguchi methods are now considered the ultimate in engineering chic.

The Taguchi system requires very little of the experimenter. No hypothesis is established to be validated or invalidated by the experiment. After selecting a handful of factors that seem to have considerable influence to the outcome of the process, the Taguchi-style experimenter simply turns to the Taguchi "cookbook" table for that number of variables and is told what experimental design to use. The nature of the fac-

[29]Taguchi himself never speaks of fractional factorial experiments, but there is no question his methods are indeed a form of the traditional fractional factorial approach.

tors does not affect the structure of the experiment. The experimenter makes no assumptions in choosing the factors and needs no understanding of the manufacturing process. Similarly, the results are interpreted without any extraneous insight into the process. Using the Taguchi approach, even the least experienced individual can instantly become a Taguchi-approved scientist.

The sudden popularity of Taguchi's experimental design procedures is easy to understand. Unlike the full factorial experiments discussed in Design of Experiments, Taguchi's methods are easy to learn and use. Whereas full factorial experiments require considerable prior knowledge and discipline on the part of the engineer-cum-experimenter, Taguchi reduces most experimental situations to a handful of experimental design tables. The most popular Taguchi-style experiment is the two-level design (in other words, each chosen factor is tested at only two different settings). With only a tiny number of factors involved and each set at only two levels, the number of experimental runs required shrinks to almost negligible levels. Moreover, the Taguchi "cookbook" of orthogonal experimental designs makes setup of the experiment foolproof once the variables have been selected.

Unfortunately, however, Taguchi's methods tend to give a false sense of security—and the conclusions can be very erroneous indeed. The four main features that make Taguchi methods so attractive also pose significant risk. Those features consist of:

1. *Eliminating any requirement for the experimenter to make informed choices when selecting factors to be investigated.* As we saw under the discussion of partial factorial experiments, the partial factorial approach offers one great advantage over full factorial experiments: the ability to test many parameters with a small number of observations. On the other hand, the determination of factors to include (and, therefore, the factors to exclude) in the partial factorial experiment requires significant guesswork on the part of the experimenter. The more the experimenter knows about the process during design of the experiment, the greater the probability that the selection will include the most significant factors and exclude the trivial. Of course, there is no way before or after the fact to prove that the experimenter understands the process factors as well as presumed.

2. *Minimizing the number of variables considered.* Taguchi not only sanctions omitting certain factors from consideration in the experiment, he insists that this is the only valid way of conducting experiments in industry. Eliminating possibly relevant factors from the experiment is known as fractionalization. Taguchi also holds the settings of each factor to a low level (usually only two). Combining those two recommen-

dations—few factors and artificially low variation in their settings—means far less work for the experimenter. It also sharply increases the probability of false "findings."

3. *Eliminating the need to be statistically proficient.* Taguchi's cookbook approach gives the user a false sense of competence. Deming may have been wrong to argue that mastery of statistical theory is more important than familiarity with the production processes, but to believe (as Taguchi does, whether explicitly or implicitly) that in-depth understanding of the process is inconsequential goes beyond naïveté into very dangerous territory.

4. *Removing the need to have informed judgment when evaluating results.* Since the experimenter is not expected to have mastered knowledge of the process, there is no knowledge by which to assess the information provided by the process. Specifically, the experimenter is unable to answer the fundamental quality of correlation: Is the relationship real or coincidental?

Over the years, when we have expressed concern about Taguchi methodology, process engineers and design personnel invariably respond that they *did* reduce their defects by using Taguchi-style experiments. While they acknowledge that the results may be suboptimal or locally optimal, they also contend that surely they are better off than had they not turned to Taguchi methods at all. That is not necessarily true, although they may be better off than had they ignored process improvement altogether.

Taguchi's favorite example of effective application of his techniques to electronics products involves a voltage regulator used in NEC television sets.[30] According to Taguchi, NEC televisions employed a voltage regulator valued at +115 V dc at 600 mA. The device was built using inexpensive components that were rated at ±5 percent; consequently, a meaningful number of the voltage regulators produced had output variance of as much as ±25 V and could not be trimmed to the required 115 V by adjusting the potentiometer. Taguchi claims to have used his orthogonal arrays to "optimize" the regulator so that variation was reduced to a maximum of ±1.5 V without resorting to more expensive components. Unfortunately, no electronic engineer has ever been able to prove Taguchi's claim.

Bob Pease, a staff scientist with National Semiconductor Corporation and columnist for *Electronic Design* magazine, checked whether the Taguchi circuit would perform to the stated tolerances. Using Taguchi's

[30][Taguchi and Wu 1979], pp. 41–47.

circuit design, Pease built the regulator and tested its performance. He found that the circuit performed exactly as described—provided the input was 138 V, the input voltage Taguchi had specified in arriving at the "robust" design. Changing the input by 10 V changed the output by 8 V. In Pease's words: "A ripple of 20 V on the main filter gives a totally unacceptable ripple of 16 V on the output."[31]

Pease concluded that "Taguchi wished to have the output invariant to any resistor changes—but he neglected to say he still wanted it to REGULATE! He forgot to check to see if the 'optimal' circuit would still keep a constant output vs. changes in [input voltage]. So he got what he asked for. And he did not check the answer."

Pease reported his findings to Taguchi, who responded with the unusual statement, "We are not interested in any actual results because quality engineering deals with only optimization." But, as Pease concluded, a system that doesn't work can hardly be considered optimized.

Over the years, many engineers besides Taguchi himself have obtained quite bizarre "answers" to process problems by applying Taguchi methods. However, the experimenters concerned have failed to recognize that their "answers" are very wrong. Their only concern is whether the mathematics work properly; in fact, they should be most concerned about whether the findings can be verified using an effect-cause-effect analysis.

14.10.2 When Experimentation Is Not the Best Road to a Solution

Design of Experiments, partial factorial experiments, and Taguchi methods all involve great effort, cost, and risk. Certainly there are times when the output of resources and acceptance of risk are warranted. All too often, however, statistical experimentation is employed needlessly. Before embarking on such a statistical exercise, it is advisable to recall the fundamental rule of experimentation found in Charles R. Hicks's classic text *Fundamental Concepts in the Design of Experiments.*[32] "Research is a systematic quest for undiscovered truth," Hicks wrote. "The truth sought should be something that is not already known, which implies a thorough literature search (first) to ascertain that the answer is not available in previous studies." Those words may be the most important statistical lesson ever published.

---

[31] Pease, Bob, "What's All This Taguchi Stuff, Anyhow? (Part II)," *Electronic Design,* June 10, 1993, pp. 85–92.

[32] [Hicks 1982], p. 1.

14.11 Summary

The principal points raised in this chapter included:

- Every process contains variation.
- Statistical quality control (SQC) uses mathematical and graphical techniques to identify sources of excessive variation and the need for corrective action to the process.
- Collection and analysis of statistical data involves considerable effort and expense; scientific quality management requires balancing the costs of the data collection and analysis against the benefits.
- Deming and Goldratt both point out that statistics may have considerably less value than often perceived; to have any value, the statistician must be assisted by an individual with thorough knowledge of the process interactions.
- A process master can typically learn statistical methods much faster and more easily than a statistician can acquire competence in process matters.
- Statistical "data" have most value in highlighting correlation between variables; however, correlation may only be coincidence, a possibility that requires development of a cause-and-effect hypothesis that is subjected to validation testing.
- The most useful statistical tools for application in the electronics assembly industry tend to be Pareto analysis, mapping diagrams, and product failure rates; they are simple and inexpensive to employ but provide meaningful data in a very short time.
- The effectiveness of statistics depends on choosing the correct answers to several difficult questions, among them what quality characteristic(s) to measure, how to measure, and where to measure.
- Control charts can be useful in providing information about stability of process with respect to a given variable or attribute. But very large amounts of undependable product can be produced before control charts show that the process requires attention.
- The "experimentation" done in the electronics assembly industry typically is nothing more than a search for correlation; true experimentation starts with a hypothesis of cause-and-effect and leads to discovery of other effects related to the same cause.
- Fractional (or "partial factorial") experiments are risky because they require the experimenter to guess which factors have little effect on the process. Taguchi methods are particularly dangerous because the experimenters make no attempt to develop and test a cause-and-effect hypothesis.

15 Pseudoquality

15.1 Chapter Objectives

More often than not, quality improvement programs produce unsatisfactory results. This has been the case even when programs have been designed and undertaken with the best of intentions following the latest conventional wisdom of how a quality program should be structured. Unfortunately, all too many "quality" programs are not intended to improve quality at all. Rather, their goals are to increase efficiency and impress customers; in these cases, the programs use "quality" in their names for questionable purposes such as obscuring the true objectives from employees who would be disinclined to work diligently on projects to reduce jobs or from customers who wrongly believe they will be receiving more dependable products.

"Quality" programs that do not improve quality can be termed "pseudoquality." Pseudoquality programs are extremely dangerous. They can easily undermine the quality department's credibility and extinguish any enthusiasm within the company for further quality efforts. They draw away resources that would otherwise be employed in real quality improvement. And, in many cases, they are dishonest.

The company that is genuinely committed to quality improvement needs to know what pseudoquality is, its consequences, how pseudoquality can be identified, and how to avoid the pseudoquality traps. This chapter addresses those issues.

15.2 Some Obstacles to Quality Improvement

The quality revolution has not been a universally positive experience for America's electronics assembly industry. In fact, the results of many quality projects can only be described as tragic. Opportunities for substantive improvements in the company's health have been lost, enormous amounts of money and time wasted, and perhaps most tragic of all, the company caught up in confusion between the goal and means of achieving it. In the words of *Business Week* magazine: "[A]t too many companies, it turns out, the push for quality can be as badly misguided as it is well-intended...it's wasted effort and expense."[1]

Harvard Business School professor David Garvin, an early and prominent proponent of the total quality approach, expressed concern in 1995 that the quality movement has largely failed. "[A]s a hot management trend, quality has been pushed aside—by reengineering and a host of other ideas," Garvin said. "Companies once gung-ho about quality have become disillusioned and discouraged and have dropped out.

"...There's a set of companies...where quality is part of the fabric....There's another set of companies where quality was the fad of the month. They've moved on to other things....The quality movement was oversold. As soon as it didn't solve all their problems, some companies dropped out." [2]

Previous chapters have shown some of the ways in which quality improvement efforts can fail badly, among them:

1. *Lack of an operational definition for quality.* Without companywide agreement on a precise and accurate definition of "quality"—the sort of workable terminology that Deming called an "operational definition"—unsatisfactory quality improvement strategies will inevitably be adopted. Apparent consensus on actions may be achieved, but there will always be serious underlying misunderstandings. Each individual operates according to a personal agenda that will be unknown to other personnel. The hidden nature of the misunderstandings make them considerably more dangerous to the company's well-being than are clearly articulated disagreements.

2. *Failure to work in a scientific manner.* In the electronics assembly industry, key processes dominated by natural forces determine the dependability of output. Therefore, real quality can be achieved only by scientific process management. Without science there can be no control; without control, there can be no effective quality management.

What is science? *The American Heritage Dictionary* defines science as "the observation, identification, description, experimental investiga-

---

[1]"Quality: How to Make It Pay," *Business Week,* Aug. 8, 1994, pp. 54–59.

[2]"Is TQM Dead?" *USA Today,* Oct. 17, 1995, p. 1A.

tion, and theoretical explanation of phenomena." Or, using Goldratt's particularly useful three-step way of looking at the scientific process, science is achieved when the following three steps have been fulfilled: Classification, Correlation, and Effect-Cause-Effect.[3]

3. *Unreasonable expectations of benefits from quality improvement.* Quality improvement programs too often are undertaken in the belief that higher quality must translate into greater demand and profits. As Chap. 4 explained, this frequently is not true; improved quality many times does not lead to increased demand.

Quality is not necessarily free. It often involves costs as well as benefits. When long-term costs exceed long-term benefits, spending on quality improvement should end. If long-term benefits exceed long-term costs but the quality improvement program causes short-term cash flow to fall below survival levels, the program cannot be considered at all.

Benefits from quality improvement in the electronics assembly industry tend to result from production and quality evaluation efficiencies rather than greater demand. Dependability of electronics products varies inversely with the amount of handling; therefore, reducing the need for rework and repair is a form of quality improvement that can deliver substantial cost reductions. Typically, the benefits of quality improvement are more pronounced on the supply side (i.e., cost reductions) than on the demand side.

4. *Excessive demands on production personnel.* The employees closest to the actual manufacturing and quality evaluation operations are best positioned to identify problems related to quality and productivity. Under certain conditions, those same personnel are also well suited to implement corrective actions. In all cases, however, the value of employee input is constrained by their knowledge. As explained in Chap. 12, workers are not truly "empowered" until they have been thoroughly educated and provided with extensive hands-on experience in the processes and the corporate organization. Ideally, the necessary education would contain a quasi-apprenticeship for every employee, including design engineers and potential managers. The cost in dollars, time, and effort of such an undertaking is so substantial that it can be feasible only where employees remain with the same company for many years. Since few manufacturing companies today display such high levels of employment stability, adequate investment in employee development is rare and most employee participation programs end in disappointment.

---

[3]See Sec. 14.3.4.2.

5. *Improper use of statistics.* Although Deming disagrees, anything that cannot be measured cannot be effectively managed. Measurement first requires the ability to identify those output characteristics that will assist managers if measured, a requirement shown in Chap. 14. Moreover, the "information" provided by those measurements can be highly misleading, particularly where coincidence is mistaken for correlation. Effective quality management requires skillful use of statistics, but most "statistics" are actually meaningless groupings of digits. The improper use of statistics is closely tied to the lack of an operational definition in that missing or weak definitions also lead to massive volumes of meaningless statistics.

6. *Absence of urgency.* The value of quality improvement resides in the improvements themselves, not in the improvement process. Though the need to emphasize the results rather than the means would seem self-evident, companies regularly endorse programs that deliver the necessary improvements far too late. By the time the improvements are made, the competition is likely to have moved on to even better performance levels and the company's competitive position may actually be worse than at the start of the program.

Orthodox Demingism believes that a company requires 5 to 10 years to change its quality culture. While that belief may or may not be valid,[4] it has been wrongly used to justify sluggish delivery of benefits from individual departments. Substantial benefits should be seen within weeks from any properly conceived quality improvement program.

15.3 Pseudoquality and Real Quality

Considering the numerous formidable challenges facing the company that hopes merely to understand quality, it should not be surprising that little "quality improvement" activity today actually causes higher product dependability. "Quality improvement" strategies that will have minimal or no effect on product dependability can best be described as "pseudoquality," where "pseudo" means false or pretended. Pseudoquality is simply a parody of real quality.

Pseudoquality comes in two basic forms:

---

[4]In our experiences, any program with a time horizon of years is doomed. Personnel changes in the intervening period will sap the program's strength and obscure the original vision. In the modern world, the luxury of leisurely change is too costly.

1. Some pseudoquality programs do not see quality improvement as their objective at all. They are designed and approved as schemes for becoming more efficient or attracting additional customers. The word "quality" is included in their titles for public relations purposes (rather like the "new and improved" formula for selling soap) or to distract employees who might be unwilling to support a program intended to reduce jobs.

2. Programs that truly want to bring about higher quality but are constructed in ways that ensure failure.

Both true quality and pseudoquality programs can have overlapping benefits, among them:

- Cost reductions
- Elimination of unnecessary activities
- Increased profits

The distinction between pseudoquality and real quality lies less in what benefits may result than in priorities and methodologies. Pseudoquality focuses on building demand, eliminating waste, and seeking other efficiencies to reduce costs. Real quality, of course, concentrates on ways to enhance product dependability. The ways to improve dependability *may* be more efficient, *may* save money, and/or *may* attract more customers,[5] but those are by-products of the quest for higher quality rather than the primary goals per se.

In other words, while pseudoquality and real quality programs can at times produce similar results, their goals are different. While this may seem like an exercise in semantical hair-splitting, the distinction between goals and results is extremely important. Many employees, for example, have turned against real quality improvement programs because their experience with pseudoquality has led them—often with cause—to conclude that "quality improvement" is just code for finding ways to eliminate jobs. While real quality improvement may also reduce jobs, it does so only as a means of achieving greater product dependability.

Pseudoquality programs are not necessarily bad if judged only on the basis of what they can accomplish. For example, becoming more efficient and acquiring additional customers (pseudoquality objectives) are

---

[5]Indeed, if some or all of these ancillary benefits are not achieved, there will be no advantage to the company in improving the quality.

laudable missions that companies would be foolish to ignore. At the same time, however, those objectives are not related to quality. By posing as activities to bring about better quality, pseudoquality activities lead companies into an erroneous sense of security that the quality level is being competently analyzed and improved. Since companies using pseudoquality tactics believe themselves to be working aggressively on quality improvement, real quality programs which might deliver substantial benefits are overlooked. The companies end up in a suboptimal position.

Although the gap between pseudoquality and real quality is enormous, pseudoquality programs can easily be mistaken for real quality efforts because both pseudoquality and real quality generally employ the same tools. For example, statistical methods figure prominently in both pseudoquality and real quality. The vital distinction is that, although they may superficially employ the same tools, those tools are used in very different ways. As an example, while statistical measures are used in both cases, pseudoquality typically measures by proxy variables while real quality concentrates on changes in the actual dependability of output. Similarly, pseudoquality would employ a defective statistical method like Taguchi methods without hesitation while real quality programs would not.

For reasons presented in the following chapters, Total Quality Management, the Malcolm Baldrige National Quality Award, and Six Sigma™ all fall into the pseudoquality category. ISO 9000, though originally designed as a quality improvement tool, has also acquired many of the attributes of pseudoquality. It is a reflection of how little industry comprehends quality that few people today can distinguish between pseudoquality and real quality.

15.3.1 Ten Attributes of Pseudoquality

Pseudoquality programs possess some or all of the following attributes:

1. *Originate with consultants, business school faculties, or other groups lacking actual operations experience.* Pseudoquality consulting has been very profitable for many inexperienced theorists and general consulting practitioners, few of them with any hands-on plant experience. Just as Crosby convinced corporate executives that they could make money by *actually* improving quality, consultant Tom Peters's enormous financial success with his book *In Search of Excellence* demonstrated to the consulting and business school establishments that considerable money could be made by selling programs *claiming* to improve quality. Equally

important, Peters showed that pseudoquality consulting could be practiced successfully by individuals completely lacking any hands-on experience in quality.

2. *Apply statistical tools to interpret and manage processes without in-depth firsthand knowledge of the process and variables.* Deming (see Sec. 14.3.4.1) contended that knowledge of statistical practices alone was not sufficient to ensure that any process could be accurately diagnosed. "No one should teach the theory and use of control charts without knowledge of statistical theory...supplemented by experience under a master," Deming argued.[6] Scientific use of statistics begins with a hypothesis and conducts tests on what happens if the hypothesis holds or fails; it is exceedingly difficult—generally impossible—to formulate a meaningful hypothesis without also possessing some a priori knowledge of the process to be examined. The history of Taguchi partial factorial experimental methods shows how easily the naive theorist can seriously misinterpret the nature of a process and the effects of variables.

3. *Do not define "quality"—or define it in subjective terms that make substantive program planning and management impossible.* Although the meaning of quality is dependability, pseudoquality programs emphasize the number of identified defects, customer opinions, process documentation, and other proxy factors. A proxy factor is any attribute believed to reflect quality but is not directly related to product dependability. For example, Six Sigma™ measures quality by defect levels during production; the Malcolm Baldrige National Quality Award stresses customer satisfaction and employee participation; and ISO 9000 places considerable emphasis on paperwork. Higher real quality can result as a by-product of pseudoquality programs, but any changes will be purely accidental rather than inevitable.

4. *Encourage bureaucracy.* Pseudoquality programs often contain the very bureaucratic characteristics that businesses properly disparage in government. A few such characteristics include:

1. Endless meetings
2. Enormous paperwork
3. Majority rule
4. Lack of accountability (or, more accurately, self-evaluation)

5. *Defy attempts at quantitative analysis.* Popular pseudoquality yardsticks such as "customer satisfaction" cannot be employed in a homogeneous, objective fashion. If the definition of "world class qual-

---

[6][Deming 1986], p. 131.

ity" is—as so many authorities claim—"meeting the customer's expectations," one way to achieve world class status is to seek out customers with low expectations. The customer who is impossible to satisfy cannot be considered an accurate reflection of the manufacturer's performance, yet a company afflicted by unreasonable customers can destroy itself searching for ways to satisfy the unsatisfiable.

6. *Insist on constant improvement activity without consideration of relative costs and benefits.* Pseudoquality orthodoxy contends that a company can never have too much quality. Higher quality is always preferable to less *if the price is right*—but excess quality may raise costs and drive away customers. For example, the additional quality may be achieved by using more expensive components that increase costs to the point where customers cannot afford to be in the market. Quality improvement that does not help the company prosper is no achievement at all.

7. *Involve high costs.* Bureaucracy involves greater overhead and, therefore, greater costs. The supporting documentation required for consideration in a competition such as the Malcolm Baldrige National Quality Award is very costly. One medium-sized U.K. company with which we are well acquainted spent approximately £150,000 ($240,000) in preparation costs prior to being audited by one of the major U.S. aircraft manufacturers for compliance with the American firm's own version of the Baldrige; to further put the cost in context, the American firm requiring the implementation and subsequent audit was only one of the company's many customers. Similarly, paying for regular audits by ISO 9000 assessment bodies also adds to operating costs.

8. *Work slowly if at all.* According to Deming, "Companies with good management will require five years to remove the barriers that make it impossible for the hourly worker to take pride in his work. Many companies will require ten years."[7] Deming disciples do not see any problem in such extended time to attain acceptable performance levels. "We are a monumentally impatient people who want everything finished yesterday, and if it can't be done quickly, Americans question whether it is worth doing at all," one pair of reporters turned quality pundits wrote in defense of Demingism.[8] "Quality doesn't work that way. The Deming management system takes years because it is a philosophy, not a technique. Thinking is always more difficult and time-consuming in a quality management system than doing....Americans

[7][Deming 1982], p. 90.

[8][Dobyns and Crawford-Mason 1994], p. 24.

believe that everything can be done quickly—ipso facto, everything that isn't quick must be bad. That's [a] myth that has to go."

Looking for more immediate results is not necessarily indicative of intemperance or immaturity, however; it may be pragmatic recognition that long-term turnarounds require short-term survival. In a world where a product generation may be as short as 6 months, turning to solutions that will not bear fruit for a decade could mean the company will die before the cure can be effected.

Much can happen in 5 years; the normal tenure of a CEO at major companies is shorter—often much shorter—than that. In 10 years, the entire world can change. The possibility is very real that the company will not survive until a 5- to 10-year turnaround is complete. Although opportunities for further incremental improvements in quality will often appear long after a program to increase quality has been undertaken, any program that cannot produce quantum gains in a matter of months should not even be considered. The myth that must be changed is that which claims all short-term thinking is destructive.

9. *Are primarily image-driven.* In real quality programs, enhanced quality is its own reward; demand may rise and costs will likely fall. Pseudoquality depends on public recognition for its allure. Companies spend much more time and money by entering a Baldrige National Quality Award or Deming Prize competition than are needed to implement and manage a highly successful program that does not involve competition. The extra spending is purely marketing driven and has nothing to do with quality improvement; if the company wins an award for its progress, it has leveraged its pseudoquality investment into a promotional tool.

10. *Thrive on jargon.* Paradigm, paradigm shift, visioning documents, empowerment, world class, Total Quality Management, Total Quality Leadership, and Total Quality Control are only a few of the terms that have crept into the pseudoquality language. These are not phrases of inclusion; they are used in the worst tradition of jargon—to exclude anyone who is not a member of the inner circle. Jargon impedes communication.

15.3.2 The Forces Enabling Pseudoquality

To properly understand the nature and dangers of pseudoquality, it is necessary to understand why it so readily infiltrates companies run by highly intelligent personnel. Companies commit to pseudoquality programs—as opposed to true quality improvement strategies—for one or more of a relatively small number of reasons. The most common reasons include:

- Idealism
- Opportunism
- Peer influence
- Coercion
- Inexperience

15.3.2.1 Idealism. The quality department may be one of the last pockets of idealism in present-day businesses. The idealist believes the function of business is to improve society, not just by providing dependable products at a reasonable price but also by providing happiness to customers ("exceeding customer expectations") and greater job satisfaction to employees. The pure idealist has no doubt that happier customers and contented employees will make for a more prosperous company in addition to a better world and a healthier company. At times, the idealists are right; other times, they are wrong. At no point, however, are the idealists operating from an objective point of view. Management by ideology is never an acceptable substitute for management by objective information.

15.3.2.2 Opportunism. Opportunists are polar opposites of the idealists. They take a short-term view of business and strive to maximize income today rather than caring for the company's long-term prosperity. They will take on any program—quality improvement included—that promises to boost short-term earnings. Opportunists are notorious for turning programs on and off without considering the longer-term effects; these, in other words, are the individuals who create employee cynicism about "flavor of the month" programs.

Image building is a form of opportunism. A depressingly large number of companies blatantly exploit "quality" themes in their messages to customers, shareholders, and employees without having any knowledge of what quality means—or whether it can be found in their products.

15.3.2.3 Peer Influence. Senior executives listen to what their counterparts at other companies say. If a respected or well-known CEO, for example, tells the corporate world (through press releases, advertising, speeches, and interviews) that a particular activity helped his company enormously, other CEOs can be converted. For example, John Akers, while CEO of IBM, announced that company would be following Motorola's lead and adopting a Six Sigma™ strategy; his interests soon shifted to other matters, and IBM's flirtation with Six Sigma™ was very brief.

15.3.2.4 Coercion. Major customers can coerce suppliers into taking actions that the supplier would not have chosen—and may well have worked diligently to avoid—on its own. Numerous companies invested in ISO 9000 certification because of customer requirements, even though their management does not believe that the program improved their companies in concrete ways. We have seen customers require ISO 9000 certification for suppliers as small as single-person businesses! Other large companies have required their suppliers to implement programs based on the Malcolm Baldrige National Quality Award criteria.

15.3.2.5 Inexperience. Wisdom cannot be learned from books or lectures; it comes only from the lessons of life. Increasingly, the quality profession has come to rely on individuals who are well prepared in theory but lack the first-hand operations experiences that lead to fully informed decisions. Three primary causes for this dwindling store of experience can be identified:

1. The quality profession has acquired much of the glamour and career potential previously found only in the highly visible corporate activities such as finance and sales. As a quality career offered better career prospects, its ranks have swelled with bright, aggressive individuals who lack seasoning. In time, the new generation of quality professionals will be better equipped to plan and execute quality programs than earlier generations who were not exposed to the same extensive academic grounding in quality theory. In the meantime, the natural—and, generally, costly—tendency is to apply academic principles to real world situations.

2. Many companies in recent years have forced experienced veterans—primarily those aged 50 to 60—into early retirement. Younger (and lower-paid as well as less knowledgeable) workers have been hired as replacements for the veteran staff. Inevitably, the steadying influence brought to planning and decisions by managers who possess considerable first-hand experience in operations has been seriously diluted. As they depart, the veterans take with them the real world experiences that prevent repetition of mistakes.

3. Lack of experience greatly increases the probability of overlooking the flaws in new systems. In the early 1980s, American companies rushed into quality circles in the incorrect belief that worker groups were an important factor in Japan's industrial competitive strength. Being unfamiliar with Japanese culture and industrial organization, few American managers realized that quality circles would fail without substantial corporate support in the form of education, guidance in how to organize and conduct meetings, strong management to assess the

broader implications of improvement suggestions before they would be implemented, and much more. American quality circles proved very costly to the companies that employed them. For example, wages were paid for time spent in meetings rather than production. Poorly conceived "improvement" plans put into action without proper scrutiny from managers tended to be more harmful than helpful. At the same time, circle members found they could not meet the challenges and became disillusioned; this outcome was a predictable consequence of failing to provide the members of the circles with appropriate education and experience in the techniques of:

- How to run effective meetings
- Setting priorities among alternative activities
- Communication (with other employees noted represented in the circle and with management) and reporting
- Leadership
- Systems design
- Implementation
- Measurement

Supervisors and managers were also denied the education and executive support that would have allowed them to function more effectively in the new environment. Some companies assembled teams to develop improvement suggestions but had no systems for following through on any suggestions that emerged. Other companies distorted the original concept of employee involvement to total consensus management in which every voice was equal; rather than recognizing that someone needs to make a final decision after all the opinions have been aired, those companies undercut the authority of managers and paralyzed decision-making.

Valuable lessons were learned from the mistakes of the quality circle experiments, and few who participated in those experiments would make the same errors again. Those lessons included how to make effective use of teams as well as how not to operate quality circles. However, most veterans of the quality circles have moved on to other things, taking their expertise with them. The errors of the quality circle days are being repeated today under the name of employee teams.

15.4 Summary

True quality-oriented systems are less common than the large number of companies operating "quality improvement" programs suggests. Often, a "quality improvement" project is directed toward results that are not related to greater dependability—such as productivity increases, winning awards, reducing defects, and so on. We refer to those mislabeled "quality" programs as pseudoquality. Some of the most popular activities in contemporary quality are actually pseudoquality. Those pseudoquality systems include the various forms of Total Quality, the Malcolm Baldrige National Quality Award, ISO 9000, and Six Sigma™.

Many of pseudoquality's objectives are desirable. Quality improvement, after all, is only one of many means for a company to progress toward the ultimate long-term goal of greatest profitability. Other pseudoquality aspects—the quest for bragging rights through competitions, for example—are at best questionable. However, it is certain that the outcome of any pseudoquality project will include enhanced dependability only by chance as a second-order effect.

Pseudoquality systems should dismay advocates of real quality improvement for two reasons. First, adoption of pseudoquality causes company executives to wrongly believe quality improvement is being addressed forcefully and effectively. Second, funds wasted on inefficient pseudoquality activities are not available for real quality initiatives.

The company as a whole should be distressed by pseudoquality because the return on investment in pseudoquality programs is low. Unfortunately, because avoidance of quantifiable objectives is a prominent feature of pseudoquality, the poor performance of these programs is rarely recognized.

16 Examples of Pseudoquality

16.1 Chapter Objectives

Chapter 15 described pseudoquality in generic terms. The generic description is given life in this chapter by examination of several well-known "quality improvement" programs that will turn out to be largely or wholly pseudoquality rather than real quality. In addition to warning about some highly publicized but dangerously flawed approaches to quality, the exercise will demonstrate how easily pseudoquality can be confused for real quality.

16.2 Pseudoquality Programs Examined in This Chapter

The quality world abounds with pseudoquality programs. The broad variety of perils that lie in wait for the unwary company can be illustrated here by reference to three programs that have been widely implemented in American industry within the past decade: (1) Total Quality Management (TQM), (2) the Malcolm Baldrige National Quality Award, and (3) Six Sigma™.

Total Quality Management (or, as it is known in some organizations, Total Quality Leadership) epitomizes pseudoquality. The favorable notice that TQM has received in recent years has been substantially overstated in all but a very few cases. The Malcolm Baldrige National Quality Award—essentially a TQM-based competition—suffers from all the TQM ailments and introduces some additional liabilities.

Inasmuch as voluminous reference material describing TQM and the Baldrige can be found in virtually any library or bookstore, this chapter will be largely restricted to pointing out some generally overlooked weaknesses in those programs.

Six Sigma™ is an example of statistical double-talk obscuring gross inaccuracies in concept and management practices. Much less has been written about this idiosyncratic approach to quality improvement, in part because it is essentially unknown outside the electronics industry. Within the electronics assembly industry, however, some powerful customers have imposed this system on suppliers. While not warranting lengthy discussion, Six Sigma™ is briefly examined in these pages to assist suppliers in their negotiations with customers who might impose the program as a condition of purchase.

ISO 9000 has taken on many attributes of pseudoquality and properly belongs in this chapter. On the other hand, while disillusion is growing among companies that have tried the other examples of pseudoquality, ISO 9000 experience remains more limited in America. Owing to the increasing reliance currently being placed by customers on ISO 9000 certification and the astonishingly large amounts of money being spent on promotion of ISO 9000 by the apparently endless numbers of consultants who profess competency in the field, a separate chapter is devoted to that topic.

16.3 Total Quality Management

"What do we do after Total Quality Management?" Crosby says in the introduction to *Completeness.*[1] "People ask me that question continually.... They forget, or perhaps didn't know, that I have never recommended they fool around with TQM in the first place. They are just a set of initials without definition or formulation that have been used by organizations in order to avoid the hard work of really managing quality. People are so busy working on techniques such as 'building teams,' and doing statistical process control, that they never get around to actu-

[1][Crosby 1992], p. xi.

ally learning how to build prevention into their organization...a lot of time has been spent chasing fairy dust instead of being real about quality."

While Crosby may not favor TQM, he has definitely been in the minority among quality consultants over the past decade. Even the Juran Institute—one of the last places we would have expected to see TQM embraced—joined the TQM movement; the institute's promotional brochure for its 1994 annual conference recommended attendance for "Professionals seeking an understanding of the challenges and issues of implementing TQM...Those new to TQM...Anyone active in or wanting to be active in a TQM effort."[2] Hundreds of books have been written, endless magazine pages filled, countless conferences held—all devoted to TQM.

While Crosby has never been our favorite quality authority, his disdain for TQM is solidly based. In most respects, TQM exemplifies pseudoquality.

16.3.1 The Many Faces of Total Quality

The first reference to "total quality" appears in a landmark *Harvard Business Review* article by Armand Feigenbaum.[3] "The underlying principle of this total quality view...is that...control must start with the design of the product and end only when the product has been placed in the hands of a customer who remains satisfied," Feigenbaum wrote. Moreover, he noted, "The first principle to be recognized is that *quality is everybody's job.*" In his *Total Quality Control* book, he states: "One essential contribution of total-quality programs today is the establishment of customer-oriented quality disciplines in the marketing and engineering functions as well as in production. Thus, every employee of an organization, from top management to the production-line worker, will be personally involved in quality control."[4]

Feigenbaum's systems came to the attention of Japanese engineers through a 1957 reprint in *Industrial Quality Control* of his *Harvard Business Review* article. They liked the idea of company-wide quality control

---

[2]Sales brochure for "IMPRO94, Rediscovering Quality: The Next Steps," a Juran Institute conference held in Orlando, Fla., Nov. 16–17, 1994.

[3]Feigenbaum, Armand V. "Total Quality Control," *Harvard Business Review*, November/December 1956, pp. 95–98. The concepts that would form the basis for the first edition of Feigenbaum's *Total Quality Control* 5 years later were contained in this article.

[4][Feigenbaum 1991], p. 13.

but apparently misunderstood Feigenbaum's admonition to bring every employee into the quality process; in their minds, Feigenbaum was advocating the domination of quality by a specialized department, and they could not accept centralized power. The concept of TQC adopted in Japan was very much like Feigenbaum's in that everyone in the company was encouraged and expected to participate in quality improvement. However, wrongly believing that their "total quality" was substantively different from Feigenbaum's TQC, the Japanese began a hunt for a new name. Ishikawa urged that the name of the Japanese system be changed to "Japanese-style total quality control," but the new name was rejected as "too cumbersome." In 1968, the Japanese engineering society JUSE settled on the term "company-wide total quality control" to distinguish their system from Feigenbaum's.[5] To this day, the Japanese apparently fail to recognize that they owe Feigenbaum a considerable ideological debt.

Subsequently, other "total quality" terms appeared. Total Quality Management entered the general vocabulary late in the 1980s. By the middle of the 1990s, Total Quality Leadership was replacing TQM in many organizations. Some people still refer to plain "total quality."

Although each professes to be different from and superior to the others, all the concepts of total quality share one characteristic; their meanings are utterly ambiguous. For example, *The St. Petersburg Times* asked Edward Popovich, a quality consultant retained by the Florida Department of Environmental Protection, to implement a controversial TQL program, to explain Total Quality Leadership. His response as reported by *The Times* reveals the lack of substantive content in total quality: "TQL can be viewed in many ways. It's an overriding philosophy, an attitude, a focus on your customer. The relentless pursuit of perfectionism....It's like saying 'Can I explain Catholicism to you?'"[6] This sort of hair-splitting is all too reminiscent of theology debates in the Middle Ages, tempting one to ask how many TQL consultants can dance on the head of a pin.

16.3.2 Defining Total Quality

The most pressing problem for TQM (or, as terminology changes, TQL) is, therefore, its lack of clear meaning. Many other TQM practitioners besides Popovich have noted the philosophical nature of their field.

---

[5][Ishikawa 1985], pp. 90–91.

[6]Olinger, David. "Quest for Quality Is Costing Quite a Bundle." *The St. Petersburg Times*, Nov. 13, 1995, pp. 1B, 4B.

Unfortunately, what they are really saying is that TQM is too abstract for unambiguous explanation. Or, as Crosby said, TQM is "just a set of initials without definition or formulation."

The TQM definitional problem, more precisely, is not a lack of definitions, although a distressingly high percentage of TQM "how to" books never bother to specify what the subject means. Rather, there are *too many* definitions and none of them meets Deming's requirement of an "operational definition" discussed in Sec. 3.2.3.[7]

The number of books and articles we have read on TQM over the years runs into the hundreds. Most publications about TQM say only its nature will become clear from the examples given. They then provide loose accounts of so-called TQM ventures that share few common elements undertaken by companies that themselves have little in common. We have visited many corporate offices and plants of companies that contend they follow TQM, inevitably coming away with the same sad realization—that TQM or (TQL) is whatever the implementing organization thinks it to be.

16.3.2.1 The Best Definition of TQM. Out of all those books, magazine articles, and company visits, the most rigorous definition of TQM we have found comes from the husband-and-wife team of Joseph and Susan Berk.[8] Their definition says merely: "Total Quality Management is a blend of technologies focused on four concepts: defect prevention, continuous improvement, focusing on the customer, and a philosophy that quality is not just the responsibility of an organization's Quality Assurance department, but rather, it is a responsibility shared by all."

While this is the clearest definition we have found, it clearly leaves much to be desired, including:

- *Improper definition of quality.* As Sec. 3.7 explained, the defect level is not in itself a good measure of quality; defects may exist only because of incorrect workmanship standards or overzealous inspection. The level of "defects" can be improved by throwing out the standards that are not dependability-related and implementing systems that will reduce the numbers of false "defects."

---

[7]It is ironic that Deming inhabits both the pseudoquality and real quality worlds.

[8][Berk and Berk 1993], p. 4.

- *Violating the fundamental principle of cost-benefit analysis.* Continuous (or continual, as Deming insisted for linguistic accuracy) improvement can readily violate the requirements of quality optimization. Improving quality is not advisable if the costs outweigh the returns.
- *The emphasis on customer-driven quality.* The impossibility of operating a typical company on the basis of customers defining quality was presented in Sec. 3.4. Some of the reasons quality cannot be led by customers include: (1) Customers are not all alike; (2) customers do not always know what they want; and (3) what the customer wants may not be what the customer needs. In the electronics assembly industry, the customer focus that matters most is ensuring product integrity.
- *Quality cannot be everybody's responsibility.* A philosophy that responsibility for quality is shared by all conveys a sense of moral superiority but, as Crosby noted more than 30 years ago, "if (quality) is everybody's job, then it is nobody's especially."[9] The Berks are, in fairness, much more restrained than other TQM advocates, most of whom contend that quality decision-making be devolved to the lowest levels of the company hierarchy.

So the Berks' definition of TQM, while one of the best available, is fatally flawed in several ways. Compared to alternative definitions, however, the Berks crafted a paragon of precision and usefulness.

16.3.2.2 Other Definitions of TQM. More typical of the "definitions" permeating TQM is one found in a 1994 book published by the Juran Institute:[10] "TQM is tough because it is not just a way of delivering better products and services; it is also a way of changing how we think, work, and relate to people. It involves improving everything an organization does."

In a similar vein, one American Management Association book[11] takes a back door approach to defining TQM by arguing "Sometimes knowing what something is not helps one to gain a better understanding of what it really is." In our experience, when the only way to explain something is by saying what it is *not*, the probability is that no one knows what it is.

---

[9] Crosby, Philip B., "Z Is for Zero Defects," *Industrial Quality Control*, October 1964, p. 183.

[10][Main 1994], p. ix.

[11][Williams 1994], p. 16.

The range of opinions and lack of objective definitions on which can be based quantifiable planning, implementation, and evaluation is all too similar to the problems associated with definitions of "quality" examined in Chap. 3. As was seen in that earlier chapter, the absence of an operational definition leads inevitably to functional paralysis. Like quality, TQM can mean everything and therefore means nothing.

16.3.3 Jargon

TQM, being a pseudoquality approach, depends greatly on jargon. Jargon can be useful (a language of "inclusion") when it consists of common widely understood colloquialisms that enhance communication even with individuals who are outside the profession. In the total quality context, however, jargon is exclusionary; that is, the meaning is not obvious to anyone who is not part of the group that coined the phrase.

The St. Petersburg Times provided one example of TQM jargon in action: "In a memo assessing her department's leadership, Secretary Virginia Wetherell wrote that 'the mission of the (Florida) Department of Environmental Protection is articulated in a visioning document that serves as the beginning point for coalescing the agency. That visioning document is a valuable beginning but for the vision to reach fruition as a leadership tool it requires further refinement and a planned comprehensive agency wide communication program to explain and encourage DEPENDABLE associates to buy in....'" Recognizing that the average reader would find the argot impenetrable, *The Times* provided a translation: "Another way to say it: Some employees need help to understand and accept what the department aims to accomplish."[12]

One of the more depressing aspects of the thoughtless use of jargon is that so many authorities fail to recognize the problems it creates. For example, The American Management Association book that defined TQM according to what it is *not* also contains a defense of jargon: "Management is not an exact science. It is a field of endeavors that changes with the times, and its proponents seek out the interest and excitement of new concepts and techniques....As a result, American management tends to follow trends. *Team Building, excellence* and *empowerment* were key words of yesterday's business books, magazine articles and seminars. Now *quality* and *TQM* are management's buzzwords."[13]

[12]"Leadership Plan: It Starts with a 'Vision.'" *The St. Petersburg Times,* Nov. 13, 1995, p. 4B.

[13][Williams 1994], p. 2.

This praise of jargon ignores the insidious nature of buzzwords—they lack universal meaning and therefore impede rather than assist development and implementation of the fitting actions. Amorphous buzzwords and jargon prevent the effective communication on which effective business decisions depend. It is difficult to picture a world in which more jargon provides any advantages to management, to the company, or to industry as a whole.

16.3.4 Origins of TQM

The origins of the term Total Quality Management are as murky as its meanings. Gabor[14] and others[15] claim that Total Quality Management originated with the U.S. Navy in 1988, about the time the term began to appear regularly in journal articles. However, retired Air Force general Bill Creech claimed in a rambling 1994 book[16] that he used the term TQM in a slide presentation to private industry upon his retirement from the Air Force in early 1985. Creech's claim, if true, would make him not just the father of TQM but the grandfather as well.

The usual history of Total Quality Management's origins attributes the name to a Navy materials quality working committee under Rear Admiral John Kirkpatrick[17] in 1987. The committee concluded that a Japanese-style quality program would be the best route and, in keeping with their intentions, recommended what they believed to be the Japanese term: "Total Quality *Control*." Kirkpatrick liked the general idea but overruled the use of "control" because of its particular meaning in military terminology. A behavioral psychiatrist, Nancy Warren, recommended "Total Quality *Management*," and the term received Kirkpatrick's blessing.[18]

New regimes like to put their personal stamp on their organization's culture so, when Kirkpatrick was replaced by Admiral Frank Kelso in 1990, the new chief demanded yet another revision to the nomenclature. The name was changed to "Total Quality *Leadership*," which remains the Navy-approved jargon and prevails throughout other branches of government down to the state level.

---

[14][Gabor 1990], p. 274.

[15]See, for example, [Dobyns and Crawford-Mason 1994], p. 214.

[16][Creech 1994], pp. 6–7.

[17]Kirkpatrick had attended a Deming seminar in 1985 and decided the approach would work wonders in the Navy. For a detailed description of Kirkpatrick's epiphany, see [Dobyns and Crawford-Mason 1994], pp. 201–204.

[18][Dobyns and Crawford-Mason 1994], p. 214.

The typical presumption about TQM is that the Navy's TQM/TQL is Demingism with a new label. Deming disagreed, however. Indeed, Deming consistently derided TQM in public and denied that any approach in which his 14 Points were not explicitly at the center was useless. Ironically, most practitioners of TQM firmly believe that the 14 Points *are* at the heart of TQM. Which once again shows how jargon obstructs effective communication.

16.3.5 The Tools of TQM

Before it is possible to take deliberate action to be better, it is first necessary to want to do better. The question of whether employees wish to excel is not even asked by TQM, which begins with the presumption that all employees want to do better work but are prevented by company policies and management restrictions. This absolute acceptance of highly motivated workers is not always warranted; many workers do not want to take responsibility for decision-making and are happiest turning off their minds when they enter the plant. It is nonetheless true, however, that employees more often than not do wish to provide the best possible service to their companies.

However, desire in itself is not sufficient to achieve any improvement. It is also necessary to learn how to become better (in Deming's words, "By what method?"[19]). Further, it is essential that the employee clearly understand what constitutes "better" performance.

TQM prescribes several techniques by which to improve quality. In particular, TQM encourages the use of many mainstream real quality tools that, when competently employed in the proper context, are valuable components of an effective quality improvement program. Those tools, which other chapters of this book analyze more thoroughly outside the TQM framework, include:

- Statistical methodologies, including statistical quality control, statistical process control, and experimentation
- Greater involvement of the company's human resources
- Personnel development
- Establishment of closer and mutually supportive ties to suppliers

Individually, each tool generally has much of consequence to offer. But the sum of TQM is less than its parts, in large measure because the right tools are employed in the wrong fashion. A worker knows that

[19]p. 110.

applying the wrong tool to the task at hand—like using a hammer where a screwdriver is required—renders the tool useless and mars the outcome of the job. Even when the appropriate tool is chosen for the job, employing it incorrectly will also produce unacceptable results; for example, holding a hammer by its head to swing the handle is a highly unsatisfactory way to drive a nail.

16.3.6 The Need for Moderation

Ultimately, all of TQM's problems—including the lack of a workable definition—can be attributed to a single fatal flaw: the search for perfection in all aspects of the company.

Perfection in all activities is rarely a reasonable expectation. In our experiences, not even the most adamant TQM proponent claims that company-wide perfection will ever be achieved. Yet, despite the unreasonably ambitious nature of the expectations, all total quality programs have as their stated goal relentless pursuit of perfection in all company activities through a continual series of incremental improvements.

TQM exerts such enormous emotional appeal that rationality is easily overlooked. To appreciate the possible repercussions of striving always for perfection, value judgments must be set aside so the analysis can be carried out dispassionately and without bias.

16.3.6.1 Costs vs. Benefits. The necessity of subjecting every business activity to the most rigorous possible cost-benefit analysis has been noted in previous chapters. The requirements and methodology of ongoing cost-benefit analysis are examined at length in later chapters. The objective of TQM needs to be restated as "to pursue quality improvement insofar as such improvements result in long-term benefits that exceed long-term costs without reducing short-term cash flow below survival levels."

16.3.6.2 Diseconomies of Scale. TQM advocates changing every company operation *simultaneously*. Such a universal approach could make sense if significant economies of scale could be realized relative to the costs of designing and implementing quality improvement one sector at a time. Far from delivering economies of scale, however, company-wide quality improvement entails enormous *dis*economies. The diseconomies are so great that TQM almost guarantees failure by overloading the company's personnel, financial, and intellectual resources.

The problem lies in the fact that diverse functional groups in any company have little or nothing in common when it comes to the skills and processes they need to master. Even within similar groups of personnel, there may be geographic dispersion around different sites or temporal dispersion where the company operates more than one shift. Consequently there will be no synergy in developing the knowledge bases, educational materials, communication skills, implementation systems, and so forth. The absence of synergy in large-scale projects has serious implications since development of effective educational materials for a single process can easily tax the abilities of the most talented and professional program management team. Attempting to develop the necessary materials for all groups within the company will destroy even the best-funded and most knowledgeable education organization.

Consider, for example, just two prominent departments—production and sales—in the same company. TQM requires that quality improvement activities be undertaken simultaneously in both departments. At a minimum, putting substantive improvement programs in place simultaneously in just those two departments requires:

- *Defining an operational definition and tangible measuring approach to quality for each department.* This is not a trivial undertaking, as should be apparent from the difficulties already encountered in Chap. 3 when developing a definition of quality for the manufactured product.
- *Preparing educational materials* (or, if they exist, buying the materials from third parties). Educational materials have value only when they provide information that was previously not known to the students. Therefore, the person(s) preparing the educational materials must themselves possess thorough understanding of the scientific and humanistic parameters of each process.
- *Developing company teachers who are qualified to use the educational materials.* The best results are achieved by using personnel from the department where an activity is taking place. The all-purpose company "trainer" will be useful only in a small number of departments. It may be reasonable for an assembly trainer to teach other production-related courses in which the trainer has been thoroughly schooled; it is less realistic to expect the same person to achieve wholly satisfactory results when teaching both assembly techniques and proper sales practices.
- *Holding classes.* Classes cannot be held until it has been determined who will teach, what will be taught, who will attend, how the development of students will be assessed, and who will conduct the assessments.

- *Determining the management structure best suited to the department.* TQM advocates turning decision making over to the average workers organized into teams. The company must decide to what degree non-management personnel will be permitted to make and implement decisions as well as what authorities and responsibilities will be assigned to managers and executives. The optimal allocation of responsibilities often varies from department to department.
- *Forming teams.* Effective teams do not develop without careful planning and administration. Some of the decisions that must be reached about any team include its size, composition, and mandate; whether it will be temporary or permanent; the scope of the issues any team can address; who will lead and what form that leadership will take. In those companies where piecework still enters into the operator's compensation—a pay scheme that, by encouraging volume rather than dependability of output, violates fundamental tenets of effective quality and should be strongly discouraged—it is also necessary to devise a means of compensating the operators for time spent in meetings.
- *Selecting tasks to be improved.* Solutions come only after it has been decided what issues to study. No employee action team can deal with more than one or two issues at a time. Therefore, procedures must exist specifying (1) the manner in which potential tasks for employee teams will be identified, (2) how the possible tasks will be prioritized, and (3) who will set the priorities (generally a choice between team members and management).
- *Holding regular team meetings.* A few of the matters to be established in the beginning include (1) how often the team will meet, (2) at what time meetings will be held, (3) where the meetings occur, and (4) what amount and type of supervision from management will be involved.
- *Evaluating and implementing suggestions.* Team suggestions have no value unless action is taken and positive results obtained. Proper action can be taken only when the following management decisions have been made: (1) the method by which the team will report, (2) how often reports will be submitted, (3) what level of personnel (teams or higher authority) evaluates suggestions, and (4) the systems under which improvement plans will be used.
- *Measuring results.* It must be known from the beginning (1) who will measure the accomplishments and failures of a project, (2) how the

designated evaluators will measure outcomes, and (3) how the measurements will be used.

- *Recognition of achievements.* Contributions to a better plant require overt recognition from senior management and financial compensation in keeping with the consequences of the contributions. This is both an ethical obligation and a necessity if continued employee participation is desired. TQM orthodoxy believes team members should participate with no reward other than the joy of freedom to participate. For many workers, such psychic "rewards" will indeed be adequate. However, taking advantage of workers' good nature does raise moral issues. At the very least, employees who make the greatest contribution to the company's success should receive bonuses, promotions, greater job security, or perhaps most reasonable of all in the publicly traded corporation, shares.
- *Individual initiative.* Although TQM emphasizes teamwork, some people perform best on their own. A pragmatic company will create an environment in which employees can contribute in the ways that best suit their temperaments. Unfortunately, there is widespread belief that no room exists in the "progressive" company for independent thinkers, with the result that energetic, creative, productive, and loyal employees are castigated for not being "team players" and are pushed aside. Entrepreneurial spirits have long been the strength of American enterprise; when there is no room for disagreement, there is no opportunity to identify and prevent serious blunders at the conceptual stages. Generally speaking, it is much easier (and more secure) to be a "team player" hiding behind the efforts of other team personnel than to take an individual principled stand.

This list is by no means exhaustive. Nonetheless, it contains so many serious issues that any rational individual must recognize that the possibility of failure far exceeds the probability of success when improvement programs are too broadly defined. Moreover, as more departments are added to the system, the quantity and complexity of the quality improvement tasks rapidly expand. The list of difficult questions keeps growing as the program expands throughout the company. For example:

- In some departments, the numbers of employees may not warrant the costs of TQM participation; should the legal department, for example, be part of the program?
- How are shift workers to be brought into the system—and provided ongoing support? Not all shift workers are employed in production,

where substantial numbers of individuals perform very similar jobs no matter what the time of day. Security personnel, for example, typically operate around the clock but seldom make up a large group at any given hour.

- If the costs of including small specialized departments in quality improvement are excessive compared to the probable outcome, can TQM be omitted in those departments?
- If those departments can be excluded, where and on what basis is the line to be drawn between included and excluded departments?
- If some departments are omitted, is the program still *Total* Quality Management?
- Above all, how is "quality" defined in any given department? We know that the quality of an electronics assembly is its dependability, but does the same definition serve equally well in, for example, accounting, sales, training, or corporate planning?

Once these necessary questions are posed and the struggle for answers is begun, it is easy to see that a company-wide quality revolution is far more attractive in concept than in practice. By taking on too much at one time, the company loses its ability to focus on and bring about dramatic improvements in those areas offering the greatest return on investment.

TQM advocates point to reported successes as proof that the concept is viable and desirable. Their interpretation of TQM's record leaves room for doubt, however. Yes, TQM projects will normally cause improvements to some operations and may even satisfy the company involved that the resources expended were well used. But most of the satisfied electronics companies that we have seen should have been less pleased with their accomplishments; their satisfaction results from not knowing that more focused approaches exist which could have brought about much greater improvements in considerably less time and at lower expense.

It must also be stressed that the reported TQM "successes" often consist only of pilot projects staged in relatively small areas such as a cluster of closely related functions at a single site. The resources required for success in such a limited area are inconsequential compared with what will be necessary if the program is applied company-wide. Of course, a "TQM" program applied in a limited manner is not total quality at all; it is simply a properly disciplined, confined approach to localized issues.

We have seen many successful test programs become major liabilities when broadened to approximate a "total quality" range. One company with which we were associated conducted a TQM test program within a functional area of a multifunctional company. The results were extremely good and the company management projected that similar successes elsewhere in the firm would produce net benefits of some $150 million a year. Once the program was expanded to include the entire company, however, it lost its effectiveness and produced a substantial net loss. The company identified as a primary cost of failure the exponential increases in resources required as the program's scope expanded.

16.3.7 Time Lags

The amount of time needed for a TQM program to work deserves sober consideration before any decisions to proceed are taken. Companies considering adoption of TQM will be profoundly disappointed if they hope to see meaningful benefits in a few months or even years. TQM consultants, echoing Deming, always argue that it takes 5 to 10 years to get a total quality program working properly in the normal company.

This lengthy lag between start-up and receipt of benefits creates two serious and often fatal problems:

1. At some point between the decision to start a TQM program and the initial flow of benefits (if any), the implementing management team may be criticized or even punished for committing the company to systems where the cost flow begins on or before the program start-up but benefits will be hard to measure until years later.
2. The average tenure of a corporate CEO in this country is less than 6 years. Therefore, at the very time when some gain may finally be visible after years of financial pain, the typical company will experience a massive turnover of executive leadership. This change in leadership typically leads to one of two possible outcomes:
   a. The new executives allow themselves to be indoctrinated in the company "total quality" thinking or, more likely,
   b. The new executives impose their own ideologies, and the company starts all over again with new operating systems and targets.

An environment in which new leaders sweep out existing programs and force the company down a new path of systems familiar only to the new executive regime rarely equals good management or even good logic—but it is life. History shows that approximately every decade brings with it a new management philosophy that displaces whatever practices existed before. As evidence that management ideas change quickly and radically, a quick list of some operational philosophies that prevailed at different times in twentieth-century American business includes:

- Specialization of labor (the production line) introduced by Ford between 1910 and 1920
- Frederick Taylor's "scientific management" of the 1920s
- Humanistic management in the 1930s
- Statistical quality control in the 1940s (particularly during the Second World War years)
- Quantity and planned obsolescence in the 1950s
- Zero defects in the 1960s
- Management by objectives in the 1970s
- "Japanese" management in the 1980s
- Pseudoquality thus far in the 1990s

16.3.8 Final Observations on TQM

TQM can perhaps best be summarized as good intentions wrongly applied. Certainly, it is difficult to disagree with a desire for "perfection." At the same time, however, it is impossible to know what perfection means. The drain on a company attempting to transform its entire way of life is so severe that the company's ability to focus on performing its regular activities suffers. Most of the other failings of pseudoquality identified in Chap. 15 also afflict TQM.

Williams[20] complained that there is too much criticism of TQM and not enough support, saying "...for every unsuccessful example [of TQM], there are dozens of success stories. We don't hear enough about the success stories, unfortunately, and we hear too much about the failures." But he is wrong about the ratio of TQM successes to failures; there is a natural human tendency to publicize our successes and bury our failures. The real scenario—and certainly the scenario most consistent with our own observations when visiting companies where TQM has been considered successful—is that the successes have been overstated and the failure rate greatly under-reported.

[20][Williams 1994], p. 4.

Effective quality improvement programs must not be generic or broadly based. With breadth comes loss of focus, dilution of leadership, breakdown of administrative systems, and poorly measured results giving rise to false senses of accomplishment. Evidence supporting this position can be found in the ranks of Baldrige award winners whose prizes go to divisions, not entire companies except in the "small business" category. For the fastest and most substantive returns on quality investment by an electronics assembly company, it is always best to put the product side of the company—design, materials, production, and product integrity verification—in order and to treat each plant as a separate division.

In the final analysis, it is useful to return to Ishikawa's misinterpretation of Feigenbaum's position with respect to the amount of employee participation that is desirable in quality. Ishikawa believes that Feigenbaum claimed quality which is everybody's job ends up being nobody's. The thesis is beautifully succinct and meritorious. The only problem is that Feigenbaum himself never said anything of the sort. In fact, Feigenbaum believed strongly that every member of the company must participate in quality management.[21]

We very much wish that Ishikawa had quoted Feigenbaum correctly because the misinterpretation is considerably more valid than the thesis that quality is everyone's responsibility. Unquestionably personnel from outside the quality department must be drafted in the crusade to improve quality. At the same time, however, it should be apparent by now that a little quality knowledge is a dangerous thing (unlike fuller knowledge, which can be put to beneficial use). Quality management is too important to be left to amateurs; quality is best served by the experienced, informed quality professional.

16.4 The Malcolm Baldrige National Quality Award

Though TQM comes in many forms, the best-known framework is provided by the Malcolm Baldrige National Quality Award administered by a division of the federal Commerce Department.

The interaction between the Baldrige and TQM is stressed by one guide to applying Baldrige philosophies and winning the award itself. The Malcolm Baldrige National Quality Award, the guides states, "is

[21]See [Ishikawa 1985], p. 90; [Feigenbaum 1991], p. 13. The Japanese misinterpretation of Feigenbaum's total quality was discussed in Sec. 2.4.3.

conferred...for *total quality management*...[it] represents an approach that more and more companies are turning to in order to create new jobs, new markets, new customers and new profits."[22]

The statement itself is true. Unfortunately, too many seriously erroneous implications can be—and generally are—read into the words. The Baldrige does exemplify TQM. It is also fair to say that many companies enter the Baldrige competition or adopt its guidelines for internal quality operations in the *hopes* of acquiring new jobs, markets, customers, and profits. However, the statement fails to point out that the hopes for new jobs, markets, customers, and profits are seldom fulfilled.

16.4.1 Origins of the Malcolm Baldrige National Quality Award

The Malcolm Baldrige National Quality Award was created by Congress in 1987 with the expectation that it would help American companies in four ways:

1. By stimulating them to improve quality and productivity for the pride of recognition while obtaining a competitive edge through increased profits.
2. By recognizing the achievements of those companies that improve the quality of their goods and services and providing an example to others.
3. By establishing guidelines and criteria that can be used by businesses, industrial, governmental, and other organizations in evaluating their own quality improvement efforts.
4. By providing specific guidance for other American organizations that wish to learn how to manage for high quality by making available detailed information on how winning organizations were able to change their cultures and achieve eminence.[23]

The legislation provided for a maximum of two winners per year in each of three categories: manufacturing, services, and small business (defined as a company having fewer than 500 employees). Only passing reference was made to the criteria by which Congress believed applicants should be judged—and no reference was made to the evaluation methodology. Responsibility for implementing the whole nebulous

[22][Hart and Bogan 1992], p. 4; italicized emphasis is by Hart and Bogan.

[23]p. 13.

bundle was assigned to the National Bureau of Standards (now called the National Institute of Standards and Technology, generally abbreviated to NIST), a division of the federal Department of Commerce.

16.4.1.1 The Evaluation Standards. From the criteria mentioned in the enabling legislation—the familiar quality jargon of statistical process control, management quality leadership, worker involvement, strategic quality planning, and customer orientation—NIST devised a system of seven weighted categories and a trilevel system of evaluation. The seven categories of criteria (and their relative importances in the judging) on which applicants were to be evaluated consisted of leadership (100 points), information and analysis (70 points), strategic quality planning (60 points), human resource utilization (150 points), quality assurance of products and services (140 points), quality results (180 points), and customer satisfaction (300 points).[24] The maximum number of points, therefore, is 1000. Conspicuously absent from consideration were any measures of the company's financial performance, an omission that proved quite embarrassing when one 1990 winner, Wallace Company, filed for bankruptcy less than 2 years later.

16.4.1.2 Entrants. For its first annual contest, in 1988, the Baldrige attracted 66 entrants[25] from which three winners[26] were selected. The following year, applications fell to only 40, but 97 companies submitted applications for the 1990 contest, and in 1991 the number of applicants (106) exceeded 100 for the first and only time. Each subsequent year saw the number of applicants drop off until only 47 companies—the number of applications from small businesses alone in 1991—entered the contest in 1995.[27] *The Wall Street Journal* attributed the decline in applications to

---

[24]These were the categories and weightings at the time of publication. However, the criteria and their relative importances change periodically.

[25]Although only 66 applications were submitted, 12,000 application forms were requested. The number of applications requested continues to be immense when compared to the number submitted. In 1991, when a record 106 completed applications were submitted, 205,000 forms were requested! Though the NIST argues that the large number of applications requested shows that the Baldrige approach is widely used by companies that do not enter the formal competition, there is no evidence to support that contention. It is very possible that most companies reject the viability of the Baldrige approach once they become familiar with the system by reading the application forms.

[26]Motorola (Schaumberg, Ill.) and the Westinghouse Electric Commercial Nuclear Fuel Division (Pittsburgh, Pa.) were selected for manufacturing, and Globe Metallurgical (Beverly, Ohio) won the small business award.

[27]"Is TQM Dead?" *USA Today*, Oct. 17, 1995, p. 1B.

"growing skepticism about the Baldrige Awards and diminishing interest in quality-management, which thousands of companies had embraced in a quest for better results."[28]

The decline in prestige of the Baldrige was reflected by the almost invisible report on 1994 winners in *The New York Times.* Where winners in past years could expect to see their names on the front page of the business section, the 1994 winners were buried in a small news item at the bottom of a middle page. The 1995 results could not be found in either *The New York Times* or *The Wall Street Journal.*

16.4.2 The Baldrige's Defense

Commerce Department and NIST officials have long contended that the Baldrige's success should not be judged by the number of companies that formally enter the competition. Instead, they emphasize that tens of thousands of companies request application guidelines which they then use to structure TQM programs that will never be submitted for judging. The actual number of companies operating TQM programs based on Baldrige criteria is, of course, unknowable but is certainly substantial; many large companies required their suppliers to implement Baldrige-style programs to achieve or retain qualified vendor status. Whether the programs are helping the companies where they operate is altogether less certain.

16.4.3 The Baldrige and Quality

Nominally, the Baldrige is about quality improvement. The judges' stated reasons for decisions in the 1995 competition, however, indicate that quality as we understand it does not really enter into the equation.[29]

16.4.3.1 The Armstrong Building Products Example. The Armstrong Building Products division of Armstrong World Industries (Lancaster, Pa.), a manufacturer of acoustic ceilings and wall panels, was one 1995 winner in the manufacturing category. Among the examples cited of its better "quality":

[28]"Baldrige Award Gets Fewer Applicants from Small Business," *The Wall Street Journal,* Oct. 13, 1994, p. B2. The article provides no evidence to back up the contention that companies were losing interest in quality management. There is no reason to believe the declining number of participants reflects anything other than lack of interest in the Baldrige itself.

[29]"How Manufacturers Focused to Win First Awards," *USA Today,* Oct. 17, 1995, p. 4B.

- Operating profit—$86.8 million—increased 500 percent from 1993 to 1994. Revenues increased 7.4 percent during the same period.
- Two of seven division plants have more than 3 million consecutive hours of work without serious injury or accident, an industry record.
- Annual sales per manufacturing employee increased 40 percent since the company introduced Baldrige criteria as a management tool in 1991; output per manufacturing employee rose 39 percent in the same period.
- Product was delivered within 30 min of promised time 97.3 percent of the time; the comparable figure was 93 percent within 4 hours in 1991.
- The cost of "doing things poorly"—described as scrap, downtime, customer claims, obsolescence, waste removal and disposal, occupational sickness and injury, and administrative and sales problems—was 6.8 percent of revenue in 1994 against 10.8 percent in 1991.

Clearly, the division performed much better in terms of earnings, safety, efficiency, and production planning in 1994 than in 1993. Conspicuously absent from this list of achievements, however, is any indication that *quality* improved during that period. On the basis of the items singled out by the judges for special commendation, it appears that the national quality award was assigned without reference to true quality factors. Moreover, the results that impressed the judges so greatly do not even reflect the specified evaluation criteria for the competition (see Sec. 16.4.1.1)!

16.4.3.2 The Corning Telecommunications Products Example. In a similar vein, the judges praised the other 1995 manufacturing winner, optical fiber producer Corning Telecommunications Products (Corning, N.Y.), for many accomplishments that do not relate to quality. Among the improvements cited were:

- Output per employee doubled between 1987 and 1994.
- Returns from unhappy customers declined from 10,000 reels per million shipped in 1984 to 250 per million in 1994.
- Every customer comment—complaints, compliments, or other details—to any Corning employee is stored in a computer data base and can be accessed by any employee.
- Purchasing criteria were changed from lowest bid suppliers to including quality and delivery times in decisions (the vice president

of business systems said the company has been unable to measure benefits of the new vendor qualifications but contended "we have a gut feeling" about long-term savings).

- 800 steps involved in the order fulfillment process, from taking an order to issuing credit, are now monitored to identify ways of increasing productivity.

The only item on this list that can be considered to reflect quality improvement is the reduction in customer complaints. The data base of customer contacts is *potentially* useful for quality improvement, but there is no evidence from the judges' remarks that the potential has been tapped. Once again, the judges apparently ignored the specified evaluation criteria.

16.4.4 Costs of Entry

Applying for the Baldrige is not an inexpensive matter. Ames Rubber Corp., a winner in the 1993 small business category, estimated that participation fees alone—consisting of application fees, printing, and duplicating of documents, "plus the costs of an extended visit by six quality experts who verified details of the company's application"[30]—cost the company between $25,000 and $40,000.[31] However, the application costs represent only a small portion of the total resources invested in the competition; the real cost would be vastly greater when personnel time and output reductions during the application preparation were taken into account.

The Ames Rubber effort was not unusually costly. For its successful entry in the 1995 Baldrige competition, for example, Armstrong Building Products submitted a 70-page application and underwent a week-long audit by eight Baldrige examiners. Xerox reportedly spent $800,000 for its successful 1989 application.[32]

Typically, the costs of entry for any given company in a single year significantly understate the costs of seeking the Baldrige. Few companies are successful the first time they apply. Indeed, it is not unusual for

[30]"Baldrige Award Gets Fewer Applicants from Small Business," *The Wall Street Journal*, Oct. 13, 1994, p. B2.

[31]The inability of the company to state its costs more precisely is itself troubling, inasmuch as its fiscal responsibility is of secondary importance where competing for quality recognition is concerned.

[32][Hart and Bogan 1992], p. 193.

companies to submit applications several years before winning an award. The costs of the unsuccessful applications must be taken into account along with the costs of the successful application.

Consultants' fees are also a factor. Although not all Baldrige contestants employ consultants to guide their efforts, many do. In 1993, for example, most of the major general-purpose consulting companies offered one-day seminars outlining the basic requirements for a successful Baldrige strategy. The average fee for those seminars, which were designed for the client company's most senior management, was $5000. However, the introductory workshop was only the first step, designed to sign up clients for long-term consultant-intensive programs aimed at reaching the Baldrige finals.[33]

Baldrige winners are expected to incur further expenses in the year following their award by providing advice to other companies who may ask for guidance. When the Baldrige was designed, it was anticipated that the advice would be provided free, and that was the case until 1992. (Marlow Industries, a 1990 Baldrige winner that later declared bankruptcy, placed some of the blame for its collapse on the time its personnel spent fulfilling the commitment to inform other companies on quality management.) But when Granite Rock Co. won a Baldrige in the small business category in 1992, it decided that the burden of responding to inquiries about their procedures from other companies was too onerous. So the company began charging $300 per participant for attendance at one-day seminars about the Baldrige. In 1994, Granite claimed that it was losing money on the seminars, which it held once or twice a month.[34]

16.4.5 The Baldrige Epitomizes the Failings of Pseudoquality

From its earliest days, the Baldrige has embodied the dark side of modern quality. Among the more troubling aspects:

- Entrants are attracted by the prospect of obtaining validation that can be exploited for marketing gains.
- Self-styled "experts" from outside the company—generally individuals with no knowledge of the competing company's industry—them-

[33]Smith, Jim, and Oliver, Mark. "The Baldrige Boondoggle," *Machine Design*, Aug. 6, 1992.

[34]"Baldrige Award Gets Fewer Applicants from Small Business," *The Wall Street Journal*, Oct. 13, 1994, p. B2.

selves compete for the personal ego gratification (and possible commercial gains) of serving as contest judges and examiners.

- Use of specified tools—particularly employee teams and high-visibility statistical exercises—command more respect from Baldrige personnel than do results.
- No operable definition of quality exists.
- Resources that should be employed in strengthening the company's operations are diverted to preparation of paperwork and auditioning for judges.
- A basic template for "quality" activities is applied to every entrant.
- Large companies oblige smaller suppliers to pursue Baldrige-based "quality" programs without regard for the real state of the supplier's quality and with no concern that the costs of such programs can destroy the supplier's competitiveness.

Meanwhile, as the characteristics of winners singled out for praise by examiners in recent years demonstrates, the official national "quality" award is now assigned on the basis of criteria that have nothing whatever to do with quality. The Malcolm Baldrige National Quality Award is the clearest possible example of how the term "quality improvement" has been taken hostage for company purposes utterly divorced from quality and often not compatible with the company's only legitimate goal.

16.5 Six Sigma™

Perhaps no American electronics company has based its marketing on the quality theme as has Motorola, Inc. After several years of advertising itself as a paragon of the quality-driven company, Motorola acted as one of the principal proponents of and financial contributors to the Malcolm Baldrige National Quality Award. After winning one of the three awards issued in the first Baldrige contest, Motorola announced that it had set itself a new 5-year performance objective that it named and trademarked Six Sigma™.

Six Sigma™ is loosely based on a process capability specification—generally known as "3σ" quality—developed in the automotive industry several decades earlier. Simply stated, 3σ quality means that:

1. Output is normally distributed.
2. The process mean coincides with the optimal target value.

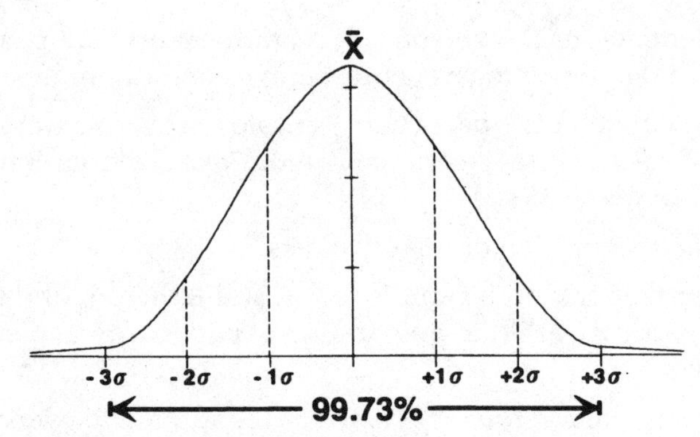

Figure 16.1. 3 sigma quality.

3. The range of acceptable product values ranges from a low of -3σ (i.e., the mean minus three standard deviations) to $+3\sigma$ (the mean plus three standard deviations).

The graphical representation of 3σ quality is shown in Fig. 16.1, where $\bar{x}$ is the output mean. When the output mean coincides with the target value, 99.73 percent of output will fall within the range $\bar{x} \pm 3\sigma$. In other words, 2700 out of every million units falls outside the acceptance range and must be rejected.

16.5.1 Six Sigma™ vs. 3σ

Motorola modified 3σ quality in several important respects:

1. The original concern with failures found in the 3σ model was replaced with "defects."

2. Rather than basing the process specification on each output unit, all possible defect opportunities in the unit would be taken into account. If an electronics assembly included 10,000 solder connections, for example, 100 units would provide 1 million possible defects. With a "3σ" rate of 2700 parts per million (ppm), the defect-free first-pass yield for such a population would be an unacceptable 0 percent.

3. The process capability would be improved to the point where the acceptable product range would extend from $\bar{x} - 6\sigma$ to $\bar{x} + 6\sigma$. Motorola claimed that a process conforming to such a tolerance range would generate only 3.4 defects per million opportunities. Effectively, the company defined its Six Sigma™ program on the basis of that stated defect rate of 3.4 ppm or less.

16.5.2 Deviant Statistical Use

For any process where the output falls within a range of six standard deviations on either side of a constant mean that equals the target value, the portion of the output population lying outside that range will not equal 3.4 ppm. In fact, the defect rate will be several orders of magnitude smaller—so small that statistics tables consider the *entire* population to lie within $\pm 6\sigma$ of the mean. The discrepancy between Motorola's figure of 3.4 defects per million and the statistician's figure of 0 parts per million lying farther than six standard deviations from the mean is attributable to a bizarre assumption on Motorola's part: that the mean will not be constant but can deviate from the target value by as much as 1.5 standard deviations in either direction. In other words, Motorola's "Six" Sigma™ is actually a deviant form of 4.5 sigma.

The image of a process so robust that the tolerance range extends from -6σ to $+6\sigma$—a truly spectacular feat—but so feeble that the mean can vary by ± 1.5 standard deviations can only be described as bizarre. The only rational conclusion for the construct is that Motorola liked the sound of "six sigma" for marketing purposes but realized that pure six sigma output is impossible. Unfortunately, a result of Motorola's extensive promotion of its Six Sigma™ has been that many people now erroneously believe that 3.4 members of a normally distributed population fall beyond the range of $\bar{x} \pm 6\sigma$.

16.5.3 Other Failings in Six Sigma™

Inappropriate use of statistics is only one of several serious flaws in the Six Sigma™ system. Of the many other failings, the most serious include:

- Violation of motivational management fundamentals
- Confusing defect numbers with real quality
- Failure to recognize differences in production challenges
- No financial linkage

16.5.3.1 Violation of Motivational Management Fundamentals. Effective motivation of personnel requires that objectives be set for relatively short time spans and be attainable within that time. Excessive time horizons will cause personnel to lose interest, while unattainable targets can only result in a sense of defeat. In today's corporate environment, employees regard 6 months as long-term; yet Motorola set up its Six

Sigma™ as a 5-year program and other companies such as IBM that emulated Motorola made the same mistake. Employees would have been motivated—and encouraged—by a series of incremental programs (each lasting no more than a few months) with specified targets that are reasonable in such a time span. With each successful completion of an incremental program, the personnel become conditioned to expect success and unwilling to settle for less. The most senior executives may have a 5-year objective in mind, but they never burden the workforce with long-range thinking.

16.5.3.2 Confusing Defect Numbers with Real Quality. Defects in electronics products are often highly subjective and not related to dependability. Whenever subjective criteria enter the evaluation, the outcomes will be erratic even for a single product type produced within the same plant. When the evaluation spectrum broadens to incorporate diverse products in diverse plants and sundry companies, differences of opinions (i.e., varying levels of subjectivity) will make "defect" levels utterly meaningless for benchmarking purposes. Any given plant can meaningfully measure its progress on defect reductions only against its own performance for a specified product. Therefore, setting a "quality" target expressed in units of defects is a senseless exercise.

16.5.3.3 Variable Production Challenges. Not all products are equally manufacturable. Aerospace systems, for example, require dense concentrations of components on printed circuits boards that may be strangely shaped to fit into tight spaces. The same sort of density is not required on many consumer products. Moreover, the aerospace product is likely to be produced in small lots (often one at a time) while consumer products are often produced in massive numbers on dedicated lines. The opportunities to achieve low defect rates and stabilize the production processes are vastly greater for the mass-produced consumer product than for the densely populated low-production aerospace system. Setting numerical targets across all product lines is irrational.

16.5.3.4 No Financial Linkage. No business activity can be rationally managed where no assessment of costs and benefits occurs. To set a numerical objective for a program without knowing the costs that will be involved to attain that objective is management folly. To compound that folly by having no understanding of what financial benefits will result from attaining the objective is irresponsible.

16.5.4 Six Sigma™ as Pseudoquality

Six Sigma™ presents some different facets of pseudoquality than those already seen in total quality. Attributes of pseudoquality that are readily seen in Six Sigma™ include:

- Improper use of statistics
- Use of proxy characteristics (in this case, defects) to measure quality
- Failure to understand employee psychology
- Ignoring the long-term company prosperity goal

16.5.5 Final Observations on Six Sigma™

For the general quality world, Six Sigma™ is not now and has never been a factor. In the electronics assembly industry, however, the scheme has attained visible usage, even if much of that usage has been grudging acquiescence by smaller suppliers to the demands of major customers such as Motorola for implementation of such programs. The failure of any company—or even any division within a company—to achieve defect levels of the Six Sigma™ standard in an economical fashion has convinced many one-time supporters that this is not a program that can be justified by any business logic. By the end of the decade, Six Sigma™ is likely to be nothing more than a footnote in the history of electronics assembly quality.

16.6 Summary

The purpose of any business activity—including quality improvement—is to enhance the company's long-term financial performance without jeopardizing its short-term survival by excessively depleting cash flow. However, many well-known quality programs are incompatible with that goal. Worse yet, these "quality" programs have little or no positive effect on real quality. We call such "quality" programs pseudoquality. Three of the most dangerous pseudoquality programs are the "total quality" family, the Malcolm Baldrige National Quality Award, and Six Sigma™. A variety of dangers are associated with each of these programs, and the programs should therefore be reviewed in detail by reading this chapter fully.

17
ISO 9000

17.1 Chapter Objectives

Most electronics assembly companies will find themselves forced to incur the costs (financial and human) of obtaining ISO 9000 certification by the end of the decade. Many have already undergone that exercise.

Supporters of ISO 9000 claim that it helps companies where it is implemented achieve higher quality, lower costs, happier customers, and greater market share. Detractors argue that ISO 9000 is time-consuming, costly, and bureaucratic, and may prevent the company from attaining its optimum quality level.

The truth lies somewhere between the two extreme positions. In this chapter, strengths and weaknesses of ISO 9000 are examined so that readers—whether employed by supplier companies contemplating ISO 9000 certification or of customer companies considering imposing certification on their vendors—can better determine whether the system helps, harms, or is neutral in its effects on their employers.

17.2 ISO 9000: The Dominant Quality Issue of the 1990s

For almost two decades in the United Kingdom, a decade in western Europe, and 5 years in North America, the most visible quality improvement activity has been an international certification scheme known as ISO 9000. Indeed, ISO 9000 has become so pervasive that it can reasonably be considered the dominant quality issue of the decade. Under ISO 9000, the formal quality management methods of goods and services suppliers are certified as acceptable by commercial bodies in return for a fee. The certified company can then use that fact in its marketing materials when soliciting customers, many of whom (particularly in Europe) accept such certification as sufficient evidence of the certified company's competence to serve as a supplier.

According to one ISO 9000 auditor,[1]

> Some of the advantages of a[n ISO 9000] certified quality management system are:
>
> - Optimized company structure and operational integration
> - Improved awareness of company objectives
> - Improved communications and quality of information
> - Responsibilities and authorities clearly defined
> - Improved traceability to "root causes" of quality problems
> - Improved utilization of time and materials
> - Formalized systems ensure consistent quality and punctual delivery
> - Documented system provides useful reference and training tool
> - Fewer rejects, therefore, less repeated work and warranty costs
> - Errors rectified at the earliest stage, and not repeated
> - Improved relationships with customers and suppliers
> - Improved control during periods of change or growth
> - Use of recognized logo on stationery and advertisements
> - Improved corporate quality image
> - Ability to bid for "ISO 9000" contracts at home and abroad
> - Continuous quality assessment by experienced professionals
> - Reduced number of customer audits
> - Improved records in case of litigation

Some of those claims are unquestionably true and important to the welfare of both suppliers and customers. Others are occasionally true and may or may not be important to either party. Still others are wishful thinking. And some of the most important consequences of ISO 9000 are not found anywhere on the list. The following sections attempt to provide a useful overview of ISO 9000's merits and demerits from the perspective of both the supplier company (which pays for the certification process) and customers.

These comments should be read with the knowledge that both authors have considerable experience working with ISO 9000 in North

---

[1]Ronald Thomas of the ISO 9000 Group, available on the Internet at www.commerce.associates.com.

America and Europe as well as the United Kingdom predecessor to ISO 9000, BS 5750. Further, one of the authors[2] is governor of a certification body in the United Kingdom. Our joint opinion is that considerable merit exists in the fundamental goal and methodology of ISO 9000 but the system as a whole suffers serious flaws and has been vastly oversold by certain vested interests to the point where it, too, comes dangerously close to pseudoquality.

17.3 The Need for Standardization

In an era of craftsmen, when the master builder made as well as assembled parts, standardization of components and materials was of no great consequence. Even where components and materials are custom manufactured to the unique requirements of a final assembler, it matters little whether those components and materials are interchangeable with those of other manufacturers.

When mass production enters the picture and one supplier may provide components or materials for many customers in several countries, uniformity offers many advantages. Just a few characteristics for which standardization is valuable are dimensions, flammability, durability, or even color. Equally—or, often, even more—important is that the characteristics be measured according to common standards.

Wherever common standards exist—for individual industries or manufacturers in general—there are clear benefits for all parties. Customers benefit by being able to design on the basis of common standardized components and materials while suppliers can anticipate their customers' needs. Both suppliers and customers are able to reduce the diversity of parts and therefore minimize inventory and setup costs. In other words, standardization eliminates risk and waste on both sides of the market.

For more than 100 years, each of the world's major industrialized nations has maintained an organization with responsibility for setting the country's standards. In some countries, the standards bodies operate without any government ties while the government role in other countries is substantial. The world's leading national standards bodies have historically been the American National Standards Institute (ANSI), the British Standards Institution (BSI), and Germany's Deutsche Industrie Norm (DIN).

---

[2]Whitehall.

17.4 A Brief History of the International Organization for Standardization

National bodies work well for internal industrial standards. In the United Kingdom, for example, the first national product standard was for tram rails; since vehicles from other countries would not use the British tramway system, there was no need to harmonize with the rest of the world. However, when international trade becomes important, common standards are equally necessary across borders as well as within them. So, in 1926, standards bodies of major industrial countries established a cooperative international standardization council known as the International Federation of the National Standardizing Associations (ISA). ISA set common standards—primarily in mechanical engineering—that member countries adopted voluntarily.

ISA dissolved during World War II, but the need for a replacement body was seen as a priority when the war ended. In 1946, 25 national standards organizations agreed to create an international body "the object of which would be to facilitate the international coordination and unification of industrial standards."[3] The Geneva-based alliance was titled the International Organization for Standardization but adopted the distorted acronym "ISO"[4] for marketing purposes. In English, the prefix "iso-" means "equal" (as in an isosceles triangle, which has three sides of equal length), and the organizers considered "equal" to be a good synonym for "standard." ISO officially opened its doors in early 1947 but did not publish its first standard—Standard Reference Temperature for Industrial Length Measurement—until 1951.

From the beginning, ISO had no involvement in electrical and electronic engineering; those industries fall under the jurisdiction of the International Electrotechnical Commission (IEC). Ultimately, however, ISO would be significantly more influential than ISA, involve many more countries (more than 100 today), and extend its authority far beyond mechanical standards. The rise of ISO 9000 extended ISO's authority into electronics assembly as well.

17.4.1 The International Organization for Standardization Prior to ISO 9000

To properly appreciate the magnitude of the increase in ISO's influence that ISO 9000 represents, it is necessary first to realize the limited scope

[3]"Introduction to ISO" from the World Wide Web site (www.iso.ch) of the International Organization for Standardization.

[4]The true acronym would, of course, be IOS.

of the organization's activities prior to the introduction of the quality management standard. Before ISO 9000, ISO restricted itself to obtaining consensus agreements on new standards for individual industries from the primary participants—customers, suppliers, and occasionally governments—in those industries. The standards typically involved specifications and criteria for the choice and classification of materials, the manufacture of products, and the provision of services. Participation—by customers and suppliers—was voluntary; ISO, having no enforcement powers, depends on agreement by its members that the organization's standards are good for all concerned.

The standards prior to ISO 9000 were purely technical. The first ISO standard—Standard Reference Temperature for Industrial Length Measurements—is typical of the pure engineering nature of the organization's work.

With the arrival of ISO 9000, the organization's role and influence on international business practices quickly rose to previously unimaginable levels.

17.4.2 The Arrival of ISO 9000

The origins of ISO 9000 lie in British defense procurement, which likely explains the inspection-oriented nature of the system.

In 1968, the British Ministry of Technology—the government body that, at the time, handled defense materiel specifications and procurement—produced one of the most innovative quality system specifications in modern military standards history. Known as AvP92, it purported to cover the entire spectrum of the contractor's operation from design to delivery. Particularly with respect to design quality, it has never been surpassed. Although it turned out to be relatively short-lived, AvP92 set the precedent for evaluating suppliers by quality management processes as well as actual product.

AvP92 was replaced in 1973 by the DEF STAN 05/20 series issued by the department now known as the Ministry of Defence (MoD). DEF STAN 05/21 covered the design, production, and service of hardware functions. This new standard was broadly equivalent to the 1969 NATO quality management specifications known as AQAP. For reasons unknown today, the United States (which, as the most influential NATO member at the time, undoubtedly exercised considerable influence over the AQAP contents) never invoked AQAP specifications for American defense contractors. MIL-Q-9858 (later MIL-Q-9858A) was introduced in its place. The conflict in standards between NATO and the U.S. Department of Defense caused considerable resentment among European defense contractors, who found themselves required to follow several divergent sets of standards. The ultimate consequence of the DoD decision, however, was that American industry would be caught unprepared when Western Europe chose to harmonize on quality management standards strongly related to the NATO and MoD requirements.

In 1979, the British Standards Institution (BSI) issued a standard (BS 5750) for quality management procedures. Under BS 5750, much like under AvP92 and AQAP, supplier quality would be judged not just by the dependability of the product (or, eventually, service) but also by its official quality assurance systems, the formal documentation of those systems, and the correlation between documented procedures and actual practices.

17.4.2.1 Underlying Decisions in BS 5750. The reasoning behind the new standard was still apparent in the introduction to the 1987 version of BS 5750: "Most organizations—industrial, commercial or governmental—produce a product or service intended to satisfy a user's needs or requirements. Such requirements are often incorporated in 'specifications.' However, technical specifications may not in themselves guarantee that a customer's requirements will be consistently met, if there happen to be any deficiencies in the specifications or in the organizational system to design and produce the product or service. Consequently, this has led to the development of quality system standards and guidelines that complement relevant product or service requirements given in the technical specifications. The series of International Standards (ISO 9000 to ISO 9004 inclusive) embodies a rationalization of the many and various national approaches in this sphere."[5] Although this rationale contains serious logical and operational flaws that will be discussed in Sec. 17.7, it does provide clear insight into the original reason for development of an international quality management standard.

BS 5750 was initially seen as a toothless curiosity by the defense contracting community, where rigorous adherence to customer-mandated processes and record keeping was as important as actual product performance in satisfying contracts. However, defense contractors are often owned by general-purpose conglomerates which also have substantial holdings in civilian industry. To those conglomerates, even a diluted military approach to quality management was better than none at all. Combined with urging by the British government to support the scheme, United Kingdom companies saw real incentive to participate in BS 5750.

---

[5]BS 5750, Part 0, Section 0.1, 1987, p. 1.

17.4.2.2 Self-Regulation. BS 5750 began as a self-regulatory system. Customers would issue contracts in which compliance with BS 5750 provisions would be required, and suppliers were expected to honor those requirements. Thus, suppliers followed the guidelines set out by the BSI and judged whether they were in compliance. Customers could inspect vendor operations to assure the accuracy of vendors' self-assessments, but no formal regulatory body existed. At the same time, BSI offered a third-party assessment service for use where suppliers or customers wanted an "official" certificate.

Inevitably, some companies graded themselves as BS 5750–compliant even though their actual procedures violated significant requirements. This caused customers seeking BS 5750 compliant suppliers to conduct audits of each new supplier. Some of those major corporate customers reaudited suppliers at regular intervals. Customers who emphasized supplier process more than product dependability began urging establishment of a national body to carry out audits and issue approvals.

17.4.2.3 The Rise of Third-Party Regulation. In 1984, the British government's Department of Trade and Industry gave BSI authority to regulate certification companies through a subsidiary organization called the National Accreditation Council for Certification Bodies (NACCB). Though the NACCB—now part of the United Kingdom Accreditation Service (UKAS)—in itself did not prevent self-assessment, its creation in reality marked the demise of the two-party (i.e., customer and supplier) certification system. The transition resulted from wording of BS 5750, Part 0, Section 0.1, 1987. "Assessments of a supplier's quality system are utilized prior to a contract to determine the supplier's ability to satisfy the requirements of ISO 9001, ISO 9002 or ISO 9003 and, when appropriate, supplementary requirements. In many cases, assessments are performed directly by the purchaser. By agreement between purchaser and supplier, pre-contract assessment may be delegated to an organization independent of both contracting parties. The number or the extent of assessments can be minimized by using ISO 9001, ISO 9002 or ISO 9003 and by recognizing previous assessments carried out in accordance with these International Standards by the purchaser or by an agreed independent assessing organization."[6]

The "agreed independent assessing organization(s)" were for-hire certification companies that sold their services to both customers and suppliers. Customers who lacked sufficient staff to audit suppliers could enlist these independent auditing bodies. Obversely, suppliers

[6]Section 8.4, Pre-contract Assessment.

seeking accreditation carrying more weight than their own claims of compliance could also retain the same auditing bodies to vouch for their systems.

The appeal of third-party certification to customers was very great. Rather than bearing the burden of auditing suppliers, the customers could pass along the costs to suppliers. Some suppliers also welcomed the shift in emphasis to third-party certification, since that certification greatly reduced the time spent in their facilities by various customer representatives. But for those companies that had not been subjected to frequent customer audits, the cost of BS 5750 compliance rose significantly as it became necessary to pay for third-party auditing. Moreover, the potential for abuse by the certification bodies was enormous, and not all such organizations resisted the temptation to squeeze payments from supplier companies in exchange for certification.

17.4.2.4 BS 5750 Becomes ISO 9000. BSI's close ties to ISO ensured that BS 5750 came to the attention of the international standards body. At the same time, movement toward uniformity of industrial and social standards throughout the European Community ensured that leaders of other European countries found BS 5750 highly attractive. In 1987, a quality management format identical to BS 5750 was adopted by the International Organization for Standardization under the designation ISO 9000.

ISO 9000 was strongly promoted by most governments in the European Community. (Germany was the most resistant but eventually enlisted as well.) That was not the case in the United States, where federal interest was focused on the newly created (and government-run) Malcolm Baldrige National Quality Award. Consequently, for its first 5 years of existence, ISO 9000 was almost exclusively a European standard generically known as EN 29000,[7] although each country tended to have its own national designation (as, for example, the United Kingdom, which retained the BS 5750 label). Only in recent years have national designations been largely replaced by reference to the EN series.

17.4.2.5 ISO 9000 Arrives in America. As American manufacturers became increasingly export-oriented in the 1990s, they discovered that the low value of the dollar relative to major European currencies made

[7]European Standards are administered by the European Committee for Standardization (CEN) in Brussels. CEN's membership consists of the national standards organizations of Austria, Belgium, Denmark, Finland, France, Germany, Greece, Ireland, Italy, the Netherlands, Norway, Portugal, Spain, Sweden, Switzerland, and the United Kingdom.

U.S. companies attractive suppliers for European companies. However, the European customers expected ISO 9000 compliance from suppliers abroad as well as at home. Meanwhile, European multinational corporations began implementing ISO 9000 in their American subsidiaries. In 1992, the American National Standards Institute (ANSI) and the American Society for Quality Control (ASQC) jointly established an ISO 9000 program officially titled ANSI/ASQC Q-90.

Once the same standard was endorsed on both sides of the Atlantic, the national terminologies became a hindrance. Reference to ISO 9000 rather than an EN or ANSI/ASQC designation became the norm. (In 1994 America's three major auto manufacturers modified a few aspects of the ISO 9000 model and introduced yet another standard: QS-9000. Suppliers to Chrysler, Ford, and General Motors are therefore required to obtain QS-9000 certification in addition to needing ISO 9000 certification if they sell to other customers.)

17.5 How ISO 9000 Works

Although the basics of ISO 9000 have remained fairly constant, modifications to requirements do appear from time to time. The last major revision at this time was released in 1994. It is best to review the most current standard by purchasing an official copy.[8] Several reasonably good manuals are available from bookstores and professional associations, although a preponderance of the available books are disappointingly superficial.[9] It is best to research the subject thoroughly *before* soliciting advice from consultants.

The requirements and procedures can be summarized as follows:

1. *There are no formal requirements for ISO 9000 certification.* The only requirement for ISO 9000 certification is insistence by customers. Not all large customers—even in Europe—require ISO 9000 certification from their suppliers.

2. *The basic concept is simple.* Reduced to its essentials, ISO 9000 only requires that a company document what it does and act according to its

[8]The U.S. edition is available from: American National Standards Institute (ANSI), 11 West 42d Street, New York, NY 10018; (212) 642-4900, fax (212) 302-1286; or American Society for Quality Control (ASQC), P.O. Box 3005, Milwaukee, WI 53201-3005; (414) 272-8575, fax (414) 272-1734.

[9][Clements 1993] is reasonably thorough and readable.

documentation. There are only two ways to fail: (1) failure to document processes and procedures thoroughly and (2) failure to follow the documentation in actual practices.

3. *Third-party certification is required.* The concept of two-party (supplier and customer) agreement envisioned in the original version of BS 5750 no longer applies. A relatively small number of companies—known as variously as *registrars, third party certification bodies* (the usual U.K. terminology), or *certification agents*—have been approved by the International Organization for Standardization affiliate of one or more countries to certify companies as ISO 9000 compliant. *Only* those registrar companies can issue ISO 9000 certification.

A registrar may not be recognized by all ISO member countries. Certification by a registrar not approved in a country where your customers are located will have no official status, although registrars with accreditation in only a small number of countries often have cross-certification agreements with similarly restricted registrars abroad. The cross-certifications are not always recognized by potential customers, however. European customers in particular tend to reject certifications from registrars possessing only North American accreditation. Retaining a registrar directly accredited in the countries where your clients are located is always the best policy.

Registrar's fees can be very high. Prices—and the packages of services they provide in addition to certification—can vary substantially among registrars. Obtaining bids from several registrars is normally a cost-effective strategy.

Consultants are available, either independently or through the registrar. In theory, registrars do not extend preferential treatment to applicant companies which retain consultants affiliated with the registrar. This is a highly contentious issue because of the potential for abuse by registrars selling both the procedures and evaluation of the procedures. As noted below, there is no formal requirement for applicant companies to hire consultants.

The final element in the certification procedure is the *assessor.* Where the applicant is a medium to larger-sized company (roughly 500 or more employees), several assessors (reporting to one lead assessor) will be assigned to the audit. The assessors are employed by the registrar, although the assessor is often an independent contractor rather than a salaried employee. The assessors' evaluation determines whether certification will be issued by the registrar.

To sum up, therefore, a registrar and assessor(s) cannot be avoided and fees will be involved. Consultants are optional and cost extra.

4. *Several types of certification exist.* The scope of certification can vary widely—from a single product, department, or plant to an entire company. Further, the term "ISO 9000 certified" is erroneous, inasmuch as ISO 9000 actually consists of three different levels of compliance—none of which is numbered ISO 9000. Of those three levels of compliance, only one will be right for any particular organization.

The most comprehensive level (ISO 9001) applies to companies that develop as well as manufacture products.[10] ISO 9002, a less comprehensive and stringent subset of ISO 9001, applies to companies or plants that manufacture but do not design or develop products. Finally, ISO 9003, the least comprehensive and stringent standard, relates to detection and management of problems discovered during final inspection and test; it is also employed by nonmanufacturing operations such as distributors.

The actual ISO 9000 document itself is simply a guide for determining which level of certification is suitable for the reader's operation. Another guide, ISO 9004, is only a reference for terminology and themes found in ISO 9001 through ISO 9003.

5. *The standards are not rigorous.* Recognize that the standards are amorphous and should be approached as "models" rather than inflexible requirements. The wording is often open to interpretation. Considerable flexibility is available for tailoring the final system to the company's unique characteristics.

6. *Much of the system may already exist in your plant.* Determine the extent to which your existing quality management practices fulfill the requirements for the level of compliance relevant to your company, division, or plant. If you already produce dependable products, much of what is required is probably already in place.

7. *Consultants may not be necessary.* An outside consultant can be retained for guidance, or most of the work can be carried out internally. An experienced consultant can reduce the uncertainty and needless work that anyone undertaking the exercise for the first time is likely to consider necessary. Beware of consultants who bring predefined all-purpose "templates"; generic documentation is generally irrelevant to electronics assembly companies, and what value there is in the ISO 9000 exercise comes primarily from lessons learned in defining and documenting the processes.

---

[10]These members of the ISO 9000 family of standards also apply to service companies, an area that is not relevant to this particular book.

8. *ISO 9000 certification does not ensure high quality.* As discussed in Chap. 6, three separate process states exist simultaneously: the natural process, the specified process, and the actual process. Dependable quality is achieved only when the actual process (what is done) conforms to the natural process (the scientific requirements specified by natural forces).

ISO 9000 concerns only the relationship between the specified process and the real process, ignoring the requirements of the natural process. If the specified process is comprehensive and the real process corresponds to the specified process, the operation is ISO 9000 compliant. However, since there is no assurance that the specified process equal the natural process, an ISO 9000 compliant operation can be completely incapable of delivering dependable product.

17.6 Strengths of ISO 9000

As subsequent sections of this chapter will show, ISO 9000 suffers several serious flaws. Fundamentally, however, ISO 9000 conforms to three sound principles:

1. Formal process definition is essential for consistency in output.
2. Lack of consistency is incompatible with cost control.
3. Documentation is the only way to identify opportunities for improvement.

17.6.1 Documentation and Consistency

Consistency in performance of output is a fundamental requirement of quality. Whenever variations in performance are found, not all units can have equal dependability and some units are likely to fall short of the minimum dependability requirements.[11]

While several forces can introduce variation into output, the single most important cause—and the factor that is most readily controlled by the company itself—is lack of constancy in production processes. Inconsistency in processes is almost inevitable whenever the steps of

[11]Variation in output performance will not constitute a quality problem if every unit provides at least the minimum promised dependability. However, variation typically means that some units fall short of the promised dependability and thus represents a serious quality concern.

those processes are not documented and carefully monitored for conformance to the documentation. Without such documentation and oversight, personnel work to what they *believe* or *remember* about the desired process parameters, not to the defined specifications. ISO 9000's emphasis on documentation and verification of repeatability is, therefore, entirely consistent with the quality department's goal of ensuring dependability.

17.6.2 Consistency and Cost Minimization

Lack of explicitly stated operation-wide methods and standards leads to disagreement about what constitutes "acceptable" quality. Some personnel will attempt to "improve" product by negative-value-added activities such as touchup, which leads to higher failure rates, repairs, and retesting. Other employees may seek to ship undependable product that they feel is "good enough." Endless hours of management time—in addition to the wasted direct labor—will be spent arguing over conformance issues. Scrap costs also soar. These are all costs of misunderstanding inflicted by lack of formal standards and procedures. They will typically be reduced—generally by substantial amounts—when ISO 9000–type documentation and control procedures are in place.

17.6.3 Identifying Opportunities for Improvement

Determining where processes are deficient requires first knowing the steps in the existing processes. The assumption is often made that identifying the process steps is a relatively straightforward, even simple, act. More often than not, accurately stating the processes *as they are truly conducted* requires months of painstaking investigation and documentation. Even after so much effort, it is common to find that the stated procedures differ from the real processes; the ISO 9000 requirements take the probability of such documentation errors into account as well.

Once accurate records of the procedures are in place, troubleshooting can begin by reviewing the various operations step by step to ascertain where the real processes may deviate from the natural processes. There is, of course, no assurance that this exercise will reveal the reasons for quality problems, but it is almost certain that the necessary answers will not be found in the absence of properly documented procedures.

17.6.4 Documentation Meets Self-Interest

The importance of documenting all company activities—and ensuring that the documentation coincides with what actually takes place in the plant—cannot be overemphasized. Long before the International Organization for Standardization entered the quality management world, our first step with customers was putting on paper the real processes and setting up systems to ensure stability in those real processes. Although this task—which can be termed "baselining"—does not in itself improve the company's performance in any respect, it lays the foundation on which solid improvements can be based. The necessity for setting up formal documentation that factually reflects the company's real operations applies to every company, not just those seeking the ISO 9000 seal of approval.

17.7 Weaknesses of ISO 9000

As noted earlier, the International Organization for Standardization's function until the introduction of ISO 9000 was to set common technical standards that could be objectively measured using unambiguous rules. Management systems—notably including quality management—do not allow for such rigorous objectivity. Therefore, it seems fair to say that, by entering the management appraisal business, the International Organization for Standardization undertook responsibilities for which it and its national affiliates were unprepared. At the same time, ISO 9000 has brought the International Organization for Standardization prominence, authority, and income unprecedented in its history.

As noted earlier, the fundamental principle of ISO 9000 is sound. *Effective* process control requires that procedures be documented and that the documentation accurately reflect the actual operations. This is often stated as "say what you do and do what you say." Beyond meeting the need for more rigorous management of operations, however, the positive attributes of ISO 9000 are few and the concerns numerous.

17.7.1 Distorted Perceptions

An advertisement placed by a microprocessor manufacturer summed up the fundamental misunderstanding of ISO 9000: "We've got ISO 9000 certification—in plain English, that means world class manufacturing facilities,"[12] the advertisement claimed. But the ISO 9000 certification means nothing of the sort.

In fact, any level of ISO 9000 certification means only that the company has demonstrated to the satisfaction of a registrar that (1) its operations (most commonly the manufacturing processes) are the same as the company's documented procedures and (2) the documented procedures comply with the requirements of the particular level of ISO 9000 being sought. In other words, the real procedures coincide with the specified and mandated procedures.

17.7.2 Hazards to the Naive Customer

ISO 9000 certification is too often wrongly perceived as security for customers, a fact reflected in the following statement from *The New York Times:* "The certificates...provide assurance of a company's ability to deliver a product or service that consistently satisfies customer requirements."[13] This statement is valid only if the customer's sole requirement is supplier certification to ISO 9000 levels.

Since there is nothing in the ISO 9000 structure that ensures the real processes correspond to what we have termed the natural processes, it is entirely possible that the operations—documented and real—are harmful to quality rather than helpful. For example, the use of highly activated fluxes during soldering would increase the probability of product failures due to ionic contamination but not affect the company's ISO 9000 status. Thus, customers who rely on ISO 9000 certification in selecting their suppliers frequently discover they have been terribly naive. One of the more common mistakes involves companies that believe ISO 9000 certification of suppliers reduces their need for vigilant and aggressive purchasing department staffs. Thinking that the ISO 9000 system protects them from unreliable suppliers, those companies cut back on their purchasing staffs, a decision that typically leads to costs far in excess of the savings. Customers that rely on ISO 9000 certification rather than first-hand knowledge that a supplier's

---

[12] *PC Week,* Dec. 26, 1994/Jan. 2, 1994, p. 103.

[13] "New Quality Standards Selling Brazilian Exports."

products work are managing their procurement badly. ISO 9000 assessors, after all, need not know the science of the processes involved; they need only be familiar with the procedures for writing and maintaining documentation. For customers seeking assurance of dependable products, the only answer lies in rigorous analysis of supplier products and timeliness.

17.7.3 Twisted Logic

A core justification for centralized third-party regulation of company quality management systems was stated in the 1987 edition of BS 5750 and cited in Sec. 17.4.2.1: "Technical specifications may not in themselves guarantee that a customer's requirements will be consistently met...." This cannot be valid unless the technical specifications are themselves erroneous or incomplete, provided the customer exercises responsible judgment in selecting the supplier. If the specifications are erroneous or incomplete, the supplier cannot meet the customer's real needs except by (1) violating the official specification and/or (2) reading the customer's mind. No party other than the customer itself—not the supplier and certainly not some third party without any knowledge of the technical requirements inherent in the customer's business—can make the ultimate decision about what the specifications should be and whether the supplier meets those needs.

17.7.4 Participating for the Wrong Reasons

The intent of BS 5750 was for the initial documentation and certification to be the preliminary step in ongoing quality improvement. Sadly, this is not the usual outcome of ISO 9000. The very nature of ISO 9000 as a requirement to become a supplier to many companies ensures that suppliers will seek certification for the wrong reason—as a marketing requirement—and ignore the need for ongoing improvements after certification. Suppliers receive their certificates and, in the face of myriad other business pressures, tend to believe that they have won the process lottery rather than simply made a down payment on a ticket.

On the other side of the equation, customers are encouraged to become complacent. Rather than remaining vigilant about supplier performance, many customers wrongly assume that the ISO 9000 supervision eliminates the potential for supplier failure.

17.7.5 Rigorous Documentation as a Barrier to Progress

ISO 9000 documentation makes process improvements more difficult. The very formality of documenting processes tends to lock in the status quo, which need not be perfect or even largely free of errors. As previous chapters have illustrated, many crucial electronics assembly processes are much more difficult and subtle than commonly realized. Because the component damage that results from those imperfect processes is largely invisible, companies frequently fail to recognize that their own procedures and standards constitute their most serious quality threat. Thus, documentation that locks the plant into maintaining those destructive practices is certainly not beneficial to the company's welfare; following those documented but undesirable procedures and processes will only ensure that the output continues to suffer serious damage. No certificate vouching for the plant's rigorous adherence to those documents will alter the fact that the product reliability has been compromised.

17.7.6 Bureaucracy

There is no small amount of irony in the fact that ISO 9000 imposes a DoD-type level of bureaucracy at the same time as the DoD is seeking ways to reduce the compliance hardships it creates for its suppliers. The registrar and assessors are completely analogous to the field inspectors used by the DoD to regulate defense contractors. And the national standards bodies are themselves the equivalents of the DoD standards committees. Bureaucracy, of course, is the primary reason for high costs and dubious dependability of defense electronics products. Many of the corporate leaders who support ISO 9000 have also been outspoken about the need for less government interference in business affairs—but what makes a private sector bureaucrat less harmful to a company's health than a government bureaucrat?

17.7.7 Overstated Benefits to Applicants

The primary argument in support of ISO 9000 is that it opens doors to new markets. This claim has been endorsed by companies that got early certification and therefore gained status as preferred suppliers to customers which demand the ISO certificate. In many—but far from all—cases of early certification, the costs of becoming certified were offset by greater market share. As more and more competitors become certified, however, the value of certification decreases but the costs of document maintenance and regular third-party reassessments remain. If most companies become certified, there will be no competitive benefits, only additional costs. Thus, ISO 9000 becomes nothing more than the price of admission to the market.

That admission price is generally quite high. For example, the direct labor cost alone for a small to medium-sized company to document and implement its systems in 1996 was estimated at between $100,000 and $150,000 while a 3-year agreement with an ISO-approved registrar would cost between $12,000 and $45,000.[14]

For one group of companies in particular, the cost of admission to ISO 9000 can be punitive and prevent the company from selling to those customers that demand the ISO certificate. The companies that bear this burden are the small businesses. While the absolute cost of gaining and maintaining the ISO seal of approval is usually less for smaller businesses, the relative cost is much greater because of the smaller sales volume over which to amortize the expense. Any increase in the costs of doing business—and the amount of paperwork involved—can hardly be considered an advance in the company's competitiveness or efficiency.

Meanwhile, Japanese companies have been slow to embrace ISO 9000. Although their European plants generally have little option but to obtain certification if they wish to sell into many of the European markets, the parent operations in Japan have largely avoided the scheme. As the Western world acquires another layer of expense and bureaucracy, the Japanese will have become more competitive.

17.7.8 Documentation Can Hamper Performance

The greater the amount of documentation in a company, the less likely will be widespread knowledge of the documents' contents by employees. This has been proved over half a century in defense contracting facilities where massive documentation exists but personnel rarely refer to the process or workmanship specifications.

---

[14]Hilary, Rachel. "Quality in the USA: Behind the Stars and Stripes," *QW*, January 1996, p. 22.

17.7.9 Inflexibility

The entire ISO 9000 system fails to account for the differences between very small companies and larger firms. Sole proprietorships have been forced into ISO 9000 paperwork, a task that has no more value to the small enterprise than audited financial statements for internal consumption. (Though the United Kingdom has responded to this criticism by creating a "small firms scheme" with assessments better suited to the characteristics of businesses with fewer than 10 employees, the wisdom of imposing any bureaucracy on small firms remains questionable.) The limitations on small companies that could function fairly, competently, and effectively as registrars are another reflection of innate biases in the ISO bureaucracy. When it comes to company size, ISO does not mean equal.

17.7.10 Certification Is History

Certification may reflect a company's past performance but cannot predict future performance. It is not unusual for companies to lose their passion for perfection once the immediate goal of certification has been realized. The regular reassessments by the registrar's representatives are intended to prevent any backsliding by the certified company, but no outside observers of an organization can easily spot cracks in the organization's structure. The difficulty in identifying problem areas is compounded when the assessors are not authorities in the organization's industry.

17.8 The Money Factor

One of the more troubling aspects of the ISO 9000 system is the rampant commercialism. The International Organization for Standardization derives substantial income from the scheme. The national standards affiliates prosper from selling standards guidelines (or collecting royalties from other publishers) and fees charged for accreditation of registrars and assessors. The normally dispassionate nature of quality professionals can be distorted by the power and income. There are even internal political squabbles—among various international standards bodies as well as between standards bodies and registrars—that would dismay seasoned public officials. The line between service and tyranny is perilously fine.

17.9 QS-9000

To find evidence that ISO 9000 will not meet the claimed objective of simplifying life for suppliers, it is necessary to look no further than the requirements currently being imposed on suppliers to the three major American auto manufacturers.[15] Generically known as QS-9000, the auto standard "is a superset of ISO 9001."[16] QS-9000 is most easily described as a merger of Ford Motor Company's Q-101 "Worldwide Quality System Standard" with ISO 9000. However, since the final document was the product of a committee drawn from Chrysler, General Motors, and varied smaller vehicle manufacturers in addition to Ford, QS-9000 involves requirements not found in either ISO 9000 or Q-101. Moreover, each of the automakers is applying the standard in its own fashion. Ford, for example, has not set rigorous compliance deadlines for its suppliers, but Chrysler and GM are requiring that all suppliers be certified to QS-9000 status by the end of 1997.

QS-9000 consists of seven volumes, sold as a set by the Automotive Industry Action Group (AIAG), an industry association whose membership includes the North American vehicle manufacturers and parts suppliers.[17] As with the International Organization for Standardization, AIAG is historically a technical standards group that evidently fails to recognize that its background in objective standards has not prepared it for the less tangible world of management systems.

For automotive parts suppliers whose clients include foreign and domestic automakers, QS-9000 is a significant additional administrative problem compounding the burden of ISO 9000. Yet another set of audits and reinspections must be endured. In a very real sense, the auto parts industry is taking on an uncomfortable resemblance to the defense contracting industry of decades past when different contracts involved different standards. The purchasing departments of the automakers involved may see QS-9000 as advantageous to them, and the auto parts suppliers who will bear the costs of the additional requirements have accepted the new obligations without much public complaint; however,

---

[15]Assorted smaller motor vehicle manufacturers—principally producers of trucks—are also following the lead of the Detroit "big three" automakers.

[16]Philip Stein, chair-elect, ASQC Measurement Quality Division, in a description filed on Dec. 5, 1995, with QUALITY.ORG on the Internet.

[17]The seven-volume set of QS-9000 standards—officially known as the Chrysler, Ford, GM Supplier Quality Requirements "7 Pack"—can be purchased from Automotive Industry Action Group, 26200 Lahser Road, Suite 200, Southfield, MI 48034.

it is difficult to see how increased administrative costs among suppliers can be advantageous to the North American auto industry as a whole.

The cost to the automotive parts industry aside, the short time available to Chrysler and GM suppliers to obtain QS-9000 certification presents a serious personnel problem: There are not enough qualified auditors to carry out the necessary certifications, particularly with the current high demand for ISO 9000 certification audits. The rush to certification may result in admission of many unqualified field auditors. A rapid collapse in demand for such services after 1997 would then throw the assessing business into serious recession, possibly depressing the desire of energetic new prospects from entering the quality profession for many years to come.

17.10 ISO 9000 Compared to the Baldrige

The sudden prominence of ISO 9000 also threatens the future of the Malcolm Baldrige National Quality Award, a fact recognized several years ago by the Commerce Department's National Institute of Standards and Technology, which administers the Baldrige. In a 1993 paper,[18] Baldrige administrators contended:

> There is much confusion regarding the relationship between the Baldrige Award and ISO 9000 registration. Two common misimpressions stand out: (1) they cover the same requirements; and (2) both address improvement, both rely on high quality results, and hence, both are forms of recognition. Many conclude the Baldrige Award and ISO 9000 are equivalent, and that companies should choose one or the other.
>
> These conclusions are incorrect. The Baldrige Award and ISO 9000 registration differ fundamentally in focus, purpose and content.
>
> - The focus of the Baldrige Award is enhancing competitiveness. The Award criteria reflect two key competitiveness thrusts: (1) delivery of ever-increasing value to customers; and (2) improvement of overall operational performance. The Award's central purpose is educational—to encourage sharing of competitiveness learning and to "drive" this learning, creating an evolving body of knowledge nationally.
> - The focus of ISO 9000 registration is conformity to practices specified in the registrant's own quality systems. Its central purpose is to enhance and facilitate trade.

[18]Reimann, Curt W., and Hertz, Harry S., "The Malcolm Baldrige National Quality Award and ISO 9000 Registration: Understanding Their Many Important Differences," Office of Quality Programs, National Institute of Standards and Technology, Gaithersburg, MD 20899.

The defense is seriously flawed in at least three respects:

1. ISO 9000 registration requires conformance to the specified practices of the registrant (as the paper does not) together with meeting minimum standards imposed by the International Organization for Standardization itself (a fact not stated in the paper). The consequence of omitting the second aspect of the certification procedure is that the rigors of ISO 9000 are understated.

2. The distinction between "enhancing competitiveness" (given as the purpose of the Baldrige) and "to enhance and facilitate trade" (the stated purpose of ISO 9000) seems more semantic than real.

3. The outline shows that quality per se has very little to do with either the Baldrige Award or ISO 9000, at least in the perception of the Baldrige's administrators. Both programs use "quality improvement" for emotional appeal and moral posture, but neither, in fact, is more than pseudoquality.

In the final analysis, one feature in particular separates the Baldrige from ISO 9000. The Baldrige is a voluntary competition while implementation of ISO 9000 is rapidly becoming obligatory. While both systems are gravely flawed, spending on ISO 9000 certification is the better of two unpleasant business decisions.

17.11 Summary

ISO 9000, for better or worse, has become an almost unavoidable requirement of modern business. The system is bureaucratic, costly, and unlikely to improve real quality. Indeed, it has the effect of imposing an inefficient and capricious defense-style procurement structure on civilian industry. Moreover, by instilling a sense of undeserved complacency in many purchasing departments, it makes selection of the most competent suppliers by customers less likely. Worst of all, by conveying an unwarranted impression that it is a quality improvement initiative, ISO 9000 siphons off resources that would otherwise be used for genuine quality improvement.

In short, ISO 9000 certification should not be sought out eagerly but should be undertaken only when required by major customers. And customers confronted by supplier promotion claiming that ISO 9000 certification verifies them as high-quality suppliers would do well to reconsider the supplier's true knowledge of quality management.

18 Quality Optimization

18.1 Chapter Objectives

Quality, as we noted in Chap. 1, is not necessarily free. At some point in the evolution of a quality improvement system, the cost of further increasing quality will exceed the value of that greater quality. When the marginal cost of increasing quality equals the marginal benefit, no further investment in quality improvement is warranted. At that point, quality can be said to be optimized.

The fundamental principle of quality optimization is fairly simple. However, that apparent simplicity masks many difficult questions. Those questions will be asked and answered in this chapter so that the reader will understand how and why to implement quality optimization.

18.2 The Basic Issue Revisited

Quality has become an emotional pursuit, as reflected in book titles from *Quality or Else* to *Quality Is Personal.* Too often today, "quality" is confused with an evangelical spirit of motivation. Appearances—*what* a company is doing rather than *why*—wield excessive influence, not in all cases in a disconcertingly large number.

There is, of course, a more substantive quality tradition exemplified by Shewhart, Juran, and, less obviously but most helpfully, Goldratt. That school of thought recognizes quality as one segment of the broad business management spectrum, unquestionably vital but still subject to the same rules governing all enterprise. While companies that ignore quality—theirs and competitors alike—should not expect to survive let alone prosper, the same caution applies to companies who regard quality improvement as a panacea.

This book began with the thesis that it is possible to have "too much" quality. That proposition does not mean that lesser quality is preferable to greater quality or that the search for quality improvements should be abandoned. The thesis does mean that any investment—in quality improvement, expansion of output capacity, marketing, new product development, acquisitions of other companies, or every other possible business activity—must generate a positive long-term net increase in income greater than any other investments that must be rejected in order to free up the resources required by the chosen investment.

Unless the company possesses unlimited resources with marginal cost approaching zero, the rational business will normally find it necessary to decline some quality improvement opportunities in favor of other forms of investment promising greater net returns. In the short term, the net return on all possible forms of quality improvement may exceed the net returns of alternative investments that must be forgone; in the long term, however, it is impossible to imagine every possible quality improvement investment meeting this standard.

The theoretical principle is clear. The problem that must be addressed is how to apply the principle to everyday quality management. The answer lies in quality optimization.

18.3 Quality Optimization Defined

Quality optimization can be defined as the science of determining when, why, and how to improve quality—and when to stop. By explicitly specifying that all quality management activities be constantly assessed to determine whether they should be maintained, this methodology is incompatible with the current "total quality" fashion of seeking continual improvement. Quality optimization may not be in vogue and may even be heretical. However, with experience showing conclusively that every quality program eventually encounters negative returns,[1] the need for quality optimization must not be underestimated.

Quality is optimized when *all* the following criteria have been met:

[1]i.e., the costs of incremental improvements exceed their worth.

1. The company has a single clearly defined primary goal.
2. That goal is known, understood, and accepted by all company personnel.
3. Every proposed and existing company activity—including but not limited to quality improvement—is evaluated and maintained according to its contribution to achieving that goal.
4. The company defines "quality" clearly and unambiguously in a manner that allows changes in quality to be quantified directly rather than by proxy variables such as defect rates.
5. The value of the last unit of quality improvement achieved equals the cost to achieve that unit. This is the optimization point.

Each of these requirements is necessary for attainment of quality optimization. No single requirement, however, is sufficient in itself for achieving quality optimization.

18.3.1 The Primary Goal

With the exception of nonprofit organizations (which should judge their performance according to the quantity of the good or service they are able to deliver to their "customers" in any given time period), the company should have as its goal the maximization of long-term profits without reducing interim cash flow below survival levels.[2] Provision of jobs, creating satisfying new products, and other social objectives so often cited by modern total quality advocates are possible only when the company possesses the financial strength to make such things possible.

Long-term profit maximization is the only private sector goal consistent with development of a healthy, lasting company. But many business leaders (and investors) do not care about long-term prosperity; their priority is short-term results. (As economists are fond of saying, in the long term everyone is dead.) Such short-term thinking generally results in

[2]Psychologists might argue that some companies controlled by a very small number of shareholders (usually one) would rationally violate this objective if the owner(s) derive their satisfaction from the enterprise itself rather than prospering. Indeed, firms run as much for amusement as income by their proprietor(s) are not uncommon. Such firms should be excluded from normal business analysis in which the objective is purely financial. The "hobby" business lasts only as long as its owner(s)' interest and willingness to accept financial sacrifices. In contemporary quality terms, the owner takes the place of the customer.

substantial harm to some or all parties—shareholders,[3] employees, creditors, suppliers. and customers in particular—affiliated with the company. However, where the goal is explicitly stated and the consequences realized by all parties (including employees who may wish to seek work elsewhere if long-term job security ranks among their priorities), criticism of the choice can only be based on value judgments. The advice in this book does not apply to companies deliberately pursuing a "live fast, die young" strategy.

18.3.2 Employees Know, Understand, and Accept the Goal

The standard by which the company measures its success or shortcomings has value only when all personnel are aware of the standard, understand its meaning, and accept the goal as the criterion for assessing their own performance. This is not easily accomplished, particularly in an age when company leaders must placate so many constituencies that do not realize strong earnings serve their interests as well as the shareholders'. One of the appeals of pseudoquality programs is the ease of avoiding employee resistance by camouflaging productivity improvement schemes, for example, as quality initiatives.

Comprehension of the company's goal does not in itself assure that employees will base their actions on the company's priority. Self-interest often contradicts the company's best well-being. Few employees will aggressively support efficiency programs that will eliminate their jobs. Even CEOs, when compensation is tied to short-term stock performance or annual earnings rather than building a stronger company for the long term, can lose sight of the company's welfare while maximizing their own. As quality professionals, we must always resist the temptation to use ineffective pseudoquality programs as ways of inflating our perceived importance.

18.3.3 Managing by Progress toward the Goal

Every quality professional knows the difficulties of measuring the benefits of quality improvement by the financial standards applied through-

---

[3]Shareholders must be analyzed in the aggregate over time rather than cross-sectionally at any given moment. A strategy of short-term profit maximization can be highly advantageous to transient shareholders (assuming that the stock market itself does not recognize the company's underlying problems and reduce share prices accordingly). Only if assessed over the life of the company will the total shareholder benefit be measurable.

out the rest of the company. Costs can be determined much more readily than gains, some of which may never be quantifiable. Yet our profession's credibility depends on providing the company's leaders with accurate financial data on which solid budgeting decisions can be based. The difficulty of providing such financial analysis of quality activities is no greater—and in many ways is easier—than the challenges facing sales and marketing which have long been expected to provide the same financial reports.

True quality optimization requires strong leadership from the very top of the company. Only there can the opportunity cost of capital—the benefits forgone when funds are diverted from other uses to quality improvement activities—be determined. Within the quality department itself, the best that can be achieved is cost-benefit evaluation in which projects are halted when the marginal expense exceeds the marginal gain.

18.3.4 Adoption of a Quantifiable Definition of Quality

"Quality" has no meaning until it is expressed in an "operational" definition that allows for direct quantification of changes in quality using directly relevant numbers rather than proxy variables. The meaning of quality may be different in other industries, but within the electronics assembly industry quality means the product performs as specified without failure for a specified time that can be considered reasonable. Once this definition is adopted, the only appropriate measure of quality is the actual failure rate (failure to perform as specified) during the specified time. "Defect" rates (where a defect may be some product attribute other than failure to perform to specification), touchup rates or other indirect measures of dependability are irrelevant.

18.3.5 The Value of the Last Unit of Quality Improvement Equals the Cost

As the level of quality attained by the company increases, the costs of each additional increment in quality will typically rise or, at best, remain constant; this rise in marginal cost reflects the tendency of quality improvement programs to focus first on the highly visible problems that tend to offer the greatest payback when solved. At the same time, the benefits received with each additional increment of quality would be expected to decrease.[4]

Ultimately, the point will be reached where further increases in quality will cost more than the resulting benefits. Beyond that intersection point of costs and benefits, additional spending on the quality initiative under consideration will reduce profits, thereby moving the company away from the goal of long-term profit maximization. At the intersection point—not before but certainly at the intersection—the quality improvement program must be terminated. If the cost curve exceeds the benefits curve at all levels of quality improvement, the program should not be initiated.

The cost and benefit curves need not be smooth and continuous; either or both could be characterized by abrupt shifts in values as failure rates pass certain critical points. In general, however, it is appropriate to envision a downward-sloping benefits curve and an upward-sloping cost curve where the vertical (y) axis represents the cost or value of the marginal change and the horizontal (x) axis shows the cumulative number of failures prevented. The cost curve should show the greater of (1) the marginal cost of capital or (2) the returns that would have been generated by investments forgone to free up the capital needed for the quality improvement project. Derivation of the marginal costs for the company as a whole—a necessary condition for true quality optimization—requires substantial input from the company's senior strategic planners.

18.4 Further Aspects of Quality Optimization

In most respects, quality optimization is based on cost-benefit analysis. However, simple cost-benefit analysis alone will not bring about quality optimization. The entire subject of quality is far too complex to be properly managed by superficial examination of short-term quantities.

18.4.1 The Time Horizon

Both expenditures and benefits will likely accrue over time (the goal being maximization of long-term profits). For a properly designed qual-

[4]As an example, reducing the first-pass failure rate to levels where the value of failures detected is less than the cost of test would allow elimination of that particular test. Further decreases in the failure rate would not allow that test operation to be eliminated yet again, so the marginal value of greater quality would be less.

ity improvement program, the benefits flow will continue for a very long time while the expenditures will quickly taper off after an initial high level of spending. Both flows—expenditures and benefits—must therefore be discounted to their present values. If the benefits stream is heavily weighted to the distant future while expenses are incurred earlier, the present value adjustment will discount the benefits more than the costs. Therefore, the optimal program will generally be weighted toward rapid delivery of benefits; programs that require years to produce meaningful benefits will be less attractive than those that generate benefits at the same pace or faster than expenses arise.

The fundamental question, of course, is what time period can be considered "long-term." For any individual, long-term typically means something much shorter than the time span affecting society as a whole. Fortunately, when businesses are managed rationally, the question of time horizons tends to become irrelevant. Experience shows that even 5-year plans rarely prove accurate. Projections even farther out have still less probability of accuracy. During the last decade, for example, projections were generally based on continuing price inflation; today, projections tend to assume price stability. The only point about which we can be certain is that many expectations will prove to be wrong and the magnitude of the errors will likely increase with the time period on which the expectations are based. Accordingly, while we would like to be able to manage for prosperity decades from now, uncertainty constrains most of our decision making to at most 5 years.

The issue then becomes how to gauge when a business is managed rationally. Rationality occurs when the business evaluates its investment options comparing costs and benefits based on the best available information. Game theory is probably the best approximation of rational behavior in the presence of uncertainty. In other words, investments should be based on the magnitude of the upside opportunity (the potential gain) compared to the magnitude of the downside risk (potential loss) where both upside opportunity and downside risk are weighted according to the anticipated probability of either outcome. Any choice that entails the possibility of catastrophic risk would normally be irrational.

18.4.2 Problems Quantifying Benefits

The gains induced by quality improvements may, in Deming's words, be "unknown and unknowable." Will better quality stimulate greater demand (which, in turn, means higher gross income)? Inasmuch as other changes are likely to be occurring in the market at the same time as the quality improvement, separating out the effect of changes in quality from, for example, more aggressive or persuasive advertising, changes in features, or shifts in the competition's quality will seldom be feasible. Fortunately, as explained in Chap. 6, the most important benefits of quality improvement are unlikely to involve demand. The greatest benefits almost invariably arise on the Supply Side from cost reductions resulting from less inspection, test, rework, touchup, repair, and field service. When the competition has a disproportionately lower failure rate and our own failure rate is sufficiently high to be noticeable by a meaningful percentage of customers, the Demand Side benefits of quality improvement will carry greater weight. However, enormous differences in dependability among competitors in the electronics assembly industry are far less common today than was the case a decade ago. Accordingly, the unknown Demand Side benefits can usually be left out of the equation without unduly affecting decisions based on cost-benefit analysis. The probability of undercounting costs is comparable, in our experience, to the likelihood of undervaluing benefits.

It is essential to keep in mind that quality can affect demand profoundly, quickly, and negatively when quality problems—real or imagined—come to the public's attention. Therefore, the greatest Demand Side concern of quality management should always be damage control rather than the dubious image polishing that underlies so much "quality improvement" activity. Increasing demand by improving quality often takes considerable time (one of the reasons quality management must take a long-term approach), but losing business through sudden lapses in quality can happen overnight.

18.4.3 The Opportunity Cost of Capital

The average company settles for suboptimal cost-benefit analysis by arbitrarily assigning budgets to departments which then allocate their funds in the ways that best meet their perceived needs and opportunities. With the current emphasis on very costly pseudoquality programs including TQM, the Baldrige Award, and ISO 9000, however, the size of the typical quality department's budget requires more sophisticated management. Not infrequently, companies sacrifice vital capital equipment purchases, expansion, and new product development to fund company-wide pseudoquality programs. Where the organization operates with capital constraints (and how many companies have easy access to all the capital needed to fulfill their wish list of activities?), the cost of the next dollar required after all financial resources have been exhausted is infinite, and no program can provide an adequate rate of return. Also in such companies, budget allocations among departments are standard operating procedure—but optimal allocation of funds is possible only when each department supports its budget requests with cost-benefit analysis appropriately weighted by the present value discount.

18.4.4 The Liability Factor

Many quality departments neglect to factor into their benefits the very real risk of legal liability in the event of certain forms of product failure. As with the effect of quality improvement on demand, it is impossible to know the actual level of exposure posed by a defective product. The best approximation is found by comparing the price of liability insurance with and without quality improvement. Many times, companies miss out on lower insurance premiums by failing to inform their insurers about improvements in product dependability.

Few product failures actually develop into litigation problems. Manufacturers can reimburse customers for material damages, and most failures are settled by repair or replacement of the defective unit. There are costs—often painfully high if the failures are frequent—but not punitive. Failures that cause personal injuries or deaths are not so readily or inexpensively resolved.

18.4.5 The Value of Conflict

Orthodox total quality theory emphasizes the need to avoid conflict (generally referred to in euphemistic terms as "competition") between individuals and departments. Conflict—particularly where none of the opposing positions are consistent with the company's goal—can detract from the company's performance. On the other hand, constructive rivalry in which departments compete to find the investment opportunities that best advance the company toward its goal is more likely to be beneficial than harmful to the company's development. Dynamic environments stimulate development of better products, new products, enhanced efficiencies, and all the other advances that make companies stronger; complacent environments hold back those same advances.

Competition does harm the company when the rules for evaluation are not consistent across all departments. The most prominent example of this lack of consistency has traditionally been found in the inability (or refusal) of quality departments to quantify their programs using the same financial methods applied to other departments. Where departments compete for funds using varying justification formulas (including, in quality matters especially, reliance on emotional appeals), the budget allocations will be distorted by lack of knowledge on the part of the company's financial planners. This, of course, is the inevitable outcome of embracing the TQM doctrine that quality is too important to be subject to cost-benefit analysis.

Although most quality managers have grown comfortable with their department's relative freedom from formal financial justifications, the ad hoc assessment of quality initiatives proposed and ongoing:

1. Ensures that the company will not allocate its resources optimally
2. Threatens the quality department's credibility by delivering too much or too little funding to the department. Too much funding means unacceptably low returns of marginal programs, thus pulling down the department's average performance. Too little funding will prevent implementation of some quality improvement programs that should have been undertaken.

Neither outcome can be in the quality department's best interests.

18.4.6 The Necessity of Interdepartmental Cooperation

While competition stimulates the company to better performance, interdepartmental cooperation is usually vital to maximizing the value of quality improvements. For example, even though we know that better quality seldom increases demand by meaningful amounts (at least in the short term), the amount by which demand changes will be greatly influenced by customer awareness of changes. The flaw in Emerson's better mousetrap argument, after all, is that customers will not beat a path through the woods if they are unaware that the better mousetrap exists or recognize the characteristics that make it better. This is where sales and marketing (including but not limited to advertising) become important. The sales and marketing departments inform the world of the quality department's accomplishments, accelerating the rate at which better quality translates into greater demand.

The observation that quality department achievements often require sales and marketing efforts to maximize the value of those achievements to the company may seem to contradict our earlier reproach of pseudoquality as sales programs masquerading as quality improvement. Our criticism of pseudoquality programs designed to influence sales rather than improve quality was not based on any belief that greater demand is undesirable. Rather, we are concerned that "quality improvement" programs in which the improved quality is entirely illusion to bolster the company's image will eventually sour the public perception of claims for improved quality and therefore undermine the usefulness of real quality improvement.

Similarly, cooperation between the quality and production departments will be necessary when new quality evaluation standards are adopted. If the production personnel continue to operate according to the historic standards, the Supply Side benefits to the company from the quality improvement will be compromised.

18.4.7 Quality Optimization Is Compatible with Continuous Improvement

The financial analysis of returns on quality improvement investments does not invalidate the continual improvement ideology of modern quality. It requires only that the company avoid spending for improvement initiatives whose benefits are less than the costs. Other, cost-effective, avenues of quality improvement should typically be available, so that the quality optimization evaluation exercise merely redirects use of the department's finite resources from unproductive to productive uses.

At times, opportunities for improving quality consistent with the quality optimization investment criteria may not be evident. In the dynamic world of business, however, the void is likely to be filled in short order. The desirability of taking temporary breaks from the pursuit of quality improvement when no opportunities for positive return present themselves is inherent in Deming's insistence that the quest be for continual rather than continuous improvement. "Disciplined advance" should be the company doctrine, not "advance at any cost."

Additionally, nothing in the practice of quality optimization alters the fact that many of the most impressive quality improvements are discovered utterly without cost. They do not emerge from organized quality improvement programs, consultants' advice, or any other activity involving use of resources. They come about because employees, as they acquire greater experience and knowledge, look at old practices with fresh perspective. No cost-benefit analysis will ever threaten a change that provides benefits without costs.

18.5 Summary

Use of company resources to improve quality must be carefully and continuously assessed with respect to values of both the resources used (the costs) and the benefits generated. The costs and benefits must both be expressed in the monetary terms employed throughout the rest of the company to evaluate investments. The objective of the assessment exercises is to help ensure that quality optimization will be achieved.

Quality optimization is achieved when the present value of the benefit from the last incremental improvement in quality equals the opportunity cost of achieving that improvement. The cost is always measured in terms of opportunity cost: the benefits lost as a result of other investments forgone to free up the resources for the chosen activity. Costs and benefits must always be discounted to their present values before being compared.

Quality optimization need not—and usually will not—restrict achievement of continual improvements. The practice will, however, ensure discipline in the selection of improvement targets and techniques and maximize the probability that the company will employ its finite resources most efficiently.

19 A Program for Improving Quality in Electronics Assembly

19.1 Chapter Objectives

Previous chapters have provided the scientific and social management lessons necessary to achieve success in quality management. This chapter shows how the earlier lessons can be combined to form a quality improvement program framework specifically designed for and proved in the electronics assembly industry.

19.2 Fundamentals

Over the years, our clients have enjoyed considerable success in quality improvement by following the approach described in this chapter. "Success" in this context means quality (measured in the real sense of dependability rather than proxy terms) has increased dramatically, quickly, and profitably. When properly employed, the program generates meaningful improvements within the first month and is operating at peak efficiency in not longer than 6 months.

19.3 Creating the Program

Creating a successful action program is rather like building a model airplane. Construction requires a variety of physical parts and liberal quantities of glue—all, of course, assembled in the correct order and joined in a way that ensures the presence of the necessary strength without excess glue spilling out of the joints to spoil the finish.

The physical parts of the following program consist of tangible, scientifically accepted theorems.[1] The theorems can be proved or disproved (within the existing bounds of scientific knowledge) by experimentation and other forms of analysis. Accordingly, they can be learned through hands-on exercises in classrooms or the actual shop floor environment.

The "glue," however, consists of instincts, experiences, and innate problem-solving abilities. Such characteristics are not readily transferred to new personnel nor can they be acquired as readily as the tangible physical sciences. However, the integrity of the glue determines whether the program, like a model plane, will hold together and perform satisfactorily. It is fair to say that the most significant challenge of creating an effective program lies not in the technical knowledge but in the human interactions.

Effective programs—those that cause substantial increases in product dependability and reduce costs—do not just happen. Nor can they be achieved by following generic formulas in quality manuals (including this one). They result from painstaking consideration of the details, including intensive review of often overlooked details such as how extensive the program should be, where and how to start, and how to track accomplishments. Most quality improvement programs that we have encountered believe they have implicitly taken these fundamental issues into account; however, explicit investigation generally reveals that the critical questions have been neither asked nor answered.

19.4 Where to Start

Quality, the total quality advocates repeat endlessly, must start "at the top." The advice—which is difficult to fault—is based on the logic that

[1]Scientific theorems, also known as laws or postulates, are often erroneously called "facts," a term that implies the presence of absolute truth. The reality is quite different from the implication: Nothing in science can be put into such a category. Scientific laws remain valid only until a better explanation for the behavior of nature can be found. Almost without exception, subsequent discoveries result in more valid "laws" that displace their predecessors and are themselves subject to eventual replacement. The importance of this point cannot be overstated.

better performance in any endeavor depends on the example of leadership. But the question then arises: What is the "top"?

19.4.1 Specialization

To determine what rung on the corporate ladder represents the "top," we must first identify the objective of our quality improvement program. Total quality approaches the company as a whole with the intent of overhauling every operation simultaneously. The total quality aim may be noble, but the immensity of the task also renders it ineffective. The education requirements alone make company-wide quality unrealistic. Few companies possess the financial resources to achieve such sweeping changes, and no company possesses sufficient knowledge of the myriad processes that determine outcomes of the countless activities. Without both the funds and the knowledge, development and implementation of the vast numbers of education programs is impossible.

The most effective solution we have found after years of intimate involvement with many electronics companies throughout the world is specialization. Rather than trying to change everything at once, focus on what is manageable and consequential. In the context of electronics assembly, that means concentrating on product quality, which we know to be dependability.

19.4.2 The Plant Is Pivotal

Electronics assembly takes place in plants, not in executive suites. Even design—the assembly activity most likely to be physically separated from the others—will be more closely aligned to the plant than corporate offices.[2] The plant, for all practical purposes, is a self-contained community. In larger companies, the company chairman and president rarely influences plant operations more than the President of the United States affects the average American city; the authority may exist, but it is distant and considerably less tangible to residents on a daily basis than decisions by the mayor and city council. As far as the personnel who are most essential to quality improvement are concerned, therefore, "the top" is found within the plant organization rather than the executive offices.

---

[2]In those companies where design is not tightly integrated with the production and quality evaluation departments, that integration must be established before any quality improvement program is even considered.

The "top" individual for purposes of quality improvement, then, is that person with ultimate day-to-day responsibility for running the plant. Although the title of that individual varies from company to company (and sometimes even among plants within the same company), the two most common titles for this position are plant manager and director of operations. Depending on the company's structure, the division manager or division president may also be close enough to the production activities to warrant being designated the "top" person. In smaller companies, the owner or president often functions as de facto plant manager. Rarely, however, will quality improvement efforts (or plant improvements in general) be more successful if the corporate executives overtly take part.

Basic psychology also provides support for the thesis that any production-related activities work best *without* public support from the CEO. Few CEOs have production backgrounds; they typically feel out of place when put into manufacturing environments, and their discomfort is apparent to all the personnel with real manufacturing backgrounds. Shop floor personnel have learned to be extremely cynical about pronouncements from corporate leaders; when top executives pontificate about corporate commitment to quality or employee well-being, rank-and-file personnel stop listening.

(Juran offers an astute observation on the subject of executive participation. "There is no known way," he writes, "for defining upper-management participation through use of such broad words as 'commitment' and 'involvement.'"[3])

Executive support in the form of quiet cooperation with plant management *is* necessary, especially with respect to issues such as budgeting. Public personal involvement in the form of speeches and writings is not necessary and quite often, by striking a cynical chord in various audiences, is counterproductive.

19.5 How to Start

Having identified where to start, it is necessary to know how to start. That is, we must set out some general rules for types of activities to include and the scope (i.e., how much of the plant to include) of those activities. While the program will likely expand in number of activities and scope as it matures, the initial priority is to ensure that the program will survive infancy. The probability that the program will reach maturity is maximized through planning in the early stages.

---

[3][Juran and Gryna 1988], p. 8.3.

19.5.1 Mission Statements

Total quality consultants often recommend that the company begin by writing and posting a mission statement. While it is necessary to know where we hope the program will lead the company, mission statements rarely serve any purpose other than transparent propaganda.

Mission statements are often termed "vision statements" (or "the vision thing," in the words of former President Bush). They should clearly articulate the company goal by which all operations could be measured. Unfortunately, although mission statements abound in the modern corporate world, true vision is hard to find.

Considerable cynicism (or, more charitably, naïveté) underlies these exercises. Mission statements are designed primarily for external consumption—i.e., by customers—or in the false expectation that they will motivate employees. Often termed "vision statements," mission statements tend to be generic, filled with inoffensive platitudes like making high-quality products and selling at low prices to make the customer happy. Whether customer opinions are positively affected by such insipid statements is doubtful. Internally, however, the vision statement tends to distract from the real purpose of earning profits and developing the business. This leads to erroneous choices on the part of the operations personnel since they lose sight of the important purpose that the company earn profits.

Mission statements, in other words, are perceived as window dressing attempts to cover up empty leadership. The army—which has much in common with manufacturing organizations—has a saying reserved for times when an incompetent officer sends them off with emotional appeals but no proper briefing instructions: "We don't know where we are or what we're doing—but we're making really good time." The same sentiments are expressed regularly by employees in most industries, electronics assembly included.

The pertinent point about mission statements as the groundwork is laid for presentation of a program to improve quality is that they exemplify the waste of funds through total quality methods and the productive activities characterized by real quality. As we know (see, for example, Chap. 12) and as Deming taught, employees cannot improve by being told to do better; they improve only through acquisition of knowledge. The importance of motivation must not be underestimated, but sloganeering should not be confused with motivation. The time and money spent on developing vision statements or printing banners extolling the virtues of better serving the customer would be far more useful if invested in providing the tools—especially education—that allow employees to work smarter.

19.5.2 Priorities

The effective program must be structured in such a way that the employee is better equipped to help the operation improve. Better equipped means more knowledge, not greater motivation. Therefore, priorities must include ensuring that every employee:

1. Is fully informed of the company's true goal (not the bromides about becoming "world class" and being "customer driven")—not to be confused with the "mission" statements that are mostly public relations exercises.
2. Sees the goal as attainable.
3. Feels inspired to help the company reach that goal.
4. Knows his or her function (i.e., job) within the company.
5. Understands why his or her function exists.
6. Can ascertain whether performing the function as presently defined and implemented helps the company reach its goal.
7. Possesses the knowledge and skills to properly carry out that job.
8. Has ready access to easily readable documentation about what is right and wrong (i.e., unambiguous workmanship standards and procedures) and is not embarrassed to be seen reading the documents.
9. Has the opportunity but not the obligation to help solve problems.
10. Receives informed feedback on how well management feel the employee is performing those aspects of his or her function that the employee can control.

19.5.3 Scope and Evolution

In setting up any improvement system, there is much to be learned from the techniques of new product introduction. As the idea for a new product emerges, hopes for success are generally high. Despite their expectations, however, experienced personnel know that rushing from concept to design to production without interim checks and verifications leads to failure. Designs are checked along the way through prototypes, and production is ramped up gradually. Only when dealing with small variations to a mature product are some of the interim assurance steps bypassed—and even then bringing the new model to full-scale production too quickly invites problems.

New management programs should be approached in the same manner as new product development. The desire to change the entire company—or at least the plant—will be strong. The thrill of trying something new is compelling. But attempting complete and immediate transformation is the equivalent of rushing from product concept to production without any verification steps. Attempting to implement a company-wide program is highly analogous to simultaneous introduction of many new products without prior verification.

The chances of success will be much greater if the program begins as a test program dealing with a particular problem in a portion of a plant. As experience with the abilities and limitation of the techniques is acquired, the program can be modified and expanded. Eventually, the program may prove so effective that it is extended throughout the entire plant (and, possibly, adopted in affiliated plants), but that is by no means an inevitable outcome in the program's early days.

Highly successful pilot projects regularly collapse when expanded to incorporate more of the plant. This can happen when the program is governed by diseconomies of scale. That is, as the program expands, the support structure must expand as well. The company may lack the resources to provide sufficiently broad support for a full plant-wide implementation. Implementations in several plants at the same time place even more strain on the support system.

Finally, the lesson of quality optimization—that a point may be reached where more quality improvement is not desirable—must never be forgotten. The test program may be highly successful because it solves a particularly costly problem. The benefits of application to other problems may be substantially less. In other words, the most successful plants are often those that are always experimenting with new methods of dealing with diverse problems rather than those that expect a single methodology to work equally well throughout the plant.

19.6 The Program

A step-by-step approach to improving quality and reducing costs in electronics assembly plants can now be presented. We have used the following framework for more than a decade to achieve exceptional and lasting results in a very short time. Indeed, in our experiences with many electronics assembly plants in a variety of countries and cultural conditions, we have found this framework to be almost foolproof.

Although the framework has almost never failed, several caveats must be emphasized nonetheless. First, this is a *framework* and, as such, requires fine-tuning to meet the unique needs of any given plant. Second, the results will be desirable only to companies that subscribe to the new model of quality priorities—notably including quality optimization—developed in the previous chapters of this book. Third, in quality improvement programs, as in the rest of life, good things come only to those who commit fully in deeds as well as words to turning concepts into realities.

The third caveat requires some additional blunt words. As we have observed previously, personal goals often conflict with the company goal. Furthermore, the community in which *every* member embraces change is rare. Some—and generally many—employees will be uncomfortable with the prospect of change and act to defeat it. Goldratt explains the issue well by noting that:

1. "Any improvement is a change."[4]
2. "Any change is a perceived threat to [the individual's] security."[5]

Goldratt's observations are important. Many larger companies wrongly believe that all employees fully endorse new ways of operation, of defining desirable performance, and the myriad other modifications that will certainly affect those employees in ways that they cannot completely foresee. Sweeping change therefore creates high levels of apprehension in some personnel. Almost inevitably employees with inordinate fear of change will refuse to support the new agenda and, to the extent their positions permit, may actively work against the program's success. Those committed to maintaining the status quo may speak out, but often the greatest damage will be done by individuals who are most vocal in praising the new order. When senior management becomes aware of employees who are undermining the project, authoritarian measures must be taken to make it clear that there will be no place in the company for individuals who refuse to be part of the team. Theory Z management techniques will not work when confronting the negative influences; team-oriented management methods cannot produce satisfactory outcomes when applied to personnel who have no desire to be team players.

---

[4]Though any change is not necessarily an improvement.

[5][Goldratt 1990], p. 10.

Also, be aware that some of the steps described in the following pages will often take place concurrently.

19.6.1 Step 1: Determine the Company Goal

This point has been discussed at considerable length in various parts of this book. Suffice it to say now that this framework is based on the premise that the company has set as its primary goal long-term profit maximization without reducing cash flow below survival levels in the meantime. If the company's goal is different from this, the program will not provide satisfactory results.

19.6.2 Step 2: Ensure Universal Understanding of the Goal

The goal is of use only when it is universally known. Therefore, every individual involved in designing, operating, and evaluating any quality improvement effort must be fully acquainted and comfortable with the company's goal. Every aspect of the quality improvement program must be conceived, structured, and executed in ways that move the company closer to the goal.

19.6.3 Step 3: Identify the Core Process

As noted earlier, most productivity and quality improvement programs tackle the entire plant as a single entity. In other words, they try to improve *all* processes *simultaneously.* Such a strategy can be termed a global approach—and global approaches almost guarantee failure.

The global approach presumes that manufacturing plants are homogeneous. In reality, of course, plants are highly diversified environments made up of many overlapping but distinct processes. Although the processes may interact, they are quite different one from the other. Therefore, no economies are available by addressing the processes collectively rather than individually.

Moreover, fixing all the processes at once is an impossibly big job. Management and technical resources must be spread over many processes, stretching those resources beyond the breaking point. All personnel must acquire in-depth knowledge of several processes at the same time, a requirement that leads to information overload. Decisions get made in haste, generally without adequate knowledge. Proper measurement of advances (or reverses) becomes impossible. Results, if any, come slowly. Employee enthusiasm withers. Chaos, not control, reigns.

In contrast, the approach recommended here breaks the project into bite-sized pieces. It begins with the core process (the plant's central operation) and essentially ignores the other processes until the core process is under control.

The core-based approach has many advantages:

- Since all available resources focus on a single process, management burnout is prevented.
- Personnel need master only one process initially, thereby avoiding the information overload associated with studying several processes simultaneously.
- Results come quickly and employee enthusiasm grows with each new success.

It is not necessary to wait until the core process has been perfected before exploring solutions to other processes. The only requirement is that the core process has been substantially tamed when attention turns gradually to other processes. Thus the project flows in a controlled manner from the central operation to surrounding processes.

Not all industries contain a core process. In those industries which lack such a focal point, significant improvements may be impossible even over several years. Fortunately, however, electronics assembly does contain a core process: soldering.

The following operations normally precede soldering and can be called the "upstream processes":

- Design
- Purchasing
- Receiving and incoming inspection
- Stores
- Assembly
- Presoldering inspection

The actual soldering may be carried out entirely by machine, entirely by hand, or through a combination of machine and hand. Then the following operations (which can be called the "downstream processes") typically occur:

- Postsoldering cleaning
- Visual inspection and touchup
- Hand soldering
- In-circuit test (often followed by rework)
- Functional test (again often followed by rework)
- Subassembly
- Burn-in
- Final assembly
- Final test (often followed by rework)
- Shipping

As explained in Chaps. 7 to 10, many of these activities are unnecessary and may even reduce product dependability while increasing costs.

Soldering, then, is central to the entire operation. Most other activities either lead into or flow from soldering. Controlling the soldering process will make control of all other operations easier.

This point can be illustrated by examining the upstream processes. Since these influence the quality of the soldering process, their failings will become obvious when the soldering process improves. (For example, materials deficiencies will become more visible when soldering process improvements have eliminated equipment-related causes of defects. The project team can then track these deficiencies back to purchasing or stores policies.) This knowledge greatly reduces the workload when the time comes to deal with the upstream processes. But that knowledge will not be available if we try to improve all processes simultaneously.

The downstream processes, on the other hand, are strongly influenced by the soldering itself. For example, visual inspection and touchup exist solely because of soldering process defects (except in defense contracting, in which 100 percent visual inspection forms an integral part of most contracts). Eliminating the soldering defects allows visual inspection and touchup to be discontinued. In other words, there is no need to *improve* the processes of visual inspection and touchup since bringing the core process of soldering under control eliminates the need for that downstream process. Thus gaining control over the downstream processes will be dramatically simplified by first bringing the core upstream process under control.

19.6.4 Step 4: Conduct a Literature Search

Having identified the core process, it will be tempting to undertake experiments in the hopes of acquiring more knowledge about the process. This is the time to remind ourselves of Hicks' fundamental rule for experimentation: "The truth sought should be something that *is not already known*, which implies a thorough literature search (first) to ascertain that the answer is not available in previous studies."[6]

Literature surveys almost invariably result in the discovery that different authorities hold opposing views. The existence of these disagreements often frustrates truth seekers to such an extent that they turn to experimentation (see step 5) in the hope that they will find their answers through statistical testing of hypotheses. Experimentation, however, is seldom successful. Although experiments involve considerable time and expense, they rarely lead to accurate answers (see Chap. 12).

Fortunately, such heroic measures as experimentation are rarely necessary. Applying logic while reading the conflicting "facts" in the literature will normally identify many of the arguments as being clearly nonsensical. A distressingly high percentage of soldering process literature, for example, contains "facts" that violate natural laws familiar to any high school science student (see Chap. 7). Thus anyone with basic knowledge of high school sciences can easily weed out much of the nonsense.

The individuals, groups, or companies that gain the most from literature searches possess three vital traits. First, they recognize that widespread agreement is not necessarily synonymous with truth. Second, they are constantly in search of new truths. Third, they know truth can be differentiated from fiction only by painstaking effort to learn the scientific laws. Invariably, the most successful students are those that ask the powerful question "Why?"

The literature search is the ideal time to apply the most important lesson of Japanese successes: borrow shamelessly and, if necessary, ruthlessly. Reinventing the wheel is never a good investment. Refining the basic concept, on the other hand, often turns a good concept into a brilliant accomplishment.

19.6.5 Step 5: Experiment

The literature search alone may not provide all the answers. Some issues may not be discussed in any publication, or the search may overlook crit-

[6][Hicks 1982], p. 1.

ical articles. The hunt for satisfactory educational course material may be fruitless. By now, however, the number of questions should be reduced to readily manageable levels. At this point, experimentation *may* be feasible, even at the full factorial level. However, experimentation entails substantial risks; incorrect answers are more likely to emerge than correct solutions. Coincidence may appear to be valid statistical correlation. Effect-cause-effect analysis must be employed to substantiate the findings.

19.6.6 Step 6: Formalize a Process "Recipe"

Much emphasis was placed earlier in this book on the "natural" process—the activities that make up the process and the sequence in which natural forces require that those activities be carried out. Once that natural process is known, it can be formally written as the "specified" process that the plant will follow. The specified process may be thought of as a process recipe since following the instructions will always lead to a successful outcome. The process recipe is the most important information to be taught in classes.

19.6.7 Step 7: Prepare Educational Materials

The information uncovered in the previous steps must be passed along to *all* plant personnel, including managers. Accordingly, educational material must be developed.

Education is not the same as training. Training conditions the individual to follow instructions blindly, even when the instructions are wrong. Education, in contrast, develops the individual's ability to reason. A plant whose employees do not deliberate on their actions and assist in the improvement process is doomed to failure. Many plants fail merely because no employees have sufficient knowledge or interest to question inefficient practices.

In our experience, lasting results are obtained only when the educational material includes hands-on problem-solving exercises that accurately reflect real plant conditions. Such exercises develop thinking, understanding, and teamwork (in other words, the qualities that are needed to bring change to the shop floor). Further, the lessons learned in solving a problem will be retained long after lectures have been forgotten. Above all, however, a problem based on real plant conditions gives the program relevance, ensuring that students will pay attention.

We have also found that the material must be edited and packaged in different ways for the plant's various personnel groups. Normally, four modules are necessary:

1. A senior management overview which broad-brushes technical issues
2. A course for middle managers that combines the management methods of the senior management workshop with a condensed version of the technical information provided to engineers
3. An in-depth technical presentation for engineers
4. An operators' course that emphasizes hands-on experience

In the case of developing and presenting educational materials for use throughout the plant, it will be profitable to recognize that virtually every element needed for an exceptionally effective teaching kit is already available from outside sources for easy use by in-house teachers. The choices range from complete "plug-and-play" packages of audiovisual materials, covering every essential aspect of plant activities, down to magazine articles. Generally, it is most productive and cost-effective to take advantage of third-party educators.

The options are almost endless, and choosing the best can be a daunting task. A few necessary considerations include:

1. How much work the program team is willing and able to undertake. Ready-made educational courses place far fewer demands on the teacher, whereas literature searches require substantial effort to shape the raw data into effective educational tools.

2. How long the company can wait for the necessary improvements to be made. Development of educational materials is extremely time-consuming. A good course on a single topic can easily require a year of full-time development.

3. Economics. A course developed in-house can be tailored to the company's specific product and equipment. On the other hand, production of audiovisual materials is very expensive and requires specialized knowledge of sound and video recording techniques. The economics almost always favor second-party courses whose development costs are amortized over many customers.

Educational courses are not all equal. The probability of selecting a good course rather than one that is mediocre or worse is substantially improved by asking the following questions:

- Is the course educational rather than "training"?
- Will it help personnel learn to question all operations—including operations that are outside the specific scope of that course?
- Is the content specifically designed for the electronics assembly industry?
- Does the content meet your specific needs? (Many companies spend money "teaching" personnel about SPC in much greater detail than the employees will ever be able to use. The companies then fail to empower the personnel to use those materials. Moreover, there's no point knowing SPC if you don't also know what it is important to measure.)
- Is the content accurate (does it correspond to the process recipe derived previously)?
- What mechanism ensures course content makes the transition from the classroom to shop floor action?

19.6.8 Step 8: Hold Classes

A well-designed educational program will, of course, provide the students with knowledge that they didn't know before. (While it may seem self-evident that there is little point to classes that teach what the students already know, many courses do exactly that.)

The teacher's job is not to lecture. Rather, the teacher must generate discussion and encourage questions. Some people refer to this as "facilitating" since it promotes involvement; others know it best as "Socratic" methodology. Regardless of the terminology used, it is wise to remember that many less educated employees will initially be nervous in a learning environment that expects their participation. Not all employees are comfortable with intellectual challenges.

Optimal class size is roughly 8 students, though a good teacher can achieve equally good results with as few as 5 or 6 students and as many as 10. Fewer participants put considerable strain on the students to answer the questions. Too many participants mean that some students will not participate in the discussions.

Properly guided, the students begin to look at their jobs in a new light. They prove to themselves that many of their job practices are inefficient or even detrimental to product quality. Then, most importantly, they bring these issues to the teacher's attention. The truly successful teacher is one who sufficiently inspires the trust of students that they will bring work-related issues to the teacher's attention—even after the students return to their jobs.

19.6.9 Step 9: The Executive Committee, the Supervisory Council, and Employee Action Teams

Process improvements can be implemented in either of two ways. The first is by edict. Under this autocratic system, a small group of managers and engineers decides what changes will be made and imposes its will on the rest of the plant. The second approach is employee involvement which provides every employee the opportunity (but not the *obligation*) to participate. The second approach works; the first doesn't.

Change cannot be imposed because the individuals responsible for making the changes successful (i.e., the workers) lack ownership in the changes. People support changes that they understand, and they understand best those changes which they have helped design. The supervisory council and the executive committee constitute the structure that provides employees with the opportunity to participate in plant improvements.

The supervisory council is a permanent body and acts as the program steering committee. The supervisory council is administrative; it does not actually solve the problems. Rather, the supervisory council compiles the employee suggestions, establishes employee action teams, assigns problems to the teams, and manages the teams' progress. Only a small group of employees (fewer than six) drawn from supervision and line personnel should belong to the supervisory council.

The supervisory council reports regularly to a committee of senior managers. Typically, this senior management committee includes the managers of manufacturing, quality, and engineering. Ideally, the plant manager or equivalent should also sit on this committee. The reports should be written with supplemental discussions in regular meetings. The frequency of the reporting meetings is flexible; in our experience, weekly meetings lasting not more than one hour work best. Regardless of the timing chosen, the schedule must be kept with the same rigor as board meetings. For a member of the senior management committee to miss a meeting is tantamount to declaring that the quality improvement program has low priority. Employees already suspicious of management programs will abandon ship at the first signs of waning management interest.

Employee action teams are not permanent, although creation, performance, and winding down of action teams should be a constant part of the plant's life. An action team is assembled by the supervisory council when the council becomes aware of a problem to be solved or opportunity to be addressed. Members of the team are selected by the supervisory council on the basis of their familiarity with the problem in question. The members should be drawn from the supervisory and line ranks of several departments to ensure that the "solution" does not simply move the problem from one department to another. Not every employee will want to take part in teams; even in the most progressive companies, some workers want only to do what they are told and avoid thinking. Overlapping talents should be avoided where possible.

When the team completes its assignment (or cannot produce a reasonable solution), it is dissolved and the members become available to serve on other action teams.

The structure described here is often mistaken for quality circles. There are significant differences between this system of temporary teams and the traditional quality circle, however. Particularly important, no action team is assembled until there is a task for it to carry out. Also, the team exists only until the task has been completed (or it is obvious no solution will be found). Quality circles, in contrast, are permanent teams responsible for developing their own agendas and constantly finding new activities.

19.6.10 Step 10: Collect Improvement Suggestions

Many people believe that the hardest part of any improvement program is solving problems. That is not always true. Identifying problems is often more challenging than developing solutions.

The best source of information about shop problems is the people who are closest to the problems (i.e., the employees). Many techniques exist for gathering employee input, beginning with the clichéd suggestion box. The classroom, as already noted, is a particularly fruitful environment for obtaining employee suggestions. Suggestions need not be solutions or even recommendations; questioning something that the employee does not understand often leads to discovery of important but previously unrecognized opportunities.

Whatever system is used for collecting information, suggestions flow quite freely in most plants. The problem lies not with obtaining information but with providing systems to follow through on the suggestions. Companies that judge their employee involvement programs by the number of issues raised rather than the numbers acted upon have missed the point of the exercise.

As already noted, the best time to gather that information is during classes which encourage free speech. Back in the strictly regimented hierarchy of the shop floor, employees are less likely to speak out.

Therefore, the questions and comments made by the students during the classes must not be ignored. The teacher must itemize all questions or suggestions about plant operations raised during each class. These become action items in the process improvement program.

19.6.11 Step 11: Select the First Problem to Solve

In step 3, the complexity of the quality improvement program was reduced by concentrating on just one process—the core process of soldering—first. Just as it is advantageous to limit the number of processes to be perfected, the number of problems being pursued at any one time should also be restricted. Initially, only one problem should be selected. The supervisory council selects the problem and forms a process action team to solve the problem. Further problems can be addressed as quickly as process action teams can be assembled—and the supervisory council develops the capability to manage additional teams simultaneously.

When two process action teams working on related problems come up with contradictory "solutions," the teams should be merged and assigned responsibility for reconciling the contradiction.

19.6.12 Step 12: Compare the Process with the Recipe

The process action teams' first act in resolving any process problem should be to compare the "real" process (the procedures followed on the shop floor) with the requirements of the process recipe developed earlier. Most problems arise when the real procedures do not coincide with the recipe requirements.

19.6.13 Step 13: Implement the Solutions

Solutions devised by the process action teams (having been accepted by both the supervisory council and the management committee) can now be implemented. If solutions are not implemented, the program accomplishes nothing. Of course, if the "solutions" are defective, the program has not helped, either.

Implementation, as should be obvious by now, entails considerably more than changing shop floor operations. Unless the formal standards are changed, no protection exists against falling back into old practices. If the formal processes do not conform to the actual processes, the effect is the same as having actual processes that do not conform to the formal processes. Accurate documentation is essential; on this point, at least, ISO 9000 cannot be attacked.

19.6.14 Step 14: Apply SPC

The preceding steps will create a controllable process. At this point, implementation of SPC to maintain process control is practical. Note that only now does SPC enter into the program, in stark contrast to total quality thinking that calls for use of statistical methods from the beginning. (Knowledgeable and experienced plant members will likely have been employing some statistical practices prior to this point to acquire information about the nature, location, and other characteristics of process problems.)

19.6.15 Step 15: Branch Out to Other Processes

Once the core process is under control (i.e., the process has been normalized and SPC applied), upstream or downstream processes can be refined in the same manner. Now that the operation's central problems have been corrected, many other problems not previously recognized as being attributable to the core process will have disappeared. Thus, by confining our attention initially to the core process, we have avoided needless work on problems that were not properly understood.

As application of the program spreads, however, progress must be carefully monitored to ensure that the requirements of quality optimization continue to be met.

19.7 Summary

The framework provided in these pages appears both simple and simplified. However, those who study and apply its methods will quickly find that the program framework is more complete than it seems on the printed page. Much of the scientific knowledge about the core process can be found in Chaps. 7 through 10 in this book, and the recipe can be derived fairly readily. Each step follows logically from explanations found in previous chapters.

With use, the program basics become extremely intuitive and straightforward. At the same time, we must emphasize that this program framework will never serve well when used indiscriminately. It is not intended for companies that have not set as their goal the maximization of long-term profits. Nor is it applicable in a "total quality" environment. Above all, its effectiveness depends on the user's willingness to constantly reassess the strategies and results. Dynamic environments such as manufacturing plants require dynamic solutions and dynamic administrators.

Will a program employing this framework work in your plant? Absolutely. Every outcome is determined by scientific process management and application of the quality optimization analysis. Any plant that conforms to the requirements of nature will inevitably achieve the highest possible quality and the lowest costs. The ultimate rule of quality management always is to rely on science rather than luck.

Afterword: The End of the Beginning

The end, happily, turned out very much like the beginning.

We began this book in the naive belief that decades of quality management had somehow left us qualified to advise others. The unpleasant truth quickly became obvious: We did know the conventional thinking and possessed insights into troubling aspects of trends in quality thinking. But we were humbled to realize how much remained to be learned—and how much of our existing "knowledge" was faulty. Much of what we had "learned" over the years—the mainstream lessons of quality management that every quality professional inevitably acquires with time—was seriously flawed. In some ways, we only knew enough to be dangerous.

Our collective experience was certainly not wasted. We may (in fact, we did) have lacked answers to some pivotal matters we thought we understood well. Without those decades of quality management, however, we would not likely have recognized the annoying questions—the recognition of serious contradictions between superficially complementary positions—that came to us so frequently and annoyingly. The discovery that so much remained to be learned about quality management proved wonderfully stimulating.

Did we learn it all? Of course not. We did not even unearth all the questions. Quality management contains so many thorny issues that some will never be resolved. Moreover, quality is a dynamic subject changing as rapidly as the business world in general.

So, the reader may well ask, what is gained if answers—and questions—remain hidden? To which we would respond that doubts and curiosity are in many ways more valuable than confidence and complacency. Our confidence and complacency were undermined by recognizing that:

- The subject of quality may be more popular than in decades past, but the effectiveness of quality management practices has declined as opportunities for wealth have lured glib charlatans into the field.
- Despite decades of effort by a few talented individuals and armies of unqualified fortune seekers, no adequate definition of quality itself

exists. In our opinion, the nature of quality may vary among types of businesses, but in the electronics assembly industry only dependability allows for workable planning and analysis.

- Quality improvement is not a magic bullet; it does not automatically bring well-being to a company nor can it even ensure safety from business collapse.
- Investment in quality improvement works only when the objective is known, quantifiable, and consistent with the company's ultimate goal of long-term profit maximization. Those investments will provide continuing benefits to the company only when adopted on the basis of conformance to the goal and regularly reviewed according to the same standard.
- Many of the most valuable lessons in the history of quality thinking have been shabbily distorted by latter-day interpretations. The pursuit of "quality" in far too many cases is nothing more than a game to which participants are drawn by the thrill of competition and the prospects of burnishing the corporate image.
- The Japanese international industrial success was less miracle than inevitable outcome of shortsighted American government policies, Western business practices, and financial sleight of hand.
- Ultimately, the level of quality is no better than the level of scientific understanding by the producers and evaluators. In electronics assembly, the most serious lack of scientific knowledge concerns the ubiquitous process of soldering.
- The "science" found in most modern quality prescriptions defies natural laws and logic. Even the mathematical "science" of statistics contains more pitfalls than opportunities.
- People truly are the company's most important asset—but only when intellectually nurtured and managed with realism rather than sentimental optimism.
- The quality improvement program framework that we have employed with considerable success over many years continues to perform well despite violating central tenets of conventional quality ideology. However, we now see the reasons for its success in quite a different manner.

The most important lesson of all is that this book is hardly the end of the quality improvement journey. In many respects, it may be nothing more than a tentative first step. We, however, like to believe that it stands for something more important. Not the end of the journey, perhaps, but the end of the beginning.

Bibliography

Aguayo, Rafael. 1991. *Dr. Deming: The Man Who Taught the Japanese about Quality.* New York: Fireside Books.
Bateson, John T. 1985. *In-Circuit Testing.* New York: Van Nostrand Reinhold.
Berk, Joseph, and Susan Berk. 1993. *Total Quality Management: Implementing Continuous Improvement.* New York: Sterling Publishing Co.
Bogan, Christopher E., and English, Michael J. 1994. *Benchmarking for Best Practices; Winning through Innovative Adaptation.* New York: McGraw-Hill.
Broh, Robert A. 1982. *Managing Quality for Higher Profits.* New York: McGraw-Hill.
Burstein, Daniel. 1988. *Yen!: Japan's New Financial Empire and Its Threat to America.* New York: Simon & Schuster.
Business Week, Editors, with Green, Cynthia. 1994. *The Quality Imperative.* New York: McGraw-Hill.
Byrne, John A. 1993. *The Whiz Kids.* New York: Currency Doubleday.
Champy, James. 1995. *Reengineering Management.* New York: Harper Collins.
Chatfield, Christopher. 1983. *Statistics for Technology,* 3d ed. London: Chapman & Hall.
Clements, Richard Barrett. 1993. *Quality Manager's Complete Guide to ISO 9000.* Englewood Cliffs, N.J.: Prentice-Hall.
Creech, W. L. (Bill). 1994. *The Five Pillars of TQM.* New York: Truman Tally Books/Dutton.
Crocker, Olga L., Charney, Cyril, and Chiu, Johnny Sik. 1986. *Quality Circles: A Guide to Participation and Productivity.* New York: Mentor/Penguin.
Crosby, Philip B. 1979. *Quality Is Free.* New York: McGraw-Hill.
———. 1984. *Quality without Tears: The Art of Hassle-Free Management.* New York: McGraw-Hill.
———. 1988. *The Eternally Successful Organization: The Art of Corporate Wellness.* New York: McGraw-Hill.
———. 1989. *Let's Talk Quality.* New York: McGraw-Hill.
———. 1992. *Completeness: Quality for the 21st Century.* New York: Dutton Division Penguin Books.
Deming, W. Edwards. 1951. *The Elementary Principles of the Statistical Control of Quality; A Series of Lectures.* Tokyo: Nippon Kagaku Gijutsu Remmei.
———. 1982. *Quality, Productivity and Competitive Position.* Cambridge, Mass.: Massachusetts Institute of Technology Center for Advanced Engineering Study.

______. 1986. *Out of the Crisis.* Cambridge, Mass.: Massachusetts Institute of Technology Center for Advanced Engineering Study.

______. 1993. *The New Economics for Industry, Government, Education.* Cambridge, Mass.: Massachusetts Institute of Technology Center for Advanced Engineering Study.

Dobyns, Lloyd, and Crawford-Mason, Clare. 1991. *Quality or Else.* Boston: Houghton Mifflin.

______. 1994. *Thinking about Quality.* New York: Times Books.

Eberts, Ray, and Eberts, Cindelyn. 1995. *The Myths of Japanese Quality.* Upper Saddle River, N.J.: Prentice-Hall.

Eccles, Robert G., and Nohria, Nitin with Berkley, James D. 1992. *Beyond the Hype.* Boston: Harvard Business School Press.

Emmott, Bill. 1989. *The Sun Also Sets.* London: Simon & Schuster.

Ernst & Young Quality Improvement Consulting Group, 1990. *Total Quality: An Executive's Guide for the 1990s.* Homewood, Ill.: Dow Jones-Irwin.

Feigenbaum, A. V. 1961. *Total Quality Control.* New York: McGraw-Hill.

______. 1983, 1991. *Total Quality Control.* 3d ed. New York: McGraw-Hill.

Fucini, Joseph J., and Fucini, Suzy. 1990. *Working for the Japanese.* New York: Free Press.

Fukuda, Ryuji. 1984. *Managerial Engineering: Techniques for Improving Quality and Productivity in the Workplace.* Stanford, Conn.: Productivity Press.

Gabor, Andrea. 1990. *The Man Who Discovered Quality.* New York: Times Books.

Galbraith, John Kenneth. 1984. *The Affluent Society,* 4th ed., rev. Boston: Houghton Mifflin. (First edition published 1958.)

Gitlow, Howard S., and Gitlow, Shelly J. 1987. *The Deming Guide to Quality and Competitive Position.* Englewood Cliffs, N.J.: Prentice-Hall.

Goldratt, Eliyahu M., and Cox, Jeff. 1984. *The Goal: Excellence in Manufacturing.* Great Barrington, Mass.: North River Press.

______. 1992. *The Goal,* 2d rev. ed. Great Barrington, Mass.: North River Press.

Goldratt, Eliyahu M. 1986. *The Race.* Great Barrington, Mass.: North River Press.

______. 1990. *The Haystack Syndrome.* Great Barrington, Mass.: North River Press.

______. 1990. *The Theory of Constraints.* Great Barrington, Mass.: North River Press.

______. 1994. *It's Not Luck.* Great Barrington, Mass.: North River Press.

Halberstam, David. 1986. *The Reckoning.* New York: William Morrow.

Hammer, Michael, and Champy, James. 1993. *Reengineering the Corporation.* New York: Harper Collins.

Handy, Charles. 1989. *The Age of Unreason.* London: Business Books.

Hart, Christopher W. L., and Bogan, Christopher E. 1992. *The Baldrige: What It Is, How It's Won, How to Use It to Improve Quality in Your Company.* New York: McGraw-Hill.

Hawley, Gessner G. 1981. *The Condensed Chemical Dictionary,* 10th ed. New York: Van Nostrand Reinhold.

Hiam, A. 1993. *Does Quality Work? A Review of Relevant Studies.* Report 1043. New York: The Conference Board.

Hicks, Charles R. 1982. *Fundamental Concepts in the Design of Experiments,* 3d ed. New York: CBS College Publishing.

Ishikawa, Kaoru. 1972. *Guide to Quality Control.* Tokyo: Asian Productivity Organization.

Ishikawa, Kaoru translated by Lu, David J. 1985. *What Is Total Quality Control? The Japanese Way.* Englewood Cliffs, N.J.: Prentice-Hall.

Jacoby, Jacob, and Olson, Jerry, eds. 1985. *Perceived Quality: How Consumers View Stores and Merchandise.* Lexington, Mass.: Lexington Books/D.C. Heath.

Juran, J. M. 1964. *Managerial Breakthrough.* New York: McGraw-Hill.

———. 1988. *Juran on Planning for Quality.* New York: The Free Press.

———. 1989. *Juran on Leadership for Quality.* New York: The Free Press.

Juran, J. M., Gryna, Dr. Frank M., Jr., and Bingham, R. S., Jr., eds. 1974. *Quality Control Handbook,* 3d ed. New York: McGraw-Hill.

Juran, J. M., and Gryna, Frank M. 1988. *Juran's Quality Control Handbook,* 4th ed. New York: McGraw-Hill.

Lea, Colin. 1988. *A Scientific Guide to Surface Mount Technology.* Ayr, Scotland: Electrochemical Publications Limited.

Leonida, Giovanni. 1981. *Handbook of Printed Circuit Design, Manufacture, Components and Assembly.* Ayr, Scotland: Electrochemical Publications Limited.

Lorriman, John, and Kenjo, Takashi. 1994. *Japan's Winning Margins: Management, Training and Education.* Oxford, England: Oxford University Press.

Main, Jeremy. 1994. *Quality Wars.* New York: The Free Press.

Manganelli, Raymond L., and Klein, Mark M. 1994. *The Reengineering Handbook.* New York: American Management Association.

Manko, Howard H. 1986. *Soldering Handbook for Printed Circuits and Surface Mounting.* New York: Van Nostrand Reinhold.

McGregor, Douglas N. 1960. *The Human Side of Enterprise.* New York: McGraw-Hill.

Mintzberg, Henry. 1994. *The Rise and Fall of Strategic Planning.* New York: The Free Press.

Morita, Akio, with Reingold, Edwin M., and Shimomura, Mitsuko. 1988. *Made in Japan.* New York: Signet.

Nester, William R. 1990. *The Foundation of Japanese Power.* Houndmils, England: Macmillan.

Oakland, John S. 1989. *Total Quality Management.* Oxford, England: Heinemann Professional Publishing.

Ouchi, William. 1981. *Theory Z.* Reading, Mass.: Addison-Wesley.

Persig, Robert M. 1974. *Zen and the Art of Motorcycle Maintenance.* New York: William Morrow.

Peters, Thomas J., and Waterman, Robert H., Jr. 1982. *In Search of Excellence.* New York: Harper & Row.

Rabbitt, John T., and Bergh, Peter A. 1994. *The ISO 9000 Book,* 2d ed. White Plains, N.Y.: Quality Resources.

Roberts, Harry V., and Sergesketter, Bernard F. 1993. *Quality Is Personal.* New York: The Free Press.

Sashkin, Marshall, and Kiser, Kenneth J. 1993. *Putting Total Quality Management to Work.* San Francisco: Berrett-Koehler Publishers.

Schaller, Michael. 1985. *The American Occupation of Japan: The Origins of the Cold War in Japan.* New York: Oxford University Press.

Schmidt, Warren H., and Finnigan, Jerome P. 1993. *TQManager.* San Francisco: Jossey-Bass Publishers.

Schonberger, Richard J. 1982. *Japanese Manufacturing Techniques: Nine Hidden Lessons in Simplicity.* New York: The Free Press.

______. 1986. *World Class Manufacturing: The Lessons of Simplicity Applied.* New York: The Free Press.

______. 1987. *World Class Manufacturing Handbook: Implementing JIT and TQC.* New York: The Free Press.

______. 1990. *Building a Chain of Customers.* New York: The Free Press.

Senge, Peter M. 1990. *The Fifth Discipline.* New York: Doubleday.

Shewhart, Walter A. 1931. *Economic Control of Quality of Manufactured Product.* New York: Van Nostrand.

______. 1939. *Statistical Method from the Viewpoint of Quality Control.* Lancaster, Pa.: Lancaster Press. Reprinted with new foreword by W. Edwards Deming. 1986. Mineola, N.Y.: Dover Publications.

Shina, Sammy G. 1991. *Concurrent Engineering and Design for Manufacture of Electronics Products.* New York: Van Nostrand Reinhold.

Silbiger, Steven. 1993. *The Ten Day MBA.* New York: Quill/William Morrow.

Suzaki, Kiyoshi. 1987. *The New Manufacturing Challenge: Techniques for Continuous Improvement.* New York: The Free Press.

Taguchi, Genichi, and Wu, Yuin. 1979. *Introduction to Off-Line Quality Control.* Nagaya, Japan: Central Japan Quality Control Association.

Taguchi, Genishi. 1986. *Introduction to Quality Engineering.* Tokyo: Asian Productivity Organization.

Taylor, F. W. 1947. *Scientific Management.* New York: Harper.

Tichy, Noel M., and Sherman, Stratford. 1993. *Control Your Destiny or Someone Else Will.* New York: Currency Doubleday.

Tompkins, James A. 1989. *Winning Manufacturing.* Norcross, Ga.: Institute of Industrial Engineers.

van Wolferen, Karel. 1989. *The Enigma of Japanese Power.* London: Macmillan.

Walton, Mary. 1986. *The Deming Management Method.* New York: Dodd, Mead.

Williams, Richard L. 1994. *Essentials of Total Quality Management.* New York: AMACOM.

Yamashita, Toshihiko with translation by Baldwin, Frank. 1987. *The Panasonic Way.* New York: Kodansha International.

About the Authors

JAMES ALLEN SMITH received a doctorate (A.B.D.) in economics and industrial organization from the University of Toronto in 1972 but has spent most of his career in the world of international electronics assembly. Starting as an assembly operator, he has worked as a quality control technician, operations manager, marketing executive, and venture capitalist. In 1984, he became President of International Operations for the Phoenix Group, Inc., a supplier of process management and educational services to electronics assembly plants throughout the Americas and Europe. He is currently President and Chief Executive Officer of The Phoenix Group's parent company, Cambridge Management Sciences, Inc., of St. Petersburg, Florida.

FRANK B. WHITEHALL has pursued a varied career in engineering and management spanning more than four decades and several continents. Beginning with Ferranti Defence Systems as a designer of civil and military avionic systems, he subsequently held positions in design quality, quality management, and personnel and industrial relations management. Until his retirement in 1993, he was Director of Corporate Quality. He has also served in various capacities with the Society of British Aerospace Companies (SBAC) and the Electronic Engineering Association (EEA), and was the inaugural Director of Quality for the Quality Scotland Foundation. Now an independent consultant and lecturer in Edinburgh, Scotland, he advises on quality, measurement and calibration, and airworthiness. He is also Chairman of the Board of Governors for a prominent ISO 9000 certification body in the United Kingdom.

Name Index

Subject Index